STUDENT SOl MANUAL

David S. Hage • James D. Carr
University of Nebraska, Lincoln

Analytical Chemistry
and
Quantitative Analysis

David S. Hage • James D. Carr

Prentice Hall

Boston   Columbus   Indianapolis   New York   San Francisco   Upper Saddle River
Amsterdam   Cape Town   Dubai   London   Madrid   Milan   Munich   Paris   Montréal   Toronto
Delhi   Mexico City   São Paulo   Sydney   Hong Kong   Seoul   Singapore   Taipei   Tokyo

Editor in Chief, Chemistry: Adam Jaworski
Publisher: Dan Kaveney
Marketing Manager: Erin Gardner
Associate Editor: Jessica Neumann
Managing Editor, Chemistry and Geosciences: Gina M. Cheselka
Project Manager: Wendy A. Perez
Operations Specialist: Maura Zaldivar
Supplement Cover Designer: Paul Gourhan
Cover Image Credit: Image courtesy of Rich Irish, Landsat 7 Team, NASA GSFC; Data provided by EROS Data Center

© 2011
Pearson Education, Inc.
Pearson Prentice Hall
Upper Saddle River, NJ 07458

All rights reserved. No part of this book may be reproduced, in any form or by any means, without permission in writing from the publisher.

Pearson Prentice Hall™ is a trademark of Pearson Education, Inc.

The author and publisher of this book have used their best efforts in preparing this book. These efforts include the development, research, and testing of the theories and programs to determine their effectiveness. The author and publisher make no warranty of any kind, expressed or implied, with regard to these programs or the documentation contained in this book. The author and publisher shall not be liable in any event for incidental or consequential damages in connection with, or arising out of, the furnishing, performance, or use of these programs.

> This work is protected by United States copyright laws and is provided solely for teaching courses and assessing student learning. Dissemination or sale of any part of this work (including on the World Wide Web) will destroy the integrity of the work and is not permitted. The work and materials from it should never be made available except by instructors using the accompanying text in their classes. All recipients of this work are expected to abide by these restrictions and to honor the intended pedagogical purposes and the needs of other instructors who rely on these materials.

Printed in the United States of America

10 9 8 7 6 5 4 3 2 1

ISBN-13: 978-0-321-70551-8
ISBN-10:   0-321-70551-3

Prentice Hall
is an imprint of

www.pearsonhighered.com

CONTENTS

Preface		v
Chapter 1	An Overview of Analytical Chemistry	1
Chapter 2	Good Laboratory Practices	6
Chapter 3	Mass and Volume Measurements	23
Chapter 4	Making Decisions with Data	45
Chapter 5	Characterization and Selection of Analytical Methods	74
Chapter 6	Chemical Activity and Chemical Equilibrium	92
Chapter 7	Chemical Solubility and Precipitation	106
Chapter 8	Acid–Base Reactions	127
Chapter 9	Complex Formation	159
Chapter 10	Oxidation–Reduction Reactions	173
Chapter 11	Gravimetric Analysis	195
Chapter 12	Acid–Base Titrations	213
Chapter 13	Complexometric and Precipitation Titrations	241
Chapter 14	An Introduction to Electrochemical Analysis	260
Chapter 15	Redox Titrations	273
Chapter 16	Coulometry, Voltammetry, and Related Methods	296
Chapter 17	An Introduction to Spectroscopy	305
Chapter 18	Molecular Spectroscopy	321
Chapter 19	Atomic Spectroscopy	338
Chapter 20	An Introduction to Chemical Separations	354
Chapter 21	Gas Chromatography	379

Chapter 22	Liquid Chromatography	399
Chapter 23	Electrophoresis	421

PREFACE

This Solutions Manual provides detailed information on the answers to all of the **Questions** and selected **Challenge Problems** in the text. These solutions provide instructors and students with examples of how various problems related to chemical analysis can be solved, based on approaches that are described in the text.

For students, we recommend that they first try and solve each problem on their own before consulting these answers and solutions. The process of learning how to solve such problems, and the strategies that are required, is important for the students to know and understand, as opposed to the ability to simply produce the final answer. Understanding the approach involved in each problem will help students gain the most benefit from the questions by helping each student practice and acquire the necessary skills for dealing with common calculations and choices that are encountered in modern chemical analysis.

David S. Hage
University of Nebraska, Lincoln

James D. Carr
University of Nebraska, Lincoln

CHAPTER 1: AN OVERVIEW OF ANALYTICAL CHEMISTRY

1. Analytical chemistry is the field of chemistry that deals with the use and development of tools and processes for examining and studying chemical substances. A "chemical analysis" is the general phrase used here to refer to the use of measurements to examine and study chemical substances.

2. The earliest known use of chemical testing is the analysis of precious metals like gold and silver by the fire assay. Later, methods for chemical measurements were also developed for looking at the quality of water, metals, medicines, and dyes. In modern times, these applications have continued, but chemical analysis is also now used in numerous areas that include environmental chemistry, biotechnology, and materials science, among many others.

3. Modern applications of analytical chemistry range from forensic science to biotechnology, agriculture, and materials science. Chemical analysis is also widely used in commercial applications, including the testing of foods, metals, and other manufactured products.

4. The use of analytical methods is often a two-way process because the need for more detailed chemical information promotes the development of new techniques, which then leads to more research that is made possible by the ability to obtain more data about a sample. Contributions to the field of analytical chemistry have been made by individuals with a variety of different backgrounds, including people from the fields of chemistry, physics, and medicine.

5. The portion of material taken for analysis is referred to as the "sample." The entire group of

substances that makes up the sample is called the "sample matrix" (or "matrix"). The particular substance we are interested in measuring or studying in the sample is known as the "analyte."

6. a) The sample is coal, the analyte is sulfur, and the matrix is the group of all substances that are found in the coal.

   b) The sample is the drug tablet, the analyte is the drug, and the matrix is the group of all substances that are found in the tablet.

   c) The sample is the set of fumes that were emitted by the industrial plant, the analyte is carbon monoxide, and the matrix is the group of all substances that are present in the fumes.

7. The term "major component" refers to substances that make up more than 1% of the sample. A substance present at lower levels, such as 0.01–1% of the total sample, is called a "minor component." A substance present at levels below 0.01% (100 parts-per-million) is known as a "trace component."

8. a) The protein and fat in a sample of beef are both examples of major components because they both make up more than 1% of the content of the beef (5% fat and roughly 95% protein).

   b) The aspirin in an aspirin tablet is an example of a major component because it makes up greater than 1% of the content (32% in this case).

   c) The vitamin C in an orange is an example of a minor component because it is present at a level that is 0.01–1% of the total sample content (0.05–0.06% in this example).

9. The five general steps in any chemical analysis are as follows.

1) Determine what is being asked about the sample and identify the information needed to answer this question.

2) Select an appropriate sample.

3) Prepare the sample.

4) Perform the analysis and conduct a measurement of the sample.

5) Use the results for chemical quantitation or characterization.

10. A "classical method" in analytical chemistry is a technique that produces a result by using experimentally-determined quantities, such as a mass or volume, along with atomic or molecular masses and well-defined chemical reactions. An "instrumental method" in analytical chemistry is a technique that uses an instrument-generated signal for detecting the presence of an analyte or determining the amount of an analyte in a sample. Instrumental techniques were developed long after classical methods of analysis but are used in most of today's chemical measurements. Classical methods are often performed as manual techniques, although some are found in modern laboratories as part of automated systems.

11. A separation method is often needed in a chemical analysis when the goal is to examine a chemical or group of chemicals in a complex sample. Chemical separations can be used as part of either a classical method or an instrumental method to isolate an analyte from a sample, remove interfering chemicals, or place the analyte in an appropriate matrix for analysis. Examples of separation methods are extractions, gas chromatography, and high-performance liquid chromatography.

12. In a chemical assay, a "standard" is a material that contains the analyte of interest at a known concentration. A "calibration curve" is a plot of the method's signal when the method is

used to examine a set of standards with known amounts of analyte. This calibration curve is used to determine the amount of analyte that must have been present in a sample to give a particular response.

13. a) A qualitative analysis will provide information on whether a particular analyte is present in a sample. A quantitative analysis is used to measure the analyte and provide information on the amount of analyte that is present in a sample.

    b) A structural analysis can be used to provide information on features such as the mass, composition, functional groups, or structure of an analyte. Property characterization provides information that is obtained by using a measurement of some specific chemical or physical property of the analyte, such as the chemical's ability to interact with light or electrons, its ability to react with other chemicals, or its color, crystal shape, and mechanical strength.

    c) Spatial analysis can provide information on how a particular analyte is distributed throughout a matrix. This type of analysis is valuable when you are examining a material with a composition that varies from one point to the next within its structure. Temporal analysis is used to provide information on how the amount of an analyte in a sample varies as a function of time.

14. a) Qualitative analysis would be used initially to determine if these drugs are present in the samples above an accepted cut-off limit. If any samples are found to give positive results for the drugs, they would probably be reanalyzed using quantitative analysis to determine the amount of drug that is present and confirm whether it is present above the allowed limit.

    b) A structural analysis could be used to provide information on the structure of the compound, which could then be used to identify this chemical. Property characterization might also be used for this purpose.

c) Quantitative analysis would be utilized in determining the amount of drug that is present in the product.

d) This is an example of spatial analysis because it seeks to obtain information on how a particular analyte is distributed in space along a river bed. This work would also involve quantitative analysis as the amount of pollutant is measured at different locations in the river.

CHAPTER 2: GOOD LABORATORY PRACTICES

1. A good laboratory practice (GLP) is a set of guidelines that promote proper work and conduct within the laboratory. These guidelines are important in analytical chemistry to give confidence that the final results are a valid representation of a sample and to ensure this work is carried out in a safe and effective manner.

2. A standard operating procedure (SOP) is a specific set of instructions that describes how a particular task should be performed. A SOP might be a document that describes a procedure for synthesizing a chemical, a safety protocol, or a method for calibrating an instrument. These instructions help to promote GLPs by describing how routine work should be carried out in a laboratory.

3. There are various standard operating procedures that are probably used in your laboratory, including procedures to follow in the event of a fire or chemical spil, and to follow for chemical handling and disposal. Many other types of SOPs are also possible.

4. Discuss the topics and features for each of these SOPs with your instructor or laboratory supervisor.

5. You can compare the SOPs you found on the Internet with factors that are described in this textbook and should also be part of these SOPs, as outlined in Chapter 3 (e.g., calibration of volumetric glassware, preparing solutions, and use of a volumetric pipet).

6. A chemical hygiene plan (CHP) is a set of SOPs used to promote safety in a laboratory. This plan often includes instructions on the location and use of safety equipment, such as

safety showers, eye washes, room exits, emergency phones, fire extinguishers, and first aid kits, the use of equipment for handling chemical spills, and the facilities and procedures to be utilized for the handling, storage, and disposal of chemicals.

7. Common features found in a modern laboratory include safety showers, eye washes, room exits, emergency phones, fire extinguishers, first aid kits, equipment for handling chemical spills, and facilities for the handling, storage, and disposal of chemicals. Each of these items either helps prevent your exposure to chemicals or minimizes any damage if there is such exposure. Further details can be found in Table 2.1.

8. The solution to this problem will depend on the design and facilities in your particular laboratory.

9. A chemical hazard is any chemical that is a physical or a health hazard. Chemicals that are physical hazards include those that are explosive, highly reactive, or flammable. Health hazards are chemicals that may be toxic or corrosive, produce cancer or birth defects, or cause damage to specific parts of the body, such as the lungs, skin, or eyes.

10. a) A flammable substance is a material that is easy to ignite.

    b) An explosive material is a substance that can cause a sudden, violent chemical reaction with the release of gas and heat.

    c) An oxidizer is a substance that readily yields oxygen to support the combustion or oxidation of other chemicals.

    d) A radioactive substance is a material that emits ionizing radiation.

    e) "Water-reactive" is a term used to describe a chemical that will react with water to become flammable or give off large quantities of flammable or toxic substances.

f) A compressed gas is a gas that is kept in an enclosed container at an elevated pressure.

11. a) An acute toxin is a chemical that causes a harmful effect after a single exposure.

b) A poison is a substance that can kill, injure, or impair a living organism.

c) A biohazard is a biological substance that presents a health hazard.

d) A carcinogen is a substance that causes cancer.

e) An irritant is a non-corrosive chemical that causes reversible inflammation (e.g., swelling and redness) on contact with living tissue.

f) A corrosive is a chemical that causes the destruction of living tissue at the site of contact.

g) An asphyxiant is a chemical that interferes with the transport of oxygen in the body.

h) An etiological agent is a microorganism or related toxin that can cause human disease.

i) A reproductive toxin is an agent that creates damage to the reproductive system.

j) An allergen is a substance that may produce an allergic reaction.

k) A mutagen is a substance that causes a change in DNA.

l) A teratogen is a substance that leads to the production of non-hereditary birth defects.

12. a) Hydrogen peroxide is corrosive, an oxidant, and a potential health hazard, with a National Fire Prevention Association (NFPA) health level rating of 3.

b) Potassium hydroxide is a corrosive and can be a health hazard (NFPA health level = 3).

c) Sodium chloride is a reasonably safe chemical with no appreciable hazards and only a mild health hazard (NFPA health level = 1).

d) Carbon tetrachloride is a carcinogen, toxin, and is harmful to the environment (NFPA health level = 3).

e) Acetonitrile is flammable and harmful (NFPA health level = 2).

f) Hydrogen gas has a high fire and explosion risk and can be harmful if inhaled.

13. The NFPA label is one system that is commonly used to identify chemical hazards. This label is usually drawn as a diamond with four colored areas. The blue region at the left of the diamond represents the overall health risk of the chemical. The red region at the top indicates the compound's general level of flammability. The yellow region to the right represents the chemical's ability to react with other substances. The white region at the bottom provides other information, such as the reactivity of the chemical with water or its relative ability to oxidize other compounds. The blue, red, and yellow regions each contain a number between 0 and 4, where 0 represents the safest compounds and 4 represents those with the greatest danger.

14. Sodium metal has an NFPA reactivity rating of 3 and is water reactive. A Material Safety Data Sheet (MSDS) for this chemical should further indicate that this substance should not be stored with halogens, such as chlorine gas. The NFPA label for chlorine is less clear on what effect sodium may have on this chemical, but this label does indicate that chlorine should not be stored with oxidizing agents.

15. A MSDS is a set of one or more sheets that must be sent with each chemical substance that is produced by a manufacturer or that is imported for distribution. Items provided in an MSDS include: a) a list of the chemicals found in the material and their common names, b) information on the material's chemical and physical properties, c) health hazards that are associated with the material, d) the maximum allowable limit exposure to the substance or chemical, e) an indication as to whether the chemical is a known carcinogen or has been

found to be a potential carcinogen, and f) any precautions to follow for the safe handling and use of the material.

16. Methylene chloride (CH_2Cl_2) is combustible if heated (NFPA flammability rating = 1) and is harmful if inhaled or swallowed (NFPA health rating = 2).

17. Sodium bicarbonate poses no fire hazard (NFPA flammability rating = 0) or reactivity hazard (NFPA reactivity rating = 0). It is a mild irritant (NFPA health rating = 1), but is safe to use in most circumstances.

18. The four major ways in which chemicals can enter the body are inhalation, ingestion, injection, and contact with the skin or eyes.

The inhalation of chemicals can be greatly reduced by working with volatile substances in a ventilation hood. You can also wear a mask to prevent breathing in dust or small particles.

To avoid ingestion, <u>never</u> eat or drink within a laboratory. Applying cosmetics, smoking, and using gum or chewing tobacco in a laboratory is also discouraged. For the same reason, you should always wash your hands after handling chemicals and before leaving a laboratory.

To avoid injection, carefully use sharp devices and dispose of these devices (along with broken glassware) in a sturdy container, typically labeled "sharps".

Contact with chemicals can be avoided by making sure you <u>always</u> use eye protection and have adequate clothing when you are in a laboratory. In addition, wear suitable gloves when working with hazardous chemicals and make sure to properly cover all cuts and abrasions, which can provide easy routes for chemical entry. Furthermore, clean your work area after you are finished with an experiment to avoid exposing others to substances you have been using.

19. a) This technician would be at greater risk of inhaling the volatile chemical.

b) Others in the laboratory unaware of the spill may be exposed to the acid through contact with skin or eyes.

c) The scientist and his fellow lab workers are at greater risk of being inadvertently injected with the contents of the needles.

d) The worker and others will be unaware of the potential hazards the reagent may pose or how to properly handle the reagent.

e) These graduate students may accidently ingest chemicals they have been working with in the laboratory.

20. a) Hexane should be stored in a cool, dry, well-ventilated location, away from direct sunlight and any area where a fire hazard may be present.

b) Sodium iodate is an oxidizing agent and should not be stored with reducing agents or combustible materials.

c) Zinc metal is incompatible with strong acids and bases, alkalis, and other non-metals. It is also air- and moisture-sensitive, and zinc powder can be very flammable. Storage of this material must address all of these considerations.

21. a) As indicated in the previous problem, n-hexane should be stored in a cool, dry, well-ventilated location, away from direct sunlight and any area where a fire hazard may be present. These requirements would usually cause n-hexane to be placed in a separate storage location from most of the other chemicals that are listed, but n-hexane would probably be

kept in the same general storage area as benzaldehyde.

b) Benzaldehyde has an NFPA flammability level of 2 and should be stored in an appropriate section of the laboratory for flammable organic compounds. This would probably be located in the same storage area as *n*-hexane.

c) Nitric acid is an acid, a corrosive, and an oxidizing agent. This chemical would typically be stored with other acids. It must also be kept separate from any bases or chemicals that can react with oxidizing agents.

d) Potassium metal is moisture- and air-sensitive and can spontaneously combust through the generation and ignition of hydrogen. It also reacts violently with water and acids, alcohols, and carbon monoxide. It must be stored under oil and away from other chemicals with which it might react.

e) Phosphoric acid is an acid and a corrosive. It would probably be stored in the same general area as nitric acid and must be kept separate from any bases.

f) Sodium hydroxide is a base and a corrosive. It needs to be stored in an area that is separate from any acids.

22. Laboratory waste management refers to the procedures for disposing or handling used chemicals. These procedures vary from one laboratory to the next but must always follow certain legal guidelines for safe and responsible chemical handling and disposal.

23. Consult the material safety data sheets for these chemicals and your instructor or laboratory supervisor to determine how each of these materials should be handled as part of laboratory waste management at your particular institution.

24. The laboratory notebook is a record of the procedures that were used by a scientist in an experiment, the results that were obtained, and the conclusions that were reached from the experiments. It is important to keep a thorough and up-to-date laboratory notebook because this notebook plays a vital role when these results are eventually communicated to others in articles or reports and can be used to establish when a particular experiment was conducted. A laboratory notebook also should present a record that is easy to understand by others.

25. Some recommended practices for keeping a laboratory notebook are listed in Table 2.3. Consult with your instructor or laboratory supervisor to determine which of these practices pertain to your particular laboratory.

26. An electronic laboratory notebook is a digital record of a laboratory experiment in which text can be combined directly with graphs, structures, images, and other computer-based sources of information. This system has the potential to offer much greater flexibility than a standard notebook in terms of data analysis, report generation, and the communication of results. Electronic notebooks, however, also have potential disadvantages, such as being subject to data loss due to computer viruses or computer failure. In addition, there are security issues that can result if hackers enter into the notebook.

27. A spreadsheet is a program used to record, analyze, and manipulate data. A spreadsheet is a valuable tool in automatically carrying out repetitive calculations and in analyzing data or preparing graphs for use in a laboratory notebook.

28. The SI system of measurements is a system that provides a set of uniform standards for

describing such things as mass, length, time, and other measurable quantities.

29. A fundamental SI unit is one of the basic units of measure in the SI system with which all other units of measurement can be described. The current SI base units include the meter (for length), kilogram (for mass), second (for time), mole (for the number of members of a substance), kelvin (for temperature), ampere (for current), and candela (luminescent intensity).

30. A derived SI unit is a unit of measure that can be obtained by combining the fundamental units of the SI system. Two examples are the coulomb and the volt.

    An accepted SI unit is an important or common measurement unit that is related to, but not directly derived from, fundamental SI units. Examples include the liter as a measure of volume, and minutes or hours as units of time.

31. a) (22,489 ft)(0.3048 m/ft) = 6,855 m

    b) (5.68 atm)(1.01325 × 10^5 Pa/atm) = 5.76 × 10^5 Pa

    c) (130 lb)(0.4536 kg/lb) = 59 kg (or 59.0 for 130. lb)

    d) (120 miles/hr)(1.6093 km/mile) = 193 km/hr

    e) (2200 cal)(4.184 J/cal) = 9200 J (or 9205 J for 2200. cal)

    f) (25.0 gal)(3.7854 L/gal) = 94.6 L

32. a) At a body temperature of 98.6 °F, the temperature would be as follows in °C and K:

    °C = (°F − 32 °F) · (5 °C/9 °F) = (98.6 °F − 32 °F) · (5 °C/9 °F) = 37.0 °C

    K = °C + 273.15 = 37.0 °C + 273.15 = 310.15 K = 310.2 K

    b) At absolute zero (0 K), the temperature would be as follows in °C and °F:

$$°C = K - 273.15 = 0.00\ K - 273.15 = -273.15\ °C$$

$$°F = (°C) \cdot (9\ °F/5\ °C) + 32\ °F = (-273.15\ °C) \cdot (9\ °F/5\ °C) + 32\ °F = -459.67\ °F$$

c) At a temperature of 20 °C, the temperature would be as follows in K and °F.

$$K = °C + 273.15 = 20\ °C + 273.15 = 293.\underline{15}\ K = 293\ K$$

$$°F = (°C) \cdot (9\ °F/5\ °C) + 32\ °F = (20\ °C) \cdot (9\ °F/5\ °C) + 32\ °F = 68\ °F$$

d) The temperature of a freezer (-20 °C) would be as follows in °F and K.

$$K = °C + 273.15 = -20\ °C + 273.15 = 253.\underline{15}\ K = 253\ K$$

$$°F = (°C) \cdot (9\ °F/5\ °C) + 32\ °F = (-20\ °C) \cdot (9\ °F/5\ °C) + 32\ °F = -4\ °F$$

33. a) 2.58×10^{-11} g = 25.8 pg or 0.0258 ng

    b) 125×10^{-6} L = 125 µL or 0.125 mL

    c) 150,000 g/mol = 150 kg/mol or 150 kDa

    d) 589×10^{-9} m = 589 nm or 0.589 µm

    e) 600×10^{6} Hz = 600 MHz or 0.600 GHz

    f) 25,000 V = 25 kV or 0.025 MV

34. An analog display is a display of a signal that is shown as having a continuous range of values. In recording a number from such a device, you must find the two values on the display that are just above and below the measured response. You can then estimate how far the reading lies between these two values to provide one more significant figure in the final recorded value.

    A digital display is a display of a signal that is shown as having a discrete and fixed number of possible values. To read this type of result, you would generally record all of the digits that are shown.

35. The estimated % transmittance is 53.7%, and the estimated absorbance is 0.270.

36. Figure 2.10 illustrates some GLPs for preparing a graph, plotting data points, and labeling a graph so that the final plot will be easy for others to understand.

37. Significant figures are the digits that can be used to reliably record or report a number. Using the correct number of significant figures is an important part of GLPs in determining that experimental results are used and reported in a consistent and accurate manner.

38. The significant figures are indicated in bold in each of the following numbers.

   a) $F =$ **9.64853415** $\times 10^4$ C/mol   9 significant figures

   b) $m/z =$ **183.2280** u   7 significant figures

   c) $-\log(\%T) = 1.$**238**   3 significant figures

   d) 1 km = 0.**62137** miles   5 significant figures (for miles)

   e) 1 in = **2.54** cm   3 significant figures (for cm)

   f) $[OH^-] =$ **6.00** $\times 10^{-7}$ M 3 significant figures

39. a) $[Na^+] = 1.52$ M

   b) $-\log a_{H+} = 7.463$

   c) $h = 6.63 \times 10^{-34}$ J·s

   d) $t = 5.52$ ns

   e) $10^{-pCa} = 8.37$

   f) $K_a = 0.165$

40. Five significant figures: $52.787 + 34.099 + 14.115 = 101.001$

Four significant figures: 52.79 + 34.10 + 14.11 = 101.00

Three significant figures: 52.8 + 34.1 + 14.1 = 101.0

Two significant figures: 53 + 34 + 14 = 101

41. a) 8.9×10^{-12}

    b) 1.28×10^{-9}

    c) 6.74

    d) 3.0×10^{15}

42. Three significant figures: 3.14

    Four significant figures: 3.142

    Five significant figures: 3.1416

43. The average atomic mass for C, H, and O are 12.0107, 1.00794, and 15.9994 g/mol, respectively. Using these values with the specified number of significant figures gives the following answers.

    Answer when using all significant figures for the atomic masses and final solution:

    $6 \cdot (12.0107 \text{ g/mol}) + 12 \cdot (1.00794 \text{ g/mol}) + 6 (15.9994 \text{ g/mol}) = 180.1559$ g/mol

    Answer when expressed using five significant figures: 180.16 g/mol

    Answer when expressed using four significant figures: 180.2 g/mol

    Answer when expressed using three significant figures: 180 g/mol

44. The final answers with the correct number of significant figures are shown in bold. The underlined numbers are guard digits.

    a) 107.868 + 35.4527 = 143.320_7_ = **143.321**

b) $2.5898 - 0.133 - 0.003517 = 2.453\underline{3} = \mathbf{2.453}$

c) $98.4/99.976 = 0.984\underline{2} = \mathbf{0.984}$

d) $\log(2.01 \times 10^{-6}) = -5.696\underline{8} = \mathbf{-5.697}$

e) $\text{antilog}(-2.891) = 1.28\underline{5} \times 10^{-3} = \mathbf{1.29 \times 10^{-3}}$

f) $10^{-6.82} = 1.5\underline{14} \times 10^{-7} = \mathbf{1.5 \times 10^{-7}}$

45. The final answers with the correct number of significant figures are shown in bold. The underlined numbers are guard digits.

a) $189.032 + 153.02 - 32.0861 = 309.96\underline{6} = \mathbf{309.97}$

b) $(1.053 \times 10^{-5}) \cdot (3.56 \times 10^{-8})/(0.48) = 7.8\underline{1} \times 10^{-13} = \mathbf{7.8 \times 10^{-13}}$

c) $(0.9323/0.184) + 4.8520 = 5.06\underline{7} + 4.8520 = 9.91\underline{9} = \mathbf{9.92}$

d) $0.998 \cdot (18.99840 + 12.0107) = 0.998 \cdot (31.0091\underline{0}) = 30.9\underline{5} = \mathbf{30.9}$

e) $6.82 + \log(0.1235) = 6.82 + (-0.9083\underline{3}) = 5.91\underline{2} = \mathbf{5.91}$

f) $0.238 \cdot 10^{-4.231} = 0.238 \cdot (5.87\underline{5} \times 10^{-5}) = 1.39\underline{8} \times 10^{-5} = \mathbf{1.40 \times 10^{-5}}$

46. Dimensional analysis is an approach in which you record and compare the units on numbers to make sure the final result is expressed in the desired fashion. To carry out dimensional analysis, you first need to write down the units on all the numbers in a calculation. Next, if there are any numbers being multiplied or divided by each other, you can cross out any common units that appear in both the numerators and denominators of these terms. You also need to make sure that all numbers being added or subtracted have the same units. If you have set up your equations and numbers in the right manner and used appropriate conversion factors, the final answer should have the correct units.

47. a) Mass of Cu in 5 g of $CuSO_4 \cdot H_2O$ =

$$(5.000 \text{ g } CuSO_4 \cdot H_2O) \cdot \frac{(1 \text{ mol } CuSO_4 \cdot H_2O)}{(177.63 \text{ g } CuSO_4 \cdot H_2O)} \cdot \frac{(1 \text{ mol Cu})}{(1 \text{ mol } CuSO_4 \cdot H_2O)} \cdot \frac{(63.55 \text{ g Cu})}{(1 \text{ mol Cu})}$$

$$= 1.789 \text{ g Cu}$$

b) Titrated concentration of NaOH in a sample = (95.8 mL - 25.3 mL) · (1 L/1000 mL) ·

(0.105 mol/L HCl) · (1 mol HCl/1 mol NaOH)/(0.500 L NaOH)

= 0.0148 mol/L NaOH (or 0.0148 M)

c) Molar mass of $C_{12}H_{22}O_{11}$ = (12 mol C/mol $C_{12}H_{22}O_{11}$) · (12.0107 g C/mol C) +

(22 mol H/mol $C_{12}H_{22}O_{11}$) · (1.00794 g H/mol H) +

(11 mol O/mol $C_{12}H_{22}O_{11}$) · (15.9994 g O/mol O)

= 342.2965 g/mol $C_{12}H_{22}O_{11}$

d) Density of a lead cylinder = (23.2850 g - 0.0165 g)/[(2.52 cm) · π · (0.51 cm)^2]

= 11 g/cm^3

48. a) $n = PV/RT$ = [(2.50 atm) · (3.15 L)]/[(0.0821 L · atm/(mol·K)) · (273.15 K + 25.0 K)]

= 0.321$\underline{7}$ mol = 0.322 mol

b) $\Delta G°_{AgCl}$ = -2.303 · (8.314 J/(mol·K)) · (298 K) · log(1.0 × 10^{10})

= -2.303 · (8.314 J/(mol·K)) · (298 K) · (10.00)

= -5.70$\underline{6}$ × 10^4 J/mol = -5.71 × 10^4 J/mol

c) $-\log(\gamma_{Ca2+}) = \dfrac{[0.51 \cdot (+2)^2 \cdot (0.10)^{1/2}]}{[1 + (0.10)^{1/2}]} = 0.49\underline{0} = 0.49$

d) Percent (w/w) of Sulfur in $H_2SO_4 = 100 \cdot \dfrac{(1 \text{ mol S}/1 \text{ mol } H_2SO_4) \cdot (32.066 \text{ g S}/1 \text{ mol S})}{(98.078 \text{ g } H_2SO_4/1 \text{ mol } H_2SO_4)}$

$= 32.694\underline{4}\% = 32.694\%$ (w/w)

49. In the °C = K - 273.15, the number "273.15" must have units of K to be subtracted from a temperature that is given in kelvin ("K"). To make this relationship correct, there is also an unwritten conversion factor of (1 °C/1 K) that must be present on the right-hand side. The more complete relationship would then be written as shown below.

$$°C = [K - 273.15 \text{ (in units of K)}] (1 \text{ °C}/1 \text{ K}).$$

50. $R = 8.314 \text{ J}/(\text{mol} \cdot \text{K})$

$= [8.314 \text{ J}/(\text{mol} \cdot \text{K})]/(4.184 \text{ J/cal})$

$= 1.987 \text{ cal}/(\text{mol} \cdot \text{K})$

51. A rounding error is an error that is produced when a number is rounded off too early in a calculation. One way you can avoid rounding errors is to allow each value in the calculation to carry at least one non-significant figure until the final result is obtained. These additional non-significant figures are known as "guard digits".

52. This difference was caused by a rounding error introduced by using too few significant figures in the conversion factor (1609 m/mile) for converting from miles to meters.

53. The average atomic mass for C, H, N, O, and S are 12.0107, 1.00794, 14.00674, 15.9994, and 32.066 g/mol, respectively. Using these values with the specified number of significant figures gives the following answers.

Answer when using all available significant figures for atomic masses:

$257 \cdot (12.0107 \text{ g/mol}) + 383 \cdot (1.00794 \text{ g/mol}) + 65 \cdot (14.00674 \text{ g/mol}) + 77 \cdot (15.9994 \text{ g/mol})$
$+ 6 \cdot (32.066 \text{ g/mol}) = 5807.579 \text{ g/mol}$

Answer when using six significant figures for each atomic mass:

$257 \cdot (12.0107 \text{ g/mol}) + 383 \cdot (1.00794 \text{ g/mol}) + 65 \cdot (14.0067 \text{ g/mol}) + 77 \cdot (15.9994 \text{ g/mol})$
$+ 6 \cdot (32.066 \text{ g/mol}) = 5807.576 \text{ g/mol}$

Answer when using five significant figures for each atomic mass:

$257 \cdot (12.011 \text{ g/mol}) + 383 \cdot (1.0079 \text{ g/mol}) + 65 \cdot (14.007 \text{ g/mol}) + 77 \cdot (15.999 \text{ g/mol})$
$+ 6 \cdot (32.066 \text{ g/mol}) = 5807.627 \text{ g/mol}$

Answer when using four significant figures for each atomic mass:

$257 \cdot (12.01 \text{ g/mol}) + 383 \cdot (1.008 \text{ g/mol}) + 65 \cdot (14.01 \text{ g/mol}) + 77 \cdot (16.00 \text{ g/mol})$
$+ 6 \cdot (32.07 \text{ g/mol}) = 5807.70 \text{ g/mol}$

Answer when using three significant figures for each atomic mass:

$257 \cdot (12.0 \text{ g/mol}) + 383 \cdot (1.01 \text{ g/mol}) + 65 \cdot (14.0 \text{ g/mol}) + 77 \cdot (16.0 \text{ g/mol})$
$+ 6 \cdot (32.1 \text{ g/mol}) = 5805.4 \text{ g/mol}$

These results show that as the atomic masses are rounded to fewer significant figures, the difference between the calculated mass of insulin and its true mass tends to increase.

54. Class A fire extinguishers are intended for use with ordinary solid combustibles. Class B fire extinguishers are for use with flammable solids and liquids. Class C fire extinguishers are for energized electrical equipment. Consult with your instructor or laboratory supervisor to determine which of these is present in your own laboratory.

55. a) A neurotoxin is a toxin that acts on nerve cells and the nervous system.

    b) A nephrotoxin is a toxin that affects the kidneys.

    c) A hepatotoxin is a toxin that affects the liver.

    d) A hemapoietic toxin affects cells that are related to the creation of blood and blood components.

56. a) A combustible, or inflammable, material is a substance that is relatively easy to burn.

    b) A spontaneously combustible substance is a material that can ignite spontaneously.

    c) A pyrophoric substance is a material that can ignite in the presence of air or moisture.

CHAPTER 3: MASS AND VOLUME MEASUREMENTS

1. Mass is the quantity of matter in an object and is a property that is constant regardless of the object's location. Weight is a measure of the pull of a force, such as gravity, on an object. Weight will vary depending on the gravitational attraction working on the object at its current location and the objects' buoyancy, while the mass of an object will not change as these conditions are varied. Mass is preferred for use in scientific measurements.

2. Weighing commonly refers to the process of determining either the mass or weight of a substance. The goal of this process, however, is to determine the object's mass. This goal is met when using a balance by comparing an object to a reference with a known mass.

3. Buoyancy is a force that works in opposition to gravity when you are weighing an object and depends on both the density of the material that is being weighed and the density of the medium that surrounds it. If a correction is not made for buoyancy, this force will affect the apparent weight that is measured for an object. However, buoyancy will not affect the measurement of mass.

4. Force due to gravity (for an object) = $m_{obj} \cdot g$

   Force due to buoyancy (for air as the surrounding medium) = $- m_{air} \cdot g$

   Relationship for an object and reference giving an equal response on a balance:

   $$\underline{\text{Weight of Object}} \qquad\qquad \underline{\text{Weight of Reference}}$$
   $$m_{obj} \cdot g \;-\; m_{obj} \cdot (d_{air}/d_{obj}) \cdot g \;=\; m_{ref} \cdot g \;-\; m_{ref} \cdot (d_{air}/d_{ref}) \cdot g$$

The last equation can be rearranged to show how the true mass of the object (m_{obj}) is related to its apparent mass, as represented in the following equation by the mass of reference (m_{ref}).

$$m_{obj} = m_{ref} \cdot \frac{[1 - (d_{air}/d_{ref})]}{[1 - (d_{air}/d_{obj})]}$$

5. The change in the gravitational pull from that which occurs at 340 ft above sea level to that which occurs at 5280 ft above sea level would account for this error.

6. A balance is a precision weighing instrument that is used to measure small masses. The way a balance makes this measurement is by comparing an object to a reference with a known mass. If the object and reference are placed on opposite sides of the balance, they will have the same surrounding medium, giving them similar buoyancy effects. In addition, the local gravitational field on the object and reference will be the same. Thus, the difference in force between the two sides should be directly related to their difference in mass.

7. A mechanical balance is a balance that uses a mechanical approach for determining the mass of an object. One example is the equal-arm (or two-pan) balance, in which a sample and reference weights are placed on opposite sides of a beam held across a central fulcrum and compared.

8. An electronic balance uses an electrical mechanism to determine the mass of an object, as is accomplished by attaching the sample pan to a bar held between the two ends of a permanent magnet. When a sample is placed onto the pan, the bar is pushed downward. A position sensor signals the balance to apply a current through the bar, which produces an electromagnetic force that causes the bar to again move upward. The size of the applied current that is needed to move the bar and sample pan to their original positions is then

measured, providing a value that is proportional to the mass of the sample.

9. A quartz-crystal microbalance is a device that uses a thin, oscillating quartz crystal as a sensor for amounts of chemicals in samples; if chemicals adsorb to the surface of this crystal, it will change its frequency of vibration, which provides a signal that is related to the mass of the deposited material. This device is constructed from a thin, quartz crystal. The two sides are attached to electrodes that apply an alternating current and cause the quartz to oscillate at a specific frequency. If chemicals adsorb to the crystal's surface, the mass of the crystal will change and produce a change in the frequency at which the crystal is vibrating. This change can allow the mass of deposited material to be measured.

10. A mass spectrometer first converts the atoms or molecules in a sample into gas-phase ions, which are then separated and analyzed based on their mass and charge. The result is a graph known as a "mass spectrum", in which the amount of each detected ion is plotted versus its mass-to-charge ratio (m/z). This plot can provide information on both the molecular mass of the substance and its structure. The size of the resulting mass peaks can also be used to measure the amounts of analytes in samples.

11. The "resolution" of a balance is the maximum load of a balance divided by the balance's readability, giving a measure of the number of distinct masses that can be determined by a balance. The "capacity" of a balance (or maximum load) is the largest mass that can be reliably measured by a particular balance. The "readability" of a balance is the size of the smallest division in mass that is displayed on a balance's display. The actual values for these parameters for balances in your laboratory will depend on the types of balances that are present.

12.
   a) Macroanalytical balance or semimicrobalance

   b) Precision balance

   c) Microbalance or ultramicrobalance

13. The precision balance would be used to weigh the EDTA because it has an adequate capacity and readability for determining this mass. The analytical balance has a better readability and would be used to weigh the calcium carbonate.

14. Direct weighing is a procedure in which an object is placed on a balance pan and the mass is recorded directly from the balance display. This approach is easy to use and is commonly utilized for inert and solid objects, such as reference weights. However, it cannot be used directly with chemicals.

    Weight-by-difference is a weighing procedure in which the mass of a sample is determined by taking the difference between the mass of its container and the mass of the container plus the sample. This procedure takes slightly longer than direct weighing, but can also easily be used with chemicals or samples that cannot be measured by direct weighing.

15. Taring is a procedure that involves first placing the weighing container onto the balance and having the balance electronically reset its display, so that it reads "0" when the container is present. When a chemical is placed into this container, the display can then be used as a direct reading of the amount of added substance. Resetting the tare value might affect the calibration process.

16. A buoyancy correction is the process of adjusting for the effect of buoyancy of the measured mass of an object. Buoyancy effects should be considered whenever we want to measure a

mass with four or more significant figures. This correction can be accomplished by using the following equation.

$$m_{obj} = m_{display} \cdot \frac{[1 - (d_{air}/d_{ref})]}{[1 - (d_{air}/d_{obj})]}$$

The use of this equation requires information on the apparent mass that is given on the display of the balance, the density of the sample, the density of the reference used to calibrate the balance, and the density of the surrounding medium.

17. The true mass can be found by using Equation 3.6.

$$m_{obj} = m_{display} \cdot \frac{[1 - (d_{air}/d_{ref})]}{[1 - (d_{air}/d_{obj})]}$$

$$= (5.1142 \text{ g}) \cdot \frac{[1 - (1.2 \times 10^{-3} \text{ g/cm}^3)/(8.0 \text{ g/cm}^3)]}{[1 - (1.2 \times 10^{-3} \text{ g/cm}^3)/(1.636 \text{ g/cm}^3)]}$$

$$= 5.117\underline{1}9 = 5.1172 \text{ g}$$

18. The true mass can be found by using Equation 3.6.

$$m_{obj} = m_{display} \cdot \frac{[1 - (d_{air}/d_{ref})]}{[1 - (d_{air}/d_{obj})]}$$

$$= (0.3369 \text{ g}) \cdot \frac{[1 - (1.2 \times 10^{-3} \text{ g/cm}^3)/(8.0 \text{ g/cm}^3)]}{[1 - (1.2 \times 10^{-3} \text{ g/cm}^3)/(5.25 \text{ g/cm}^3)]}$$

$$= 0.3369\underline{3} \text{ g} = 0.3369 \text{ g}$$

(Note: There is no appreciable error due to buoyancy effects in this case.)

Amount of iron in sample =

$$100 \cdot \frac{(0.33693 \text{ g Fe}_2\text{O}_3)(1 \text{ mol Fe}_2\text{O}_3/159.688 \text{ g Fe}_2\text{O}_3)(2 \text{ mol Fe}/1 \text{ mol Fe}_2\text{O}_3)(55.845 \text{ g Fe/mol Fe})}{9.85 \text{ g sample}}$$

$$= 2.39\underline{2} = 2.39\% \text{ (w/w)}$$

19. The size of the buoyancy correction can be determined by using Equation 3.6, in which the object is the stainless steel weight (density, 8.0 g/cm^3) and the reference is now the platinum-iridium cylinder (density, 21.5 g/cm^3). It is assumed in this calculation for the sake of this illustration that the displayed mass of the stainless steel weight is 1.0000 kg.

$$m_{obj} = m_{display} \cdot \frac{[1 - (d_{air}/d_{ref})]}{[1 - (d_{air}/d_{obj})]}$$

$$= (1.0000 \text{ kg}) \cdot \frac{[1 - (1.2 \times 10^{-3} \text{ g/cm}^3)/(8.0 \text{ g/cm}^3)]}{[1 - (1.2 \times 10^{-3} \text{ g/cm}^3)/(21.5 \text{ g/cm}^3)]}$$

$$= 0.9999\underline{1} \text{ kg} = 0.9999 \text{ kg}$$

The absolute size of the buoyancy correction in this case is (0.9999 – 1.0000 kg) = 0.0001 kg. The relative size of this buoyancy correction is given by the ratio of density terms that appears on the right-hand side of Equation 3.6, which is equal in this case to 0.9999.

20. The mass and volume of an object can be related to each other through the object's density. An advantage of using volume is that the volume of a sample is easier to visualize than its mass. Volumes are also more convenient to measure for liquids, where all that is required is to place the liquid in a suitably marked container. A disadvantage of using volumes is that a sample's volume, unlike mass, can vary with temperature and pressure.

21. Volumetric glassware is designed for more accurate volume measurements than can be obtained with an Erlenmeyer flask, test tube, or graduated cylinder. Volumetric glassware is also constructed with a special type of glass that is more resistant to strong acids or bases than ordinary glass and has a smaller change in size and volume with temperature.

22. Borosilicate glass is a type of glass that contains a higher percentage of the boron oxide and a lower percentage of sodium oxide than ordinary soda-lime glass. Borosilicate glass is more resistant to strong acids or bases than ordinary glass and has one-third of the change in size and volume with temperature.

23. Special plastic containers are sometimes used that are made out of teflon, polymethylpentene, or polypropylene. These materials offer good resistance to most chemical reagents and contain only trace amounts of metals. Their main disadvantage is that they melt at much lower temperatures than glass.

24. a) A volumetric flask is used to prepare solutions and to dilute them to a specific volume. The general shape of a volumetric flask consists of a long, upper neck plus a round, flat-bottomed lower region for mixing and holding solutions. The top of the neck contains an opening where a stopper can be placed for mixing the flask's contents. There is also a line etched on the neck, which indicates where the meniscus of the solution should be located when the volume of liquid in the flask is equal to its stated volume.

b) A volumetric pipet is designed to measure and deliver a single, specific volume of liquid to a separate container. Volumetric pipets are used to handle volumes that range from 0.5-100 mL and are employed when volume measurements are needed that are reliable to within a few hundredths of a milliliter. There is a mark around the neck of the pipet that indicates where this calibrated volume occurs.

c) A buret is used to accurately measure and deliver variable amounts of a liquid and consists of a graduated glass tube with an opening at the top for the addition of a liquid and a stopcock at the bottom for the precise delivery of this liquid into another container. Calibrated marks on the side make it possible to determine the change in liquid volume as some of the contents are placed into another vessel.

d) A micropipet often is used to handle small liquid volumes. These devices come with volume capacities often ranging from 0.1-5000 µL. A micropipet uses disposable tips that can easily be replaced between samples or changed to deliver liquids over different volume ranges.

e) A syringe is a volumetric device that consists of a graduated glass or plastic barrel that holds the sample of interest. An open needle allows the sample to enter or leave the barrel, while a plunger is used to push out and dispense this sample. Syringes come with volume capacities of 0.5-500 µL or larger and are used to measure and deliver small-volume samples.

f) A serological pipet is a type of pipet that has many marks on its side, allowing it to measure and deliver a variety of liquid volumes within its calibrated range. It is designed "to deliver/blow out".

g) An Ostwald-Folin pipet is used to measure and deliver a single, specific volume of liquid to a separate container. It is similar to a volumetric pipet, but an Ostwald-Folin pipet is instead used "to deliver/blow out" a liquid.

h) A Mohr pipet is a type of pipet that has many marks on its side, allowing it to measure and deliver a variety of liquid volumes within its calibrated range. Mohr pipets have maximum volumes of 0.1-25 mL and calibrated marks at 0.1, 0.01, or 0.001 mL intervals.

25. Class A and Class B are special designations given to glassware that meets specific requirements for volumetric measurements. For instance, both Class A and Class B volumetric flasks provide much better volume measurements than routine glassware, but Class A flasks have maximum errors that are only half of those for Class B flasks. Class B flasks are less expensive than Class A flasks and are often fine for use in teaching or in general purpose work. Class A flasks are the devices of choice whenever the highest quality volume measurements are desired during solution preparation.

26. "To Contain" (TC) is a marking found on some types of volumetric glassware, indicating that they are designed to contain the amount of liquid for which they have been calibrated. A volumetric flask is a device that is designed to contain the stated volume of liquid.

    "To Deliver" (TD) is a marking that indicates a volumetric device will deliver the correct measured volume when the contents of the device are released, by allowing the device to drain, without any blowing or forced delivery, into the desired receptacle. A volumetric pipet is a device that is designed to deliver that stated amount of liquid.

    "To Deliver/Blow Out" is a marking that indicates a volumetric device will deliver the measured volume only when the last bit of its contents is blown out with a pipet bulb or by some other means of forced delivery. Serological pipets and Ostwald-Folin pipets are both examples of devices that are designed to be used in this matter.

27. Four factors to consider when you are choosing a volumetric device are: 1) the general goal of your volume measurement, 2) the volume or range of volumes to be measured, 3) the degree of reliability needed for the measurement, and 4) the number of measurements that are to be made.

28.
   a) Volumetric pipet

   b) Micropipet

   c) Several devices might be used here, such as a 2.00 mL volumetric pipet, a micropipet, or a Mohr pipet.

   d) Several devices might be used here, such as a micropipet, a Mohr pipet, or a serological pipet.

   e) Syringe

   f) Buret

29. Table 3.5 gives several general rules to follow when using a volumetric device.

30. A parallax error is an error produced when reading a calibrated mark or scale when it is viewed at any non-perpendicular angle, producing readings that may be either too high or low in value. To avoid this error, you should read the meniscus with your eyes at the same level as the meniscus. You can also make the meniscus easier to see by placing a piece of paper, preferably with a dark color, behind the glassware.

31. It is important to calibrate volumetric equipment whenever you receive a new volumetric device or are using volumetric glassware at a temperature other than the one used for its original calibration. This is needed because glassware will expand or contract with a change in temperature or may change slightly in volume over time.

32. a) The displayed mass of the water must first be corrected for buoyancy effects. The density of water and the corresponding buoyancy correction at 30 °C can be found in Table 3.7.

$$m = 49.7380 \text{ g} \cdot (1.001056)$$
$$= 49.79052 \text{ g} = 49.7905 \text{ g}$$

The flask's volume can be determined by dividing this mass by the density of water, giving the true volume of the flask at 30 °C.

$$V = \frac{49.79052 \text{ g}}{0.9956502 \text{ g/cm}^3}$$

$$\therefore V = 50.00804 \text{ cm}^3 = 50.0080 \text{ mL}$$

b) The calculated volume at 20 °C would be 50.00804 mL/1.00010 = 50.0030 mL.

33. a) The density of water and the corresponding buoyancy correction at 25 °C can be found in Table 3.7.

$$\text{Mass of water} = 0.2509 \text{ g} \cdot (1.001055)$$
$$= 0.25116 \text{ g}$$

The pipet's volume can be determined by dividing this mass by the density of water, giving the true volume of the flask at 25 °C.

$$V = \frac{0.25116 \text{ g}}{0.9970479 \text{ g/cm}^3}$$

$$\therefore V = 0.25191 \text{ cm}^3 = 251.9 \text{ μL}$$

b) Mass of water = (0.25191 cm^3)(0.9982071 g/cm^3) = 0.25146 g = 0.2515 g

34. A solution is a uniform mixture of one substance (the solute) within another (the solvent). A mixture of 0.10 M NaOH in water is an example of a solution.

A solute is a substance that is dissolved within another substance (the solvent) to produce a solution. The solutes in a 0.10 M NaOH solution in water are the sodium ions and hydroxide ions that are formed as NaOH dissolves.

A solvent is a substance that dissolves another substance (the solute) to produce a solution. The solvent is water in a 0.10 M NaOH solution.

Concentration is the amount of a substance within a given volume or mass of solution. A 0.10 M NaOH solution has a concentration of 0.10 *molar*.

35. This solution contains 161 mol ethylene glycol and 278 mol water, so water is the solvent and ethylene glycol is the solute.

36. a) Weight-per-weight (w/w) is a measure of chemical content of a mixture, determined by taking the mass of the analyte or substance of interest and dividing this by the total mass of the mixture. This measure is often used for solids, but can also be used for other types of materials.

b) Volume-per-volume (v/v) is a measure of chemical content used to describe a gas or liquid mixture, as determined by taking the volume of the specific gas or liquid of interest and dividing this by the total volume of the mixture.

c) Weight-per-volume (w/v) is a measure of chemical content of a mixture, determined by taking the mass of the analyte or substance of interest and dividing this by the total volume of the mixture. This measure is often used for liquids, but it can also be used for other types of materials.

37. a) 12.5% (v/v) acetonitrile and 25.0% (v/v) methanol

b) 70.3% (w/w) iron, 17.6% (w/w) chromium, 8.03% (w/w) nickel, 2.01% (w/w) manganese, 1.01% (w/w) silicon, 0.803% (w/w) carbon, and 0.257% (w/w) other elements

c) 5.0 g/L dissolved solids

38. a) A gaseous mixture that contains 20 mL oxygen in a total volume of 3.5 L has $100 \cdot (0.020 \text{ L})/(3.5 \text{ L}) = 0.57\%$ (v/v) oxygen.

b) A 2.00 mL sample of blood that contains 12.5 µg of a drug has $(12.5 \text{ µg})/(2.00 \text{ mL}) = 6.25$ µg/mL drug.

c) A 5.00 g sample of coal that contains 4.15 g carbon has $100 \cdot (4.15 \text{ g})/(5.00 \text{ g}) = 83.0\%$ (w/w) carbon.

39. a) Parts-per-thousand is a means of expressing chemical content in which there is one part of the analyte or chemical of interest for every one thousand parts of a mixture. This and each of the following terms are often used to describe weight-per-weight, volume-per-volume and weight-per-volume ratios.

b) Parts-per-million is a means of expressing chemical content in which there is one part of the analyte or chemical of interest for every 10^6 parts of a mixture.

c) Parts-per-billion is a means of expressing chemical content in which there is one part of the analyte or chemical of interest for every 10^9 parts of a mixture.

d) Parts-per-trillion is a means of expressing chemical content in which there is one part of the analyte or chemical of interest for every 10^{12} parts of a mixture.

e) Percent is a means of expressing chemical content in which there is one part of the analyte or chemical of interest for every 100 parts of a mixture.

40. a)

$$\text{Percent Cu}^{2+} \text{ content (w/w)} = 100 \cdot \frac{(0.010 \text{ g Cu}^{2+})}{(2.0 \text{ L solution}) \cdot (1.0 \text{ g/mL solution}) \cdot (10^3 \text{ mL/L})}$$

$$= 0.00050\% \text{ (w/w)}$$

Using 10^3, 10^6, 10^9, and 10^{12} instead of 100 in the above calculation gives the following answers for this same mixture: 0.0050 parts-per-thousand, 5.0 ppm, 5.0×10^3 ppb, or 5.0×10^6 ppt.

b)

$$\text{Percent Be}^{2+} \text{ content (w/w)} = 100 \cdot \frac{(6.2 \times 10^{-3} \text{ g Be}^{2+})}{(750 \text{ mL solution}) \cdot (1.0 \text{ g/mL solution})}$$

$$= 0.00083\% \text{ (w/w)}$$

Using 10^3, 10^6, 10^9, and 10^{12} instead of 100 in the above calculation gives the following answers for this same mixture: 0.0083 parts-per-thousand, 8.3 ppm, 8.3×10^3 ppb, or 8.3×10^6 ppt.

c)

$$\text{Percent NaIO}_3 \text{ content (w/w)} = 100 \cdot \frac{(255 \text{ mg NaIO}_3) \cdot (1 \text{ g}/10^3 \text{ mg})}{(1.5 \text{ L solution}) \cdot (1.0 \text{ g/mL solution}) \cdot (10^3 \text{ mL/L})}$$

$$= 0.017\% \text{ (w/w)}$$

Using 10^3, 10^6, 10^9, and 10^{12} instead of 100 in the above calculation gives the following answers for this same mixture: 0.17 parts-per-thousand, 170 ppm, 1.7×10^5 ppb, or 1.7×10^8 ppt.

41. Maximum allowable mass of atrazine = $(3 \text{ μg/L})(0.240 \text{ L}) = 0.72$ μg in 240 mL.

42. The "molar mass" for any substance is the number of grams that are contained in one mole of that substance. For molecular compounds, the molar mass is commonly referred to as the "molecular weight" (MW). For ionic compounds and elements, the molar mass is also called the "formula weight" or "atomic mass", respectively.

43. Molality is equal to the number of moles of a solute per kilogram of solvent. Because molality is based on a ratio of masses, it is an important unit to use when changes in temperature, and thus changes in volume, are expected during an analysis.

   Molarity is the number of moles (or the number of gram molecular weights) of a substance that is present in each liter of solution. Molarity is more convenient to use than molality in describing the preparation and concentration of liquid solutions.

44. a)
$$\text{Concentration of } H_2SO_4 = \frac{(49.73 \text{ g } H_2SO_4)/(98.079 \text{ g } H_2SO_4/\text{mol})}{0.5000 \text{ L}}$$

$$= 1.014 \, M \, H_2SO_4$$

b)
$$\text{Concentration of } RuCl_3 = \frac{(4.739 \text{ g } RuCl_3)/(207.43 \text{ g } RuCl_3/\text{mol})}{1.000 \text{ L}}$$

$$= 0.02285 \, M \, RuCl_3$$

c)
$$\text{Concentration of } FeCl_3 = \frac{(5.035 \text{ g } FeCl_3)/(162.203 \text{ g } FeCl_3/\text{mol})}{0.25000 \text{ L}}$$

$$= 0.1242 \, M \, FeCl_3$$

d) Concentration of $C_{12}H_{22}O_{11}$ = $\dfrac{(27.74 \text{ g } C_{12}H_{22}O_{11})/(342.2965 \text{ g } C_{12}H_{22}O_{11}/\text{mol})}{0.7500 \text{ L}}$

= 0.1081 M $C_{12}H_{22}O_{11}$

45. Concentration (in molarity) = (5.84 g formaldehyde/0.1040 L)/(30.03 g formaldehyde/mol)

= 1.87 mol/L = 1.87 M

Concentration (in molality) = (5.84 g formaldehyde/0.1000 kg)/(30.03 g formaldehyde/mol)

= 1.94 mol/kg = 1.94 m

46. Concentration (in molarity) = (10.0 g NaOH/100 g soln)(1 mol/40.00 g NaOH)(1109 g soln/L soln)

= 2.77 mol/L (or M)

Concentration (in molality) = [(10.0 g NaOH/90.0 g H_2O)(1000 g/kg)]/(40.00 g NaOH/mol)

= 2.78 mol/kg (or m)

47. If a solute dissociates into ions or produces several forms when it is placed in solution, the number of moles of this solute that is used to calculate a concentration would be based on the number of gram formula weights instead of gram molecular weights. This approach is then used to describe the total amount of solute that was placed into solution, giving a concentration in units of moles per liter of solution and referred to as the formality (F). In this situation, formality also represents the analytical concentration of the solute in the solution.

48. Individual concentrations:

(0.00538 mol acetic acid)/(0.50000 L) = 0.0107$\underline{6}$ M = 0.0108 M acetic acid

(0.00321 mol acetate)/(0.50000 L) = 0.00642 M acetate

Analytical concentration of acetic acid plus acetate = 0.0107$\underline{6}$ M + 0.00642 M

= 0.0171$\underline{8}$ = 0.0172 M (or 0.0172 F)

49. [HA] = (0.000322 mol/0.100 L) = 3.22 × 10^{-3} M

[A$^-$] = (0.000108 mol/0.100 L) = 1.08 × 10^{-3} M

C = 3.22 × 10^{-3} M + 1.08 × 10^{-3} M = 4.30 × 10^{-3} M

50. When you are dealing with a relatively dilute aqueous solution at room temperature, the molality and molarity of the solution will have about the same numerical value because the density of water and of the solution will both be approximately 1.0 g/mL, meaning that 1 kg of water will roughly be equivalent to 1 L of solution. This relationship will not be true if you have a different solution density, as would happen when you have a non-aqueous solvent or a solution with a moderate-to-high concentration of solutes.

51. a) A surface concentration is given in units such as (mol/m^2) and is used when you must describe the amount of a chemical on a surface.

b) Normality describes the amount of a chemical that is available for a specific type of reaction, as is accomplished by using the *equivalents* of a chemical per liter of solution. This type of concentration is sometimes used in methods such as titrations (see Chapters 12, 13, and 15).

52. a) "Certified ACS Grade" refers to a chemical that meets or exceeds specifications set by the American Chemical Society (ACS), as is used in various analytical applications.

b) "Technical Grade" describes chemicals of reasonable purity for cases where no official standards exist for quality or impurity levels. These chemicals are often used in manufacturing or general laboratory applications and are also known as "Laboratory Grade".

c) "USP Grade" describes a reagent chemical that meets or exceeds specifications set by the United States Pharmacopeia (USP), as might be used in a laboratory that analyzes pharmaceutical agents.

d) "Trace Metal Grade" describes chemicals that are prepared to have low levels of trace metals, as are used for the preparation of reagents and samples for trace-metal analysis.

e) "HPLC Grade" refers to chemicals that have been purified and prepared for use in high-performance liquid chromatography (HPLC).

f) "Biotechnology Grade" refers to chemicals and solvents that have been purified and prepared for use in biotechnology; often used in molecular biology, electrophoresis assays, DNA/RNA or peptide sequencing, and synthesis.

53. A primary standard is a pure substance that is stable during storage, can be weighed accurately, and undergoes a known reaction with the solution it is used to characterize. The reagent solution that is characterized by this process is then referred to as a secondary standard.

54. Contaminants found in ordinary water include dissolved inorganic solids, dissolved inorganic gases, dissolved organics, particulates, microorganisms, and pyrogens. Specific examples are listed in Table 3.9.

55. Distillation is a relatively inexpensive method in which the water is heated to boiling, with the steam then being recondensed and used as purified water. This approach is good for obtaining water that is free of particulates, dissolved solids, microorganisms, and pyrogens.

It can also reduce some dissolved organics, but it does not help much in removing dissolved gases. Deionization makes use of cartridges that take the cations or anions in water and exchange these for hydrogen ions and hydroxide ions, which then combine to form more water. This approach is good at removing ions and dissolved gases like carbon dioxide, making it a nice complement to distillation as a means for water treatment.

56. A stock solution is as a reagent solution that is used to make other, less concentrated solutions for use in an assay. An aliquot is a portion of a stock solution or sample that is taken for the preparation of a second, less concentrated solution. A dilution (when used as a noun) is a solution prepared by adding more solvent to a reagent or sample; when used as a verb, the term "dilution" refers to the process of preparing a more dilute solution.

57. a) Weigh 5.844 g NaCl. Transfer to a 100.00 mL volumetric flask, add water, and dissolve. Dilute to the mark of the volumetric flask and mix.

    b) With a pipet, transfer 100.00 mL of the 2.5 M solution into a 250.00 mL volumetric flask, add water to nearly the mark, mix well, and add remaining water to the mark. Mix well.

    c) Use a 25.0 graduated cylinder to measure 10.4 mL of 12 M HCl. Transfer to a 250.00 mL volumetric flask, add water nearly to the mark, mix well, dilute to the mark, and mix well again. If greater accuracy is needed, titrate the HCl with a standard NaOH solution.

58. $[HNO_3]$ = (72 g HNO_3/100 g solution)(1.42 g solution/mL)(10^3 mL/L)/(63.013 g/mol)

    = 16.$\underline{2}$ M

    Volume = (1.00 M)(2.00 L)/ (16.2 mol/L) = 0.12$\underline{3}$ L = 12$\underline{3}$ mL = 120 mL

59. Concentration of creatinine = (8.5 mM)(1.00 mL)/(20.0 mL)

$$= 0.42\underline{5} \text{ m}M = 0.42 \text{ m}M$$

60. [NaOH] = (1.435 M)(25.00 mL/250.00 mL)(25.00 mL/500.00 mL)

$$= 7.175 \times 10^{-3} \, M$$

61. An increase in temperature will change the density and volume of a solution. The change in volume, in turn, will alter the concentration. An increase in volume will lower the concentration, and a decrease in volume will raise the concentration.

62. At 25.0 °C the density of water is 0.99705 g/mL.

Concentration at 4 °C = (50.0 μM)(1.00000)/0.99705) = 50.1 μM

Concentration at 45 °C = (50.0 μM)(0.99025/0.99705) = 49.7 μM

63. Concentration = (25.0 mg/mL)(d_{10}/d_{30}) = (25.0 mg/mL)(0.99970/0.99565) = 25.1 mg/mL

64. Volume of gold crown = (1000 g Au)/(1.93 g Au/cm^3) = 51.8 cm^3 (or mL)

Volume of crown with 80% gold and 20% silver

$$= \{(800 \text{ g Au})/(19.3 \text{ g Au/cm}^3) + (200 \text{ g Ag})/(10.5 \text{ g Ag/cm}^3)\}$$

$$= 41.4\underline{5} \text{ cm}^3 + 19.0\underline{5} \text{ cm}^3 = 60.5 \text{ cm}^3 \text{ (or mL)}$$

65. a) At an elevation of 1000 m and 40° latitude, g would be as follows.

g = 9.80632 − 0.02586 cos(80°) + 0.00003 cos (160°) − 0.00000293(1000)

= 9.80632 − 0.00449 − 0.00003 − 0.00293 = 9.79887

Thus, a 1.000 kg object would weigh 1.000 kg (9.79887/9.80632) = 0.99924 kg.

b) g = 9.80632 − 0.00449 − 0.00003 − 0.00000293(1020) = 9.79881

On the fifth floor, a 1.000 kg object would weigh 1.000 (9.79881/9.80632) = 0.99923 kg.

67. a)
$$\rho_{air} = 0.0012929 \cdot \frac{(273.13 \text{ K})}{(28 + 273.15)} \cdot \frac{(745 - 0.3783 \cdot 11.99)}{760}$$

$$= 0.0011\underline{4} = 0.0011 \text{ g/cm}^3$$

b)
$$m_{obj} = m_{display} \cdot \frac{[1 - (d_{air}/d_{ref})]}{[1 - (d_{air}/d_{obj})]}$$

$$= (10.000 \text{ g}) \cdot \frac{[1 - (1.1 \times 10^{-3} \text{ g/cm}^3)/(8.00 \text{ g/cm}^3)]}{[1 - (1.1 \times 10^{-3} \text{ g/cm}^3)/(0.89 \text{ g/cm}^3)]}$$

$$= 10.011 \text{ g}$$

68. Mercury's density is 13.546 g/cm^3, so a small volume would have a much greater mass, which can be measured with greater precision. The principal disadvantage is that mercury is toxic and must be handled carefully.

69. No such correction is needed for molality because neither the number of moles of solute nor the mass of solvent changes with temperature. The same is true for % (w/w). A similar correction would be necessary for % (v/v) because the volume of both solute and solvent would change with temperature.

70. a) To convert a concentration in g/L to molarity, divide (g/L) by the molar mass of the solute.

b) To convert a concentration in molarity to molality, divide (mol/L) by the density of solution (in g/L) and multiply by (g solution/kg solvent)

c) To convert a concentration for an aqueous solution that is in mg/L to ppm (w/w), multiply (mg/L) by (1 L/1000 mL)/(1.00 g/mL), where 1.00 g/mL is the density of water.

d) To convert a concentration in percent (*w/w*) to g/L, divide percent (w/w) by 100 and multiply by $(1.00 \text{ g/mL})(10^3 \text{ mL/L})$.

71. b) Molecular weight = 16 (12.011) + 19 (1.008) + 35.453 + 2 (14.007) = 274.795 g/mol

    Nominal mass = 16 (12) + 19 (1) + 35 (1) + 2 (14) = 274 g/mol

    Monoisotopic mass = 16 (12.00000) + 19 (1.007825) + 1 (34.96885) + 2 (14.00307)

    = 274.12366 g/mol

    The biggest difference in the monoisotopic mass and chemical molecular weight is in the chlorine. The major isotope, ^{35}Cl is only about 75% of natural chlorine. Chemical molecular weight is a weighted average of all the natural isotopes of each element in the molecule, whereas monoisotopic mass is based on the mass of the major isotope of each element.

CHAPTER 4: MAKING DECISIONS WITH DATA

1. A systematic error is an error that results in a constant bias of the results from the true answer. A random error is defined as an error that results from random variations in experimental data. It is important to consider these types of errors when you are making a measurement in order to properly describe the reliability of results and compare the experimental data to other values.

2. a) An electronic balance that is placed on a counter near an area with frequent air movement and vibrations would be subject to random error.

    b) The use of a 10.00 mL volumetric pipet with a piece of dust lodged in its interior would create a systematic error.

    c) The use of a reference weight density of 8.0 g/cm^3 to make a buoyancy correction when the balance has actually been calibrated with a reference that has a density of 7.8 g/cm^3 would result in a systematic error.

3. Systematic errors can be eliminated through the use of good laboratory practices and a well-kept notebook. The proper training of lab workers and regular maintenance of equipment can also help you avoid such errors. Random errors are present in all measurements and are never totally eliminated. However, it is possible to reduce the size of random errors through the proper design of experiments and the correct choice of methods for comparing data.

4. Accuracy is defined as the degree of agreement between an experimental result and its true value. Precision is the variation in individual results obtained under identical conditions. Precision is determined by random errors, but the accuracy of individual results is affected by both systematic and random errors.

5. The absolute error of an experimental result is found by calculating the difference between this result and its true value. The relative error is found by taking the difference between the true and measured values and dividing this by the true answer. Both types of values are used to evaluate the accuracy of an experimental result.

6. Result, absolute error, relative error, and % relative error:

   117 μmol/mL, 5 μmol/mL, 0.04$\underline{1}$ = 0.04, and 4%

   119 μmol/mL, 3 μmol/mL, 0.02$\underline{5}$ = 0.02, and 2%

   111 μmol/mL, 11 μmol/mL, 0.09$\underline{0}$ = 0.09, and 9%

   115 μmol/mL, 7 μmol/mL, 0.05$\underline{7}$ = 0.06, and 6%

   120 μmol/mL, 2 μmol/mL, 0.01$\underline{6}$ = 0.02, and 2%

7. Result, absolute error, relative error, and % relative error:

   25.12 mL, 0.10 mL, 0.004$\underline{0}$ = 0.004, and 0.4%

   25.15, 0.13 mL, 0.005$\underline{2}$ = 0.005, and 0.5%

   25.13 mL, 0.11 mL, 0.004$\underline{4}$ = 0.004, and 0.4%

8. a) Result, absolute error, relative error, and % relative error:

   24.96 mL, -0.03 mL, -0.001$\underline{2}$ = -0.001, and -0.1%

   25.01 mL, 0.02 mL, 0.0008$\underline{0}$ = 0.0008, and 0.08%

   25.04 mL, 0.05 mL, 0.002$\underline{0}$ = 0.002, and 0.2%

   b) Average and standard deviation for results by the first student (Problem 7): 25.13 mL (versus true answer of 25.02 mL), ± 0.02 mL (1 standard deviation)

   Average and standard deviation for results by the second student: 25.00 mL (versus true answer of 24.99 mL), ± 0.04 mL (1 standard deviation)

Student 2 had an average result that was more accurate, but Student 1 had slightly better precision.

9. The arithmetic mean, or average, is the sum of a series of values divided by the total number of values in that set. This value is used to provide a single representative number for a group of observations.

10. a) The mean for 87, 89, 90 and 91 mg is 89.$\underline{2}$ mg = 89 mg

   b) The mean for 278.8, 279.0 and 279.1 nm is 278.9$\underline{7}$ nm = 279.0 nm

   c) The mean for 135.2, 134.8, 135.4, 134.2 and 135.0 s is 134.9$\underline{2}$ s = 134.9 s

11. a) The mean for 2.32×10^5, 2.19×10^5, 2.15×10^5, 2.11×10^5, and 2.27×10^5 g/mol is $2.20\underline{8} \times 10^5$ g/mol = 2.21×10^5 g/mol.

   b) Absolute error based on the mean result = $2.20\underline{8} \times 10^5$ g/mol - 2.21×10^5 g/mol = $-0.00\underline{2} \times 10^5$ g/mol = 0.00×10^5 g/mol

   Individual result, absolute error:

   2.32×10^5 g/mol, absolute error = 0.11×10^5 g/mol

   2.19×10^5 g/mol, absolute error = -0.02×10^5 g/mol

   2.15×10^5 g/mol, absolute error = -0.06×10^5 g/mol

   2.11×10^5 g/mol, absolute error = -0.10×10^5 g/mol

   2.27×10^5 g/mol, absolute error = 0.06×10^5 g/mol

   The absolute error of the mean result is much smaller than the absolute errors obtained for the individual measured values.

   c) The mean result and absolute error of mean when using first two results: 2.26×10^5 g/mol and absolute error = 0.04×10^5 g/mol

The mean result and absolute error for mean when using the first three results: 2.22×10^5 g/mol and absolute error = 0.01×10^5 g/mol

The mean result and absolute error for mean when using the first four results: 2.19×10^5 g/mol, absolute error = -0.02×10^5 g/mol

The mean result approaches a consistent value as more points are added to the data set and the absolute error tends to approach a fixed value that is less affected by random errors in the individual results.

12. The experimental mean is the average calculated for a group of experimental values. Random errors cause the experimental mean to be only an approximation of the true mean, especially when we are dealing with small sets of numbers. It is only when we have a large group of numbers (which causes random errors to cancel out) that the experimental mean approaches the true mean in its value.

13. a) The range is the difference between the largest and smallest value in a set of data, as used to describe the variation within a group of numbers.

    b) The standard deviation (s) is a measure of the variation within a set of values, which is calculated by using the following relationship,

    $$s = [\Sigma(x_i - \overline{x})^2/(n-1)]^{1/2}$$

    where n is the number of values within the set, $\overline{x}$ is the arithmetic mean for this set of values, and x_i represents any individual value.

    c) The relative standard deviation is a measure of precision that is equal to the standard deviation for a number divided by that number's value.

      d) The variance is the square of the standard deviation for a population of results.

14. One advantage of using a range to describe a data set is that it is easy to calculate. However, the value of the range tends to increase as your number of points increases. The standard deviation is slightly more complex to calculate, but this term approaches a constant value as the number of data points increases.

15. a) The range and standard deviation for 32.8, 34.1, 33.7, 32.9, and 33.5 min:

    Range, 34.1 min − 32.8 min = 1.3 min; standard deviation, 0.55 min

    b) The range and standard deviation for 0.21, 0.24, 0.19, and 0.23% (w/w):

    Range, 0.24% − 0.19% = 0.05%; standard deviation, 0.022%

    c) The range and standard deviation for 0.01005, 0.01018, and 0.00998 M:

    Range, 0.01018 M − 0.00998 M = 0.0002 M; standard deviation, 0.00010 M

    In each of these examples, the range is larger than the standard deviation.

16. a) The relative standard deviation (%) and variance for 32.8, 34.1, 33.7, 32.9, and 33.5 min:

    Relative standard deviation, 1.6%; variance = 0.30 min^2

    b) The relative standard deviation (%) and variance for 0.21, 0.24, 0.19, and 0.23% (w/w):

    Relative standard deviation, 10.%; variance = 0.00049% (w/w)^2

    c) The relative standard deviation (%) and variance for 0.01005, 0.01018, and 0.00998 M:

    Relative standard deviation, 1.0%; variance = 1.0×10^{-8} M^2

    The size of the range, standard deviation, relative standard deviation, or variance each decrease as the precision increases for a set of data.

17. Mean, 99.8 mg/dL; range, 112 mg/dL − 88 mg/dL = 24 mg/dL; standard deviation, 5.6 mg/dL; relative standard deviation, 5.7%

18. a) Mean, 16.56%; range, 16.70% − 16.30% = 0.40%; standard deviation, 0.15%; relative standard deviation, 0.90%

    b) Range and standard deviation for first two values:

    Range, 16.54% − 16.30% = 0.24%; standard deviation, 0.17%

    Range and standard deviation for first three values:

    Range, 16.64% − 16.30% = 0.34%; standard deviation, 0.17%

    Range and standard deviation for first four values:

    Range, 16.67% − 16.30% = 0.37%; standard deviation, 0.17%

    Range and standard deviation for first five values:

    Range, 16.70% − 16.30% = 0.40%; standard deviation, 0.16%

    c) The range increases as more data points are added to the set, while the standard deviation quickly approaches a fairly consistent value.

19. Random errors make the experimental standard deviation only an estimate of the true standard deviation. The experimental standard deviation approaches the true standard deviation as the number of data points in a set is increased.

20. The "degrees of freedom" is a statistical quantity that describes the number of values in a data set that are required to define the overall population of results. There are n degrees of freedom when you are determining a mean, and $n - 1$ when you are calculating the standard deviation for a single set of results.

21. "Error propagation" is a term used to describe the way in which errors are carried through a calculation or a series of experimental steps. Error propagation can be used to predict the precision of an experimental value and to help identify the major contributions to random errors in an analysis.

22. The equations to use for error propagation in Parts (a)–(d) are listed in Table 4.1.

23. a) $0.121 (\pm 0.009) + 2.93 (\pm 0.04) = 3.05 (\pm 0.04)$

   b) $9.23 (\pm 0.03) + 4.21 (\pm 0.02) - 3.26 (\pm 0.06) = 10.18 (\pm 0.07)$

   c) $91.3 (\pm 1.0) \cdot 40.3 (\pm 0.2) \cdot 21.1 (\pm 0.2) = 77{,}600 (\pm 1{,}200)$

   d) $185 (\pm 1) \cdot 3.2 (\pm 0.3)/9.1 (\pm 0.1) = 65 (\pm 6)$

   e) $1 + 6.4 (\pm 0.2) + 36.2 (\pm 0.3) = 43.6 (\pm 0.4)$

   f) $7.53 (\pm 0.1) \times 10^5 \cdot 2.9 (\pm 0.1) \cdot \pi = 6.9 (\pm 0.3) \times 10^6$

24. The equations to use for error propagation in Parts (a)–(c) are listed in Table 4.1.

25. a) $\log[2.0164 (\pm 0.0008)] = 0.30458 (\pm 0.00017)$

   b) $\text{antilog}[-3.22 (\pm 0.02)] = 6.0 (\pm 0.3) \times 10^{-4}$

   c) $10^{2.384 (\pm 0.011)} = 242 (\pm 6)$

   d) $2 \cdot \log[7.05 (\pm 0.02)] = 1.696 (\pm 0.002)$

   e) $e^{-1.68 (\pm 0.02)} = 0.186 (\pm 0.004)$

   f) $\ln[12.6 (\pm 0.2)] = 2.53 (\pm 0.02)$

26. Results for Problem 23 when using only the number of significant figures as a guide:

   a) $0.121 + 2.93 = 3.05$

   b) $9.23 + 4.21 - 3.26 = 10.18$

c) $91.3 \cdot 40.3 \cdot 21.1 = 77{,}600$

d) $185 \cdot 3.2/9.1 = 65$

e) $1 + 6.4 + 36.2 = 43.6$ (using 1 as an integer with an infinite number of significant figures)

f) $7.53 \times 10^5 \cdot 2.9 \cdot \pi = 6.9 \times 10^6$

Results for Problem 25 when using only the number of significant figures as a guide:

a) $\log[2.0164] = 0.30458$

b) $\text{antilog}[-3.22] = 6.0 \times 10^{-4}$

c) $10^{2.384} = 242$

d) $2 \cdot \log[7.05] = 1.696$

e) $e^{-1.68} = 0.19$

f) $\ln[12.6] = 2.534$

Differences can occur when the implied precision in the last significant digit is significantly higher or lower than the actual precision that was given in Problems 23 and 25.

27. It is possible to generate formulas like those in Table 4.1 for various combinations of calculations. However, another approach is to separate these operations into a series of steps that consist of only addition or subtraction, multiplication or division, and so on. During each step, the equations in Table 4.1 can be used to examine the random errors that are carried through that particular operation. The result can then be used in the next step until the final answer is obtained.

28. a) $[4.97\,(\pm 0.05) - 1.86\,(\pm 0.01)]/21.1\,(\pm 0.2) = 0.147\,(\pm 0.003)$

b) $[1.89\,(\pm 0.03) \times 10^3 + 2.30\,(\pm 0.06) \times 10^3 - 9.8\,(\pm 0.2) \times 10^2] \cdot 5.80\,(\pm 0.06) \times 10^{-2}$

$= 186\,(\pm 4)$

c) $6 \cdot 1.00794 (\pm 0.00007) + 2 \cdot 12.0107 (\pm 0.0008) = 30.069 (\pm 0.002)$

d) $10^{[7.40 (\pm 0.02) - 3.12 (\pm 0.01)]} = 1.91 (\pm 0.01) \times 10^4$

e) $2 \cdot \log[0.107 (\pm 0.002)/0.158 (\pm 0.003)] = -0.34 (\pm 0.03)$

f) $(3/2) \cdot e^{-[2.85 (\pm 0.03)/0.103 (\pm 0.002)]} = 1.4 (\pm 0.9) \times 10^{-12}$

29. Results for Problem 28 when using only the number of significant figures as a guide:

   a) $[4.97 - 1.86]/21.1 = 0.147$

   b) $[1.89 \times 10^3 + 2.30 \times 10^3 - 9.8 \times 10^2] \cdot 5.80 \times 10^{-2} = 186$

   c) $6 \cdot 1.00794 + 2 \cdot 12.0107 = 30.0690$

   d) $10^{[7.40 - 3.12]} = 1.9 \times 10^4$

   e) $2 \cdot \log[0.107/0.158] = -0.339$

   f) $(3/2) \cdot e^{-[2.85/0.103]} = 1.4 \times 10^{-12}$

   Differences can occur when the implied precision in the last significant digit is significantly higher or lower than the actual precision that was given in Problems 23 and 25.

30. $C_{NH3} = C_{HCl} (V_{HCl})/V_{NH3}$

   $= [0.0783 (\pm 0.0003) \, M \, \text{HCl}][24.37 (\pm 0.04) \, \text{mL HCl}]/[30.00 (\pm 0.03) \, \text{mL ammonia}]$

   $= 0.0636 (\pm 0.0003) \, M$

31. $n = (P \, V)/(R \, T)$

   $= [0.492 (\pm 0.009) \, \text{J/L} \cdot 1.000 (\pm 0.005) \, \text{L}]/[8.31451 (\pm 0.00007) \, \text{J/(mol} \cdot \text{K)} \cdot 298.2 (\pm 0.5) \, \text{K}]$

   $= 1.98\underline{4} (0.03\underline{8}) \times 10^{-4} \, \text{mol}$

   $MW = m/n$

   $= [0.0500 (\pm 0.0002) \, \text{g}]/[1.98\underline{4} (\pm 0.03\underline{8}) \times 10^{-4} \, \text{mol}]$

   $= 252 (\pm 5) \, \text{g/mol}$

32. For a series of numbers that are multiplied or divided, the relative contribution of each to the overall uncertainty in the final result can be determined by comparing the relative uncertainties for all these terms. In this case, the largest contribution to the precision of the final molar mass was the uncertainty in the number used for the pressure. The smallest contribution to the final precision was the uncertainty in the value for R. Based on this information, a more precise measurement of the pressure would be the best way to improve the precision of the final result.

33. $\Delta G = -a\, b\, F\, (E°_B - E°_A)$

    $= -(1)\,(1)\,[96485.309\,(\pm 0.029)\ \text{J/(V} \cdot \text{mol)}][1.44\,(\pm 0.02)\ \text{V} - 0.771\,(\pm 0.005)\ \text{V}]$

    $= -6.5\,(\pm 0.2) \times 10^4\ \text{J/mol}$

34. A normal distribution is a common mathematical model used for describing random errors and various types of experimental measurements. This type of distribution is also known as a bell-shaped curve or Gaussian curve and is described by the following equation,

    $$y = \frac{1}{\sigma\sqrt{2\pi}} \cdot e^{-\frac{1}{2}[(x-\mu)^2/(\sigma^2)]}$$

    where x is the experimental value, y is the probability of measuring this value, μ is the true average of the data set, and σ is the true standard deviation of the data set. A normal distribution is useful for comparing experimental results and in estimating the reliability of measurements.

35. The average of the data set (μ) gives the central point for the distribution, and the standard deviation of the data set (σ), which describes the width of the distribution. Figure 4.6 shows the effects of changing each of these parameters on the shape of a normal distribution.

36. A range of the mean plus-or-minus one standard deviations corresponds to a relative area of 0.6826, or 68.26%, of the results in a normal distribution (or roughly two-thirds of all its values). A range of the mean plus-or-minus two standard deviations represents a relative area of 0.9544, or 95.44%, of all results in a normal distribution (approximately 95%, or 19 out of 20 values). These ranges are both frequently used to describe data sets and to compare experimental results.

37. The standard deviation of the mean is a factor that is used to describe the precision of an experimental mean, as calculated by dividing the standard deviation of the entire data set by the square root of n, the number of values in the data set.

38. $\bar{x} = 10.4$ mg/dL

    $s = 0.9\underline{1}$ mg/dL

    $s_{\bar{x}} = (0.9\underline{1}\text{ mg/dL})/(9)^{1/2} = 0.3$ mg/dL

39. For five replicates:

    $s_{\bar{x}} = 5\% = s/\sqrt{5}$ or $s = (5\%)(\sqrt{5}) = 11.18$

    Number of replicates needed for a standard deviation of the mean = 1%:

    $s_{\bar{x}} = 1\% = (11.18)/\sqrt{n}$

    $n = [(1\%)/(11.18)]^2 = 125$

40. A confidence interval (C.I.) is a range that is used to express the degree of certainty that can be placed in that result. The range of values in a confidence interval that follows the experimental number is called the "confidence limit".

41. The Student's t-value (t) is a mathematical factor used to correct for the greater uncertainty

that is present when working with small data sets than with larger sets. This factor is also commonly used in calculating both confidence intervals and in the comparison of the experimental values. A confidence level is the degree of probability, or certainty, that is desired in stating that two values are either the same or that a measured result falls within a specified range of values.

42. Mean = 3.0 µg/L

    90% C.I. = 3.0 ± 0.1 µg/L ($n = 5$)

43. a) Mean, 0.0992$\underline{4}$ M = 0.0992 M; standard deviation, 0.0007$\underline{6}$ M = 0.0008 M; standard deviation of the mean, 0.0003$\underline{4}$ M = 0.0003 M ($n = 5$)

    b) 90% C.I. = 0.0992 ± (2.13)(0.0003$\underline{4}$) M

    = 0.0992 ± 0.0007 M ($n = 5$)

    95% C.I. = 0.0992 ± (2.78)(0.0003$\underline{4}$) M

    = 0.0992 ± 0.0009 M ($n = 5$)

    99% C.I. = 0.0992 ± (4.60)(0.0003$\underline{4}$) M

    = 0.0992 ± 0.0016 M ($n = 5$)

    As the desired confidence level increases, the range of values covered in the confidence interval will also increase.

44. As there are more values in a data set, the size of the standard deviation of the mean will decrease, which also cases the confidence interval for the mean to become narrower.

45. Selection of the confidence level can have a big impact on the range of results we obtain for the confidence interval. Analytical chemists typically use a confidence level of 95% as a compromise between having a relatively narrow confidence interval and one that is still

broad enough to have a good chance of including the true result for a measurement.

46. a) The model is the component of a statistical technique that represents the result, method, or predicted behavior to which an experimental value is being compared.

    b) The hypothesis is a statement that describes the initial expected outcome of an experiment.

    c) The confidence level is the degree of probability, or certainty, that is desired in stating that two values are either the same or that a measured result falls within a specified range of values.

    d) The test statistic is a factor that is calculated in a statistical test to see whether two values agree or differ in their values at the desired confidence level.

47. The critical value in a statistical test is the maximum or minimum cutoff value with which a test statistic is compared to see if the model and experimental value of interest can be said to be different at the selected confidence level and given degrees of freedom.

48. The Student's *t*-test is a statistical method that uses the Student's *t*-value to compare an experimental value with a known value or compare two experimental values. For the comparison of an experimental mean with a known value, this is done by calculating the following ratio,

$$t = |\bar{x} - \mu|/s_{\bar{x}}$$

where $\bar{x}$ is the experimental mean being tested, μ is the true value for the same data set, and $s_{\bar{x}}$ is the standard deviation of the mean. If the calculated value of *t* is greater than the critical Student's *t*-value that is expected at the desired confidence level, then x and μ can be said to represent statistically different values. Two experimental means ($\bar{x}_1$ and $\bar{x}_2$) can also be compared in the Student's t-test by using a pooled standard deviation $s_{\bar{x}pool}$ to give a Student's t-value for this comparison.

$$t = |\bar{x}_1 - \bar{x}_2|/s_{\bar{x}pool}$$

A small value for this ratio would indicate our two results are close together and probably represent the same number.

49. This problem can be solved by using a Student's t-test to compare the mean value to the known true value.

    Mean for data set $(\bar{x}) = 1.52802$

    Standard deviation for data set $(s) = 0.0020$

    $s_{\bar{x}} = s/(n)^{1/2} = (0.0020)/(8)^{1/2} = 0.0007\underline{1}$

    $t = |\bar{x} - \mu|/s_{\bar{x}}$

    $= |1.52802 - 1.52810|/(0.0007\underline{1}) = 0.1\underline{1}$

    $t\ (0.11) < t_c\ (2.36)$, results are the same at the 95% confidence level

50. This problem can be solved by using a Student's t-test to compare the mean value to the known true value.

    Mean for data set $(\bar{x}) = 0.0489\underline{5}$

    Standard deviation for data set $(s) = 0.001\underline{68}$

    $s_{\bar{x}} = s/(n)^{1/2} = (0.001\underline{68})/(4)^{1/2} = 0.0008\underline{4}$

    $t = |\bar{x} - \mu|/s_{\bar{x}}$

    $= |0.0489\underline{5} - 0.0532|/(0.0008\underline{4}) = 5.\underline{1}$

    $t\ (5.\underline{1}) > t_c\ (2.35)$, results are not the same at the 90% confidence level

51. This problem can be solved by using a Student's t-test to compare the mean values for the two sets of results.

Mean for data set 1 ($\overline{x}_1$) = 0.0275$\underline{8}$ ($n_1 = 4$)

Standard deviation for data set 1 (s_1) = 0.0005$\underline{6}$

Mean for data set 2 ($\overline{x}_2$) = 0.0269$\underline{0}$ ($n_2 = 4$)

Standard deviation for data set 2 (s_2) = 0.0004$\underline{7}$

$$s_{pool} = \sqrt{\frac{(n_1 - 1)\cdot(s_1)^2 + (n_2 - 1)\cdot(s_2)^2}{(n_1 + n_2 - 2)}}$$

$$s_{pool} = \sqrt{\frac{(4 - 1)\cdot(0.0005\underline{6})^2 + (4 - 1)\cdot(0.0004\underline{7})^2}{(4 + 4 - 2)}}$$

$$= 0.0005\underline{2}$$

$$s_{\overline{x}pool} = \frac{s_{pool}}{\sqrt{(n_1\cdot n_2)/(n_1 + n_2)}}$$

$$s_{\overline{x}pool} = \frac{0.0005\underline{2}}{\sqrt{(4\cdot 4)/(4 + 4)}}$$

$$= 0.0003\underline{7}$$

$$t = |\overline{x}_1 - \overline{x}_2|/s_{\overline{x}pool}$$

$$= |0.0275\underline{8} - 0.0269\underline{0}|/(0.0003\underline{7})$$

$$= 1.\underline{8}$$

t (1.$\underline{8}$) < t_c (1.94), results are the same at the 90% confidence level

52. This problem can be solved by using a Student's t-test to compare the mean values for the two sets of results.

Mean for data set 1 ($\bar{x}_1$) = 34.4$\underline{7}$ ($n_1 = 3$)

Standard deviation for data set 1 (s_1) = 0.6$\underline{5}$

Mean for data set 2 ($\bar{x}_2$) = 32.7$\underline{3}$ ($n_2 = 3$)

Standard deviation for data set 2 (s_2) = 0.3$\underline{5}$

$$s_{pool} = \sqrt{\frac{(n_1 - 1) \cdot (s_1)^2 + (n_2 - 1) \cdot (s_2)^2}{(n_1 + n_2 - 2)}}$$

$$s_{pool} = \sqrt{\frac{(3 - 1) \cdot (0.6\underline{5})^2 + (3 - 1) \cdot (0.3\underline{5})^2}{(3 + 3 - 2)}}$$

$$= 0.5\underline{2}$$

$$s_{\bar{x}pool} = \frac{s_{pool}}{\sqrt{(n_1 \cdot n_2)/(n_1 + n_2)}}$$

$$s_{\bar{x}pool} = \frac{0.5\underline{2}}{\sqrt{(3 \cdot 3)/(3 + 3)}}$$

$$= 0.4\underline{2}$$

$$t = |\bar{x}_1 - \bar{x}_2|/s_{\bar{x}pool}$$

$$= |34.4\underline{7} - 32.7\underline{3}|/(0.4\underline{2})$$

$$= 4.\underline{1}$$

$t\,(4.\underline{1}) > t_c\,(2.78)$, results are not the same at the 95% confidence level

53. This problem can be solved by using a Student's t-test to compare the mean values for the two sets of results.

Mean for data set 1 ($\bar{x}_1$) = 24.4$\underline{9}$ (n_1 = 8)

Standard deviation for data set 1 (s_1) = 0.1$\underline{9}$

Mean for data set 2 ($\bar{x}_2$) = 23.6$\underline{8}$ (n_2 = 6)

Standard deviation for data set 2 (s_2) = 0.2$\underline{3}$

$$s_{pool} = \sqrt{\frac{(n_1 - 1) \cdot (s_1)^2 + (n_2 - 1) \cdot (s_2)^2}{(n_1 + n_2 - 2)}}$$

$$s_{pool} = \sqrt{\frac{(8 - 1) \cdot (0.1\underline{9})^2 + (6 - 1) \cdot (0.2\underline{3})^2}{(8 + 6 - 2)}}$$

$$= 0.2\underline{1}$$

$$s_{\bar{x}pool} = \frac{s_{pool}}{\sqrt{(n_1 \cdot n_2)/(n_1 + n_2)}}$$

$$s_{\bar{x}pool} = \frac{0.2\underline{1}}{\sqrt{(8 \cdot 6)/(8 + 6)}}$$

$$= 0.1\underline{2}$$

$$t = |\bar{x}_1 - \bar{x}_2|/s_{\bar{x}pool}$$

$$= |24.4\underline{9} - 23.6\underline{8}|/(0.1\underline{2})$$

$$= 6.\underline{8}$$

t (6.$\underline{8}$) > t_c (2.18), results are not the same at the 95% confidence level

54. A paired Student's *t*-test is a statistical test that is used to compare two sets of the identical samples that are analyzed by the different methods with similar standard deviations for their results. This test is carried out by calculating the average difference between each set of results and comparing this value to the standard deviation for the average difference in the individual results. The result is then compared to the critical value expected at the given confidence level to see if the results can be considered statistically to be the same.

55. This comparison can be made by using a paired Student's *t*-test.

	Mean Results (μM)		Difference in Results
Sample No.	Method 1	Method 2	$d_i = x_{Method\ 1} - x_{Method\ 2}$
1	12.8	12.1	0.7
2	35.2	34.7	0.5
3	25.1	25.2	-0.1
4	15.8	15.9	-0.1
5	31.2	29.8	1.4

$$\bar{d} = (\sum d_i)/n$$

$$= 0.4\underline{8}$$

$$s_d = 0.6\underline{3}$$

$$s_{\bar{d}} = s_d/(n)^{1/2} = (0.6\underline{3})/(5)^{1/2} = 0.2\underline{8}$$

$$t = |\bar{d}|/s_{\bar{d}}$$

$$= |0.4\underline{8}|/(0.2\underline{8}) = 1.\underline{7}$$

$t(1.\underline{7}) < t_c (2.78)$, results are the same at the 95% confidence level

56. This comparison can be made by using a paired Student's *t*-test.

Moisture Content (% w/w)

Sample No.	Lab No. 1	Lab No. 2	Difference (d_i)
1	10.2	11.0	-0.8
2	15.3	16.4	-1.1
3	21.0	21.5	-0.5
4	13.3	14.1	-0.8
5	27.8	29.3	-1.5
6	30.5	32.2	-1.7

$$\bar{d} = (\sum d_i)/n$$

$$= -1.0\underline{7}$$

$$s_d = 0.4\underline{6}$$

$$s_{\bar{d}} = s_d/(n)^{1/2} = (0.4\underline{6})/(6)^{1/2} = 0.1\underline{9}$$

$$t = |\bar{d}|/s_{\bar{d}}$$

$$= |-1.0\underline{7}|/(0.1\underline{9}) = 5.\underline{6}$$

$t\,(5.\underline{6}) > t_c\,(2.57)$, results are not the same at the 95% confidence level

57. The *F*-test is a statistical test for comparing the standard deviations or variances for experimental results as calculated by using the ratio $F = s_2^2/s_1^2$, where s_2 and s_1 are the two standard deviations being compared (with the larger of these two terms always appearing in the top of this ratio).

58. This problem can be solved by using a Student's *t*-test to compare the mean values for the two sets of results and the *F*-test to compare the precisions for the two sets of results.

Mean for data set 1 ($\bar{x}_1$) = 74.0$\underline{2}$ (n_1 = 5)

Standard deviation for data set 1 (s_1) = 2.5$\underline{0}$

Mean for data set 2 ($\bar{x}_2$) = 75.6$\underline{4}$ (n_2 = 5)

Standard deviation for data set 2 (s_2) = 0.8$\underline{6}$

Student's t-test (to compare the mean results):

$$s_{pool} = \frac{(n_1 - 1) \cdot (s_1)^2 + (n_2 - 1) \cdot (s_2)^2}{(n_1 + n_2 - 2)}$$

$$s_{pool} = \sqrt{\frac{(5 - 1) \cdot (2.5\underline{0})^2 + (5 - 1) \cdot (0.8\underline{6})^2}{(5 + 5 - 2)}}$$

$$= 1.8\underline{7}$$

$$s_{\bar{x}pool} = \frac{s_{pool}}{\sqrt{(n_1 \cdot n_2)/(n_1 + n_2)}}$$

$$s_{\bar{x}pool} = \frac{1.8\underline{7}}{\sqrt{(5 \cdot 5)/(5 + 5)}}$$

$$= 1.1\underline{8}$$

$$t = |\bar{x}_1 - \bar{x}_2|/s_{\bar{x}pool}$$

$$= |74.0\underline{2} - 75.6\underline{4}|/(1.1\underline{8})$$

$$= 1.3\underline{7}$$

t (1.3$\underline{7}$) < t_c (2.31), the mean results are the same at the 95% confidence level

F-test (to compare the precision of the two sets of data):

$$F = (s_{high})^2/(s_{low})^2$$
$$= (2.5\underline{0})^2/(0.8\underline{6})^2$$
$$= 8.\underline{5}$$

$F(8.\underline{5}) > F_c (6.39)$, the precision of the results is not the same at the 95% confidence level (Note: In this situation, a different procedure may be required to reevaluate whether the mean results are equivalent.)

59. This problem can be solved by using the *F*-test.

    Standard deviation for data set 1 (s_1) = 0.1$\underline{9}$

    Standard deviation for data set 2 (s_2) = 0.2$\underline{3}$

    $$F = (s_{high})^2/(s_{low})^2$$
    $$= (0.2\underline{3})^2/(0.1\underline{9})^2$$
    $$= 1.\underline{5}$$

    $F(1.\underline{5}) < F_c (3.97)$, the precision is the same at the 95% confidence level

60. This problem can be solved by using the *F*-test.

    Standard deviation for first method (s_1) = 0.0020 × 10^{-6} (n = 5)

    Standard deviation for second method (s_2) = 0.011 × 10^{-6} (n = 5)

    $$F = (s_{high})^2/(s_{low})^2$$
    $$= (0.011 \times 10^{-6})^2/(0.0020 \times 10^{-6})^2$$
    $$= 30.\underline{3}$$

    $F(30.\underline{3}) > F_c (6.39)$, the precision is not the same at the 95% confidence level

61. An outlier is a data point that does not fit the general trend observed for a group of results that are obtained under supposedly identical conditions. If you think a data point is an outlier, you should first recheck the results to make sure that all data were correctly recorded and that no errors were made in calculating or plotting the results. You should also determine if there were any differences in the experimental conditions for the outlier versus other data. Another item you can use in outlier detection is the precision of your analysis method. The final approach is to use a statistical method for outlier detection, such as the Q-test and T_n-test.

62. The Q-test takes the absolute difference between a suspected outlier's value and its nearest point, with this difference then being compared to the total range of values in the data set. If the difference between our suspect and its nearest neighbor is greater than a certain critical fraction of the total range at the selected confidence level, the suspected value can be said to represent a true outlier. An appealing feature of the Q-test is it involves only simple calculations. However, this test does not make use of all the available information because it employs only three values from the data set. A second problem is that this test is often misused and is based on the assumption that only a single possible outlier is in the data set.

63. The T_n-test compares the overall mean of the data set to the suspected outlier, with the absolute value of this difference being divided by the standard deviation for the entire data set. This resulting ratio is then compared to a critical value to determine if an outlier is really present at the selected confidence level. The main advantage of this test is that it utilizes all the numbers in a data set, making it more robust than the Q-test for outlier detection. However, the T_n-test still assumes there is only one outlier present, making it invalid to use

this method to reject more than one point at a time.

64. The most likely outlier in this data set is at 0.0250 M because this value has the greatest difference in value from its nearest neighbor.

$Q = |x_o - x_n|/(x_{high} - x_{low})$

$= |0.0250 - 0.0225|/(0.0250 - 0.0208) = 0.6\underline{0}$

Q (0.6$\underline{0}$) < Q^* (0.710), the outlier cannot be rejected by the Q-test at the 95% confidence level

65. The most likely outlier in this data set is at 16.30, because this value has the greatest difference in value from its nearest neighbor.

Examining this possible outlier by the T_n-test:

Mean result = 16.57%　　　　　　Standard deviation of all results = 0.16$\underline{2}$%

$T_n = |x_o - \overline{x}|/s$

$= |16.30 - 16.57|/0.16\underline{2} = 1.6\underline{7}$

T_n (1.6$\underline{7}$) < T_n^* (1.672), the outlier cannot be rejected by the T_n-test at the 90% confidence level

Examining this possible outlier by the Q-test:

$Q = |x_o - x_n|/(x_{high} - x_{low})$

$= |16.30 - 16.54|/(16.70 - 16.30) = 0.6\underline{0}$

Q (0.6$\underline{0}$) < Q^* (0.642), the outlier cannot be rejected by the Q-test at the 90% confidence level

66. a) The Q-test assumes that, at most, only that one outlier is present in a data set.

b) In the T_n-test the suspected outlier should be used to calculate both the mean and standard deviation values because it is has not yet been rejected prior to conducting this test.

c) The confidence level should be selected before conducting the Q-test to avoid creating a personal bias in the final result that is obtained.

67. Linear regression involves taking a set of (x,y) values and fitting these to a linear equation. Linear regression is useful in analytical chemistry because it shows how you can determine the best-fit line for a set of data. This procedure is often required when you are preparing a calibration curve or are comparing experimental results to a predicted response.

68. The slope is the change in the dependent variable (y) versus the change in the independent variable (x) for a graph. For a linear relationship the slope has a constant value and is represented by the term m in the following equation: $y = mx + b$. In this same equation, b is the intercept, or the point on the y-axis that is intersected by the best-fit line.

69.

x = Conc.	y = Absorbance	$x_i y_i$	x_i^2
0.0	0.002	0.0	0
10.0	0.062	0.62	100
20.0	0.125	2.50	400
30.0	0.198	5.94	900
40.0	0.244	9.76	1600
$\sum x_i = 100.0$	$\sum y_i = 0.631$	$\sum x_i y_i = 18.82$	$\sum x_i^2 = 3000$

$$m = \frac{[n(\sum x_i y_i) - (\sum x_i)(\sum y_i)]}{[n(\sum x_i^2) - (\sum x_i)^2]}$$

$$= \frac{[5(18.82) - (100.0)(0.631)]}{[5(3000) - (100.0)^2]}$$

$$= 0.0062$$

$$b = \frac{[(\sum y_i)(\sum x_i^2) - (\sum x_i y_i)(\sum x_i)]}{[n(\sum x_i^2) - (\sum x_i)^2]}$$

$$= \frac{[(0.631)(3000) - (18.82)(100.0)]}{[5(3000) - (100.0)^2]}$$

$$= 0.0022$$

70.

$x = pCa.$	$y = E\ (mV)$	$x_i y_i$	x_i^2
5.00	−53.8	−269.0	25.0
4.00	−27.7	−110.8	16.0
3.00	+2.7	8.1	9.00
2.00	+31.9	63.8	4.00
1.00	+65.1	65.1	1.00
$\sum x_i = 15.00$	$\sum y_i = 18.2$	$\sum x_i y_i = -242.8$	$\sum x_i^2 = 55.0$

$$m = \frac{[n(\sum x_i y_i) - (\sum x_i)(\sum y_i)]}{[n(\sum x_i^2) - (\sum x_i)^2]}$$

$$= \frac{[5(-242.8) - (15.00)(18.2)]}{[5(55.0) - (15.00)^2]}$$

$$= -29.7\underline{4} = -29.7$$

$$b = \frac{[(\sum y_i)(\sum x_i^2) - (\sum x_i y_i)(\sum x_i)]}{[n(\sum x_i^2) - (\sum x_i)^2]}$$

$$= \frac{[(18.2)(55.0) - (-242.8)(15.00)]}{[5(55.0) - (15.0)^2]}$$

$$= 92.8\underline{6} = 92.9$$

71. These standard deviations can be found by using the values from Problem 70 along with the following equations.

Standard deviation of all y values (s_Y):
$$s_Y = [\sum(y_i - m x_i - b)^2/(n-2)]^{1/2}$$
$$= [(13.86)/(5-2)]^{1/2}$$
$$= 2.1\underline{5}$$

Standard deviation of the slope (s_m):
$$s_m = (n/[n(\sum x_i^2) - (\sum x_i)^2])^{1/2} (s_Y)$$
$$= (5/[5(55.0) - (15.00)^2])^{1/2} (2.1\underline{5})$$
$$= 0.7\underline{1} = 0.7$$

Standard deviation of the intercept (s_b):
$$s_b = ((\sum x_i^2)/[n(\sum x_i^2) - (\sum x_i)^2])^{1/2} (s_Y)$$
$$= ((55.0)/[5(55.0) - (15.00)^2])^{1/2} (2.1\underline{5})$$
$$= 2.2\underline{55} = 2.3$$

When the appropriate x_i and y_i values are placed into the first equation, along with the known value of n, and calculated values of m and b from Problem 71, a value for s_Y of $2.1\underline{5}$ is obtained.

This value can then be used with the other two equations and the values of $\sum x_i^2$ and $\sum x_i$ from Problem 71, giving $s_m = 0.7$ and $s_b = 2.3$.

The difference between the calculated slope of -29.7 and reference value of -29.6 is much smaller than 1 standard deviation ($s_m = 0.7$), so these results are probably the same. (Note: A more formal comparison of these values could also be made by using a Student's t-test with $n - 2$ degrees of freedom.)

72. A correlation coefficient (r) is a calculated parameter that is used to judge the goodness of fit between a best-fit line and experimental data, where the numerical value of r is always

between -1 and 1. A value of 1 or -1 represents perfect agreement between the data and the best-fit line, and a value of 0 represents no correlation between the best-fit line and data. A closely related term is the coefficient of determination (r^2), which is equal to the square of the correlation coefficient and has a value between 0 and 1.

73. The correlation coefficient for the best-fit line in Problem 69 can be found by using the following equation and the various calculated values that are given in the answer to Problem 69, along with the addition of the calculated value for $\sum y_i^2$ (or 0.118$\underline{2}$).

$$S_{xx} = (\sum x_i^2) - [(\sum x_i)^2/n]$$
$$= (3000) - [(100.0)^2/5] = 1000$$

$$S_{yy} = (\sum y_i^2) - [(\sum y_i)^2/n]$$
$$= (0.118\underline{2}) - [(0.631)^2/5] = 0.038\underline{57}$$

$$S_{xy} = (\sum x_i y_i) - [(\sum x_i)(\sum y_i)/n]$$
$$= (18.82) - [(100.0)(0.631)/5] = 6.2$$

Correlation coefficient: $r = S_{xy}/(S_{xx} S_{yy})^{1/2}$
$$= (6.2)/[(1000)(0.038\underline{57})]^{1/2} = 0.998$$

74. The correlation coefficient for the best-fit line in Problem 70 can be found by using the following equation and the various calculated values that are given in the answer to Problem 70, along with the addition of the calculated value for $\sum y_i^2$ (or 8924.$\underline{6}$).

$$S_{xx} = (\sum x_i^2) - [(\sum x_i)^2/n]$$
$$= (55.5) - [(15.00)^2/5] = 10.5$$

$$s_{yy} = (\sum y_i^2) - [(\sum y_i)^2/n]$$

$$= (8924.\underline{6}) - [(18.2)^2/5] = 8858.\underline{4}$$

$$s_{xy} = (\sum x_i y_i) - [(\sum x_i)(\sum y_i)/n]$$

$$= (-242.8) - [(15.00)(18.2)/5] = -297.4$$

Correlation coefficient: $r = s_{xy}/(s_{xx} s_{yy})^{1/2}$

$$= (-297.4)/[(10.5)(8858.\underline{4})]^{1/2} = -0.975$$

According to Table 14.9, the absolute value of this correlation coefficient for five data points is large enough to represent at least a 99% probability that this line is a true relationship.

75. A residual plot is a graph for visually examining data to look for any deviations from a best-fit line. This type of graph is prepared by plotting the difference, or residual, in the experimental and calculated y values ($y_i - y_{calc}$) versus the corresponding x values.

76. A residual plot for this data shows only random variations when comparing the experimental values with those predicted by the best-fit line. Thus, this plot supports the assumption made earlier that this data set does follow a linear relationship.

77. a) Using the full range of values gives the following best-fit parameters: $m = 0.0153$, $b = 0.145$, and $r = 0.9885$. Using the data for just 10–50 mg/L gives the following best-fit parameters: $m = 0.01984$, $b = 0.008$, and $r = 0.9993$. Based on the correlation coefficients, the data from just 10–50 mg/L would appear to give a slightly better fit to the data.

b) The residual plots also support the conclusion made in Part (a) that the 10–50 mg/L data gave a better fit to a straight line than the full range of data.

78. The difference between the cutoff value of 40.0 nmol/L and the average value of 22.6 nmol/L is 17.4, which represents 17.4/6.0 = 2.9 standard deviations. By extrapolating between the results in Table 4.2 for 2.8 and 3.0 standard deviations, we can estimate that a fraction equal to 0.4980 of a normal distribution will be present between this point and the mean. The area outside of this range would be 0.5000 – 0.4980 = 0.0020, which means that 100 (0.0020) = 0.2% of all normal males will have a testosterone value at or above 40.0 nmol/L.

79. If the average of the population is 2.8 g/mL and the standard deviation of the data set is 0.8 g/mL, a range of 2.6–3.0 g/mL would correspond to (0.2 g/mL)/(0.8 g/mL) = 0.25 standard deviations above or below the mean. By extrapolating the results in Table 4.2 for 0.2 to 0.4 standard deviations, this result would represent an area of 2·(0.098) = 0.196, or about 20% of the total area under the normal distribution. Thus, there is about a 20% chance that a new sample from the same area will also fall in this range.

CHAPTER 5: CHARACTERIZATION AND SELECTION OF ANALYTICAL METHODS

1. Table 5.1 lists several sample-related questions you should ask during this process, dealing with issues like the chemical or physical nature of the analyte and matrix and whether the analyte is in an appropriate form for study. Related questions concern the amount of analyte that is to be measured and whether other interfering chemicals are present.

2. Method-related questions you should ask before conducting a chemical analysis include: "Is this technique capable of working with the desired sample?" and "Does this method have sufficient accuracy and precision for examining the analyte?" In addition, you should think about whether the method is capable of working with the expected range of analyte concentrations and if it is sufficiently fast, selective, and stable for your application. The analysis method you select should have properties that allow it to adequately address the need identified in each of these questions.

3. a) These measurements will have to be made on a short time frame, provide results within the expected range of oxygen or carbon dioxide levels in blood, and be free of any major interferences from other sample components.

    b) A simple and relatively fast method would be preferable here to keep costs to a minimum. This method must also be able to work with soil samples (or related samples prepared from soil) and to measure the acidity at the levels that are expected in the soil samples.

    c) The method will have to be relatively selective for the pesticide and must be able to measure this analyte in the concentration range that must be monitored in the vegetables.

    d) This method will have to provide structural information and be compatible with the amount of drug that is available for analysis.

4. This scientist will need a method that can selectively monitor mercury and that can measure this analyte at the levels expected in the fish.

5. Figures of merit are the properties used to characterize an analytical method. Examples include accuracy, precision, and the usable assay range.

6. Method validation is the process of characterizing an analytical technique and proving it will fulfill its intended purpose.

7. One way the accuracy of a technique can be determined is to examine reference samples with known properties or amounts of analyte, such as a certified reference material. The absolute or relative error is then calculated by comparing the measured result for the reference sample with the known value. A spiked recovery study can be used to evaluate the accuracy of an analytical method when you are working with an analyte or sample for which no good reference materials are available. Another way accuracy can be evaluated is by taking a group of several samples and measuring the amount of analyte in each by using both your method and a second, more established technique (i.e., a correlation study). You can then compare the results of these two methods by using a paired Student's t-test or a correlation chart.

8. A certified reference material (CRM) is a material that has documented values for its chemical content or physical properties. This type of material can be used to determine the accuracy of an analytical method.

9. The drug testing laboratory can obtain a CRM that contains cyclosporine. The laboratory can then measure this analyte in the reference material. The accuracy of the technique can then be determined by using this information to calculate the absolute or relative error in the measured result for the reference sample versus the known content of this sample.

10. Absolute error = 0.874 − 0.855 = 0.019% (w/w)

    Relative error = 100 · (0.874 − 0.855)/(0.855) = 2.2%

11. A spiked recovery study is a study conducted by taking a typical sample and "spiking" a portion of this sample with a known amount of the analyte. The amounts of analyte in the original and spiked samples are then measured, with the difference in these values then being compared to the amount added to the spiked sample. This method can be used to examine the accuracy of an analytical method.

12. Percent recovery = 100 · (22.7 µg/L)/[{23.0 × 10^{-9} g)/(1.0 mL)}{10^6 µg/g}{10^3 mL/L}]

    = 98.7%

13. This problem can be solved by using a Student's t-test to compare the mean value of the measured recoveries to the expected recovery of 100.00% when no systematic errors are present.

    Mean recovery for data set ($\bar{x}$) = 98.7$\underline{2}$

    Standard deviation for data set (s) = 1.6$\underline{2}$

    $s_{\bar{x}} = s/(n)^{1/2} = (1.6\underline{2})/(5)^{1/2} = 0.7\underline{2}$

$t = |\bar{x} - \mu|/s_{\bar{x}}$

$= |98.7\underline{2} - 100.00|/(0.7\underline{2}) = 1.\underline{77}$

$t\ (1.\underline{77}) < t_c\ (2.78)$, there are no significant differences in the results at the 95% confidence level, so no significant systematic errors were detected

14. A paired Student's *t*-test can be used during method validation to compare the results obtained by one method versus a reference method for a set of identical samples that are examined by each method. The individual samples can have a wide range of concentrations, but the precision of each method should be comparable throughout the range of values that are being examined.

15. This comparison can be made by using a paired Student's *t*-test (see Chapter 4).

Measured Protein Content (% w/w)

Sample No.	Old Analysis Method	New Analysis Method	Difference (d_i)
1	10.1	10.5	-0.4
2	25.8	26.9	-1.1
3	15.3	15.6	-0.3
4	30.0	32.9	-2.9
5	6.2	7.8	-1.6
6	21.7	23.1	-1.4
7	34.2	36.9	-2.7
8	40.1	44.8	-4.7

$\bar{d} = (\sum d_i)/n$

$= -1.8\underline{8}$

$s_d = 1.4\underline{8}$

$s_{\bar{d}} = s_d/(n)^{1/2} = (1.4\underline{8})/(8)^{1/2} = 0.5\underline{2}$

$$t = |\bar{d}|/s_{\bar{d}} = |-1.88|/(0.52) = 3.\underline{6}$$

$t\,(3.\underline{6}) > t_c\,(2.36)$, results are not the same at the 95% confidence

16. A correlation chart is a graph in which the results of your new method are plotted versus those of the reference technique. If the two techniques have identical or similar results, their correlation chart should give a linear relationship with a slope near 1 and an intercept near 0. If there is a systematic error in one of the methods, this will cause a deviation from the expected behavior.

17. The correlation chart is shown below. This result indicates that there is a good correlation between the two methods.

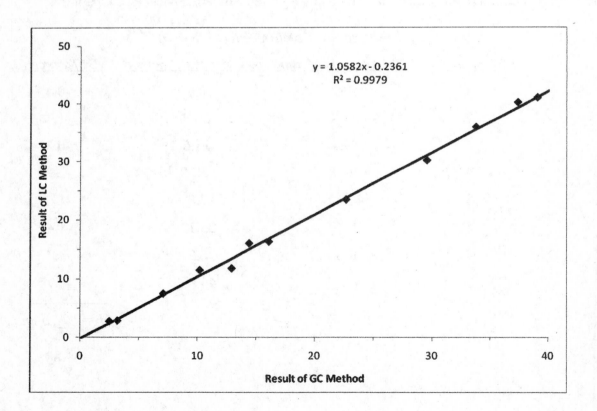

18. The correlation chart is shown below. This result indicates that there is a good correlation between the two methods, but also that the old method tends to give a slightly lower result than that of the old method.

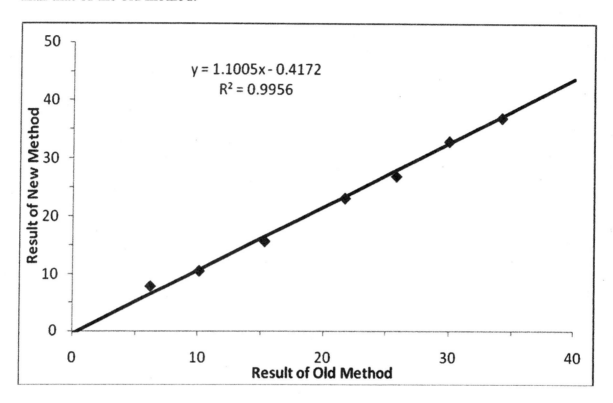

19. The precision of an analytical technique can be examined by analyzing several portions of the same reference sample. The absolute or relative standard deviation of this group of results is then calculated and used as an index of the method's precision. This calculation should be done with several samples that contain different amounts of the analyte because the precision of a method can show a large change with analyte content. This data can also be used to prepare a precision plot to give a visual representation of how the precision of a method changes as the measured property or amount of an analyte is varied.

20. Standard deviation = 0.6 mg/mL, relative standard deviation = 4%

21. Standard deviation = 1.3$\underline{5}$ ng/mL = 1.4 ng/mL, relative standard deviation = 3.7$\underline{7}$% = 3.8%

    The absolute standard deviation is larger than that measured for the lower concentration of CEA, but the relative standard deviation is similar.

22. A precision plot is a diagram that gives a visual representation of how the precision of a method changes with the concentration or measured property of an analyte. The type of graph is prepared by plotting the standard deviation or relative standard deviation measured for a set of results (each set representing a different concentration or property of the analyte) on the *y*-axis and the concentration or desired property of the analyte on the *x*-axis.

23. The precision plots based on the standard deviation and absolute standard deviation are shown on the following page. The absolute standard deviation has a minimum concentration of 30 mg/L, while the relative standard deviation has a broader minimum range between 20 and 50 mg/L.

24. The RSD from 15–35 mg/L ranges from 2.5–8%; the concentration range that gives an RSD at or below 5% is approximately 18–47 mg/L.

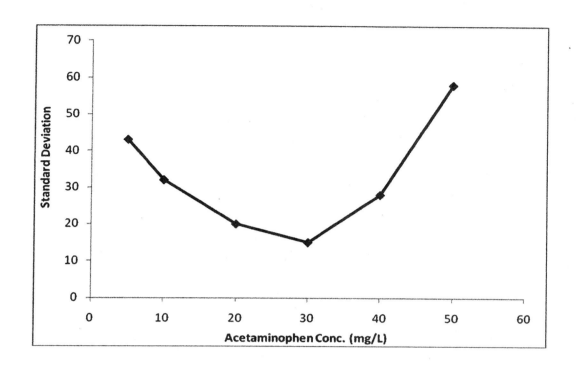

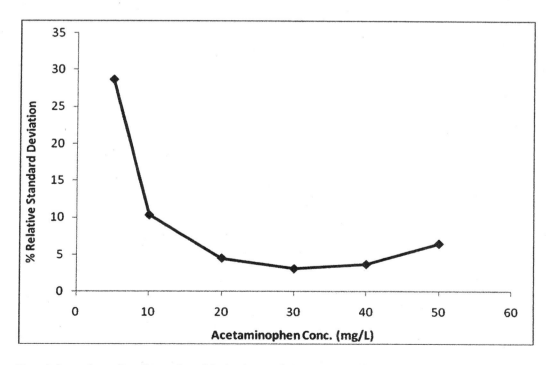

Precision plots for Question 23 (see previous page)

25. a) Within-run precision is a measure of the variation of an analysis for a single sample during one analysis session, or "run".

b) Day-to-day precision is the variation in results that is obtained over several days when using a particular analytical method.

c) Within-day precision is the variation obtained for an analytical method during a single day, usually over the course of several runs.

d) Interoperator precision is the variation obtained with a single analytical method and sample but by different analysts.

e) Interlaboratory precision is the variation obtained with a single analytical method and sample but by different laboratories.

26. a) Interoperator precision   b) Day-to-day precision   c) Within-run precision

27. Signal-to-noise ratio (S/N) is the ratio of the "noise" (or random variation in the blank signal) versus the "signal" (or the net change in response measured between the blank and the sample) for a sample. This ratio is often used to help determine the limit of detection for a chemical analysis method.

28. The limit of detection (LOD) is the lowest or highest amount of analyte that can be detected by a method. The lower LOD is often determined by using the amount of analyte that gives a particular signal-to-noise ratio in an analytical method (such as S/N = 2, 3, or 3.3).

The limit of quantitation (LOQ) is the smallest or largest amount of analyte that can be measured within a given range of accuracy and/or precision. This is sometimes also determined by using the amount of analyte that gives a particular signal-to-noise ratio in an analytical method (such as S/N = 10).

29. The best-fit line can be used to give the signal of the method at a given amount of analyte and the standard deviation of the intercept can be used to give the noise, thus providing a signal-to-noise ratio. The amount of analyte that is needed to provide a signal-to-noise ratio with a given minimum value (e.g., S/N = 3 for a lower limit of detection or S/N = 10 for a limit of quantitation) is then calculated from this information and used to find the signal and amount of analyte that results in this condition being met.

30. $LOD = 3\,(s_b/m) = 0.02$ ng/mL or $LOD = 3.3\,(s_b/m) = 0.03$ ng/mL
    $LOQ = 10\,(s_b/m) = 0.08$ ng/mL

31. $LOD = 3\,(s_b/m) = 0.3\underline{6}$ µg/mL or $LOD = 3.3\,(s_b/m) = 0.3\underline{9}$ µg/mL
    $LOQ = 10\,(s_b/m) = 1.1\underline{9}$ µg/mL

32. The linear range is the portion of a method's range that gives a linear dependence between its response and analyte's concentration or measured property. The dynamic range is the largest possible range that can be used by an analytical technique, extending from the lower limit of detection to the upper limit of detection.

33. The dynamic range extends from the lower limit of detection (insufficient data to estimate in this case) up to at least 20.00 mg/L. The linear range for data that fits within 10% of a best-fit line extends from the lower limit of detection up to approximately 12.50 mg/L.

34. Based on the best-fit line from the data from 0.00 through the 7.50 mg/L iron, the calculated concentration for a sample with an absorbance of 0.258 would be 11.5 mg/L. The result expected based on extrapolation of the standards that gives absorbances in this same range would be 12.8 mg/L. Thus, the use of the best-fit line gives an underestimation of the true concentration.

    The same effect happens for a sample with an absorbance of 0.340. The concentration based on the best-fit line at the lower concentrations would be 15.3 mg/L. The result based on extrapolation for samples with similar absorbance values is 18.6 mg/L. The best-fit line to the lower concentrations again gives an underestimation of the actual value. This situation occurs because the calibration curve deviates from a linear response at concentrations above approximately 10 mg/L.

35. Sensitivity is a measure of how the response of an analytical method changes as the amount of analyte or a sample property is varied. The limit of detection (LOD) is the lowest or highest amount of analyte that can be detected by a method. The lower LOD refers to the smallest amount of analyte that can be examined by a method, while the sensitivity refers to the smallest *change* in the amount of analyte that can be detected.

36. The student is actually reporting a lower LOD and not the sensitivity of the method.

37. The calibration sensitivity is 0.0252 for the linear range in this plot, as given by the slope of the best-fit line for this range.

38. The calibration sensitivity (or slope) at 7.50 mg/L is about 0.022 L/mg. This value is consistent throughout the linear range but decrease at concentrations above the linear range.

39. Selectivity refers to the ability of an analytical method to detect and discriminate between an analyte and other chemicals in a sample; also often referred to as "specificity". An analytical technique is said to be "specific" or "selective" if it responds to a single analyte or a small group of chemicals. Procedures that detect a wide range of compounds are often referred to as general or universal methods.

40. An interference plot is a graph in which the apparent amount of analyte that is measured by a method is plotted versus the amount of a second substance that has been added to the sample. Such a graph makes it possible to determine whether the added agent will create any problems in the assay.

41. A series of samples that contain a fixed amount of sodium should be prepared that contain different amounts of potassium. The apparent amount of sodium in the sample should then be determined for each sample and plotted versus the amount of potassium that was added to the sample. The resulting graph is an interference plot for the method, which describes the effect of potassium on the measured result for sodium.

42. A series of water samples that contain a fixed amount of vitamins, but different amounts of aromatic compounds, should be prepared. The apparent amount of the vitamins in the sample should then be determined for each sample and plotted versus the amount of aromatic compounds that were added to the sample. The resulting graph is an interference plot for the method.

43. The selectivity coefficient is a measure of an analytical method's specificity, as determined by comparing the relative signal that is produced by one substance versus another in that

method. The value can be used to characterize the specificity of an analytical method for an analyte versus other possible sample components.

44. a) The overall analysis time is the time needed for all steps in the preparation and analysis of a sample. This can be measured by keeping track of the total time that is needed for each of these steps.

b) Sample throughput refers to the number of samples that can be processed by an analytical method in a given period of time. This can be determined by calculating the number of samples that can be processed in a particular period of time by the method.

c) Robustness refers to the ability of an analytical technique to provide a consistent response when small variations are made in its experimental conditions. This can be examined by measuring the response of the method as known changes are made in the experimental conditions.

45. Total analysis time = 10 min + 10 min + 15 min

$$= 35 \text{ min}$$

46. When 10 samples can be extracted and derivatized simultaneously, the time for all 10 samples is 10 min + 10 min in the first two steps. The time required for the GC separation is still 15 min for each sample, or 150 min for all 10 samples. Thus, total time to process all 10 samples is (10 min + 10 min + 150 min) = 170 min or 170 min/10 = 17 min per sample. If two columns are used instead of one, the overall time to process a batch of 10 samples would be 20 min for the sample pretreatment and derivatization steps and 10 (35 min/2) = 175 min for the GC separation, for a total of 190 min for the 10 samples. The calculated

sample throughput would then be 190 min/10 samples = 19 min per sample. However, the time required to completely process each individual sample is still 10 min + 10 min + 15 min = 35 min.

47. Quality control refers to the process of monitoring the routine performance of a method. When you are working in an analytical laboratory, you not only need to know how to use various methods, but you must be able to determine when an assay is not working properly. Quality control allows you to obtain such information.

48. The first part of a quality control program is the control material (or "control"), which is a substance you will analyze periodically by an analytical method to determine whether the procedure is working in a consistent manner. The second part of a quality control program is a control chart, which is a graph that uses the results obtained with a control material to follow the performance of an analytical method over time. Error assessment is the third part of a quality control program and is the process of identifying all sources of errors that can occur in an analytical method and in determining how to correct these errors.

49. A "Levey-Jennings chart" (or "individual chart"), is a control chart used in situations where just one measurement per analyte is performed on a sample. This type of chart can be used as a tool in a quality control program for monitoring the performance of an analytical method.

50. The mean for this set of control results is 9.1$\underline{0}$, and the standard deviation is 0.2$\underline{4}$. The Levey-Jennings control chart that would be made based on this information is shown on the following page. (Note: This chart also includes data points from Question 51.)

Figure for Questions 50 and 51 (previous page and next page)

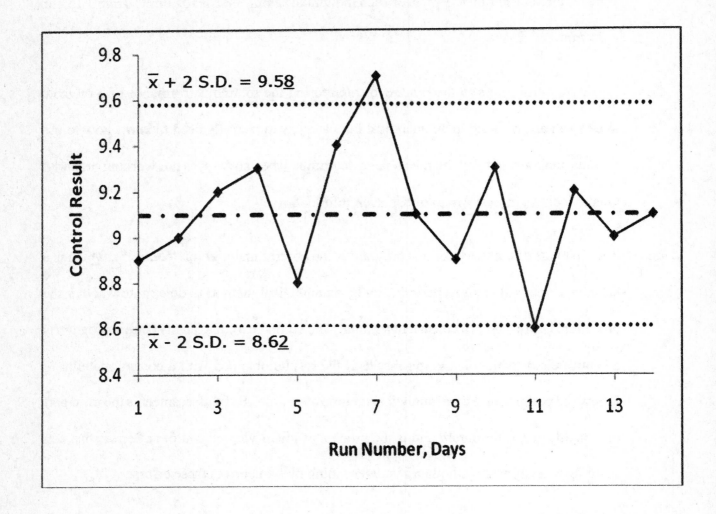

51. Using an allowable range of $\bar{x} \pm 2s$ would correspond to values of 8.6$\underline{2}$ to 9.5$\underline{8}$ in the control chart. The values that are outside of this range are those obtained on Days 7 and 11. The control samples on these days would need to be reanalyzed and possible corrective action would need to be taken before the assay could be used to examine samples and standards.

52. A Shewhart chart is a control chart used in industrial labs and in situations where enough time and material are available to perform several measurements on every sample. This differs from a Levey-Jennings chart, which instead just uses a single result for each new analysis of the control material. With the extra information that is available through replicate measurements, it is possible to determine a mean result, range, and standard deviation for each sample and to prepare separate Shewhart control charts for these various parameters. These separate plots make it easier to identify and differentiate between any systematic or random errors that might later appear in the assay.

53. A sampling plan is the specific approach used when acquiring a sample. The particular approach that will be used for a sampling plan will depend on the physical state of the material, the sample size that is needed, and the extent to which the original material is uniform in composition.

54. A sampling error is an error created when only a portion of a non-uniform substance is used for an analysis. The size of the sampling error will depend on the number of "particles" in a collected sample and the fraction of the total material that is being analyzed.

55. A random sample is a sample that is acquired by arbitrarily taking part of a material and testing this portion to obtain information on the material's entire contents. This type of

sample can be used for a homogeneous material, however, it is more difficult to obtain for non-uniform substances. A representative sample is a sample that reflects the overall composition and properties of the original material, which requires a well-defined sampling plan. Obtaining a representative sample involves more work than random sampling but provides a better picture of a material, especially when it has a heterogeneous composition.

56. The following are some reasons why pretreatment may be used on a sample before an analysis: a) the analyte is not present in a matrix suitable for the chosen method; b) there are substances in the sample that interfere with measurement of the analyte; c) the analyte needs to be placed into a chemical form that can be studied by the desired method; or d) the analyte needs to be placed at a level that is suitable for detection.

57. Discuss this question with your instructor or laboratory supervisor to determine what types of sample preparation procedures are performed in your laboratory.

58. A portion of the analyte might be lost during this process. To help correct for this loss of analyte, a fixed amount of another chemical can be added to the sample to act as an internal standard.

   A second problem that can occur during sample pretreatment is when new substances are introduced to the matrix that may interfere with analyte detection. It is possible to detect this problem by performing interference studies or by testing the specificity of the method to determine which substances may cause difficulties in the technique.

   The use of sample pretreatment will also result in the loss of some information about the sample. This situation may be acceptable in many types of analysis. However, if the analyst is unaware of such effects, these effects can lead to an improper interpretation of the final data.

CHAPTER 6: CHEMICAL ACTIVITY AND CHEMICAL EQUILIBRIUM

1. Acid–base, complexometric, precipitation, and redox reactions are commonly used in analytical chemistry, although nearly any other type of reaction can also be utilized somehow in chemical analysis. Applications of these reactions in chemical analysis include their use in titrations, gravimetry, colorimetric assays, electrochemical methods of analysis, and chromatography.

2. One must consider the products of the reaction, any side reactions, the equilibrium constant, and the speed of the reaction.

3. Chemical thermodynamics is the field of chemistry that is concerned with the changes in energy that take place during chemical reactions or phase transitions, and the overall extent to which such processes can occur. Chemical kinetics is the field of chemistry that is concerned with the rates of chemical processes. Chemical thermodynamics can be used to describe the possible extent of a chemical reaction and chemical kinetics describes the speed of the reaction.

4. Chemical potential (μ) is a measure of the amount of energy per mole that is available through the reaction of a chemical when it is present in a particular state. Chemical activity (a) is a measure of the difference in chemical potential between a substance in some given state (μ) versus a standard reference state for that same substance (μ°), where $a = e^{(\mu - \mu^\circ)/(RT)}$, R being the ideal gas law constant and T the absolute temperature.

5. A standard state is a reference form for a chemical that is said to have an amount of inherent energy per mole that is equal to the standard chemical potential. The standard state for a

chemical has an activity of exactly one and is based on either the pure form of a chemical or a particular concentration of this chemical.

6. a) Oxygen gas: The standard state is a pressure of one bar (formerly 760 torr).

   b) NaCl crystals: The standard state is the pure substance.

   c) Methanol as a solvent: The standard state is the pure liquid.

   d) An aqueous solution of NaCl: The standard state is a 1.00 M solution.

   e) An aqueous solution of methanol: The standard state is a 1.00 M solution.

   f) Helium: The standard state is this gas at a pressure of one bar.

7. If we are working with a dilute solution of NaCl, the chemical potential will generally be much less than it is for the pure form of the same compound, giving an activity that is somewhere between 0 (the lowest possible value) and 1.0. There are also some situations in which a chemical can have more energy than its standard state. For instance, the chemical potential for sodium ions and chloride ions in a highly concentrated solution of NaCl (> 1.0 M) will give a value for activity that is greater than 1.0.

8. If the concentration of solute in the solution is low, the water concentration is nearly the same as in pure water, its standard state. The activity would be assigned a value of 1.0 in this situation. If a very high concentration of solute is present, the water becomes diluted by the solute and the activity of the water will become less than 1.0.

9. An activity coefficient (γ) is a term used to relate that activity (a) of a chemical to its concentration (c) in a solution. The activity coefficient is a unitless term that is used in the relationship $a = \gamma_c (c/c^\circ)$ where c° is the concentration of the substance of interest in a

standard state. (Note: The term $c°$ is generally equal to 1 molar when molarity is the concentration unit used for the chemical of interest, so this is not always written as part of the relationship between a, γ_c, and c.)

10. Activity coefficient for each ion = $(0.0734)/(0.100\ m/1.00\ m) = 0.734$

11. Activity coefficient for each ion = $0.155/(0.20\ m/1.00\ m) = 0.77$

12. Usually, we wish to measure the concentration of a species but equilibrium constants describe the activity of these species. It is therefore the activity that dictates the completeness of a reaction. Concentration is easier to use when we are preparing a solution, but activity is what is actually measured by many analytical techniques. Failure to consider this issue can create a systematic error in the analysis.

13. $[Na^+] = [ClO_4^-] = (0.20\ g/L)/(122.44\ g/mol) = 0.00163\ M$
    Ionic strength = $0.00163\ M$

14. For 0.100 M NaCl, Ionic strength = $½\ [(1^2)(0.100\ M) + (1^2)(0.100\ M)] = 0.100\ M$
    For 0.100 M Na$_2$SO$_4$, Ionic strength = $½\ [(1^2)(0.200\ M) + (-2)^2(0.100\ M)] = 0.300\ M$
    In this example, the ionic strength for the sodium sulfate solution is higher than that for the sodium chloride solution because of the -2 charge on sulfate and the corresponding fact that 1 mole of sodium sulfate produces 2 moles of sodium ions.

15. a) Ionic strength = $½\ \{(0.10\ M)(1^2) + (0.10\ M)(-1)^2 + (0.20\ M)(1^2) + (0.20\ M)(-1)^2\}$
    $= 0.30\ M$

    b) Ionic strength = $½\ \{(0.050\ M)(2^2) + (0.050\ M)(-2)^2 + (0.100\ M)(1^2) + (0.050\ M)(-2)^2\}$

$$= 0.35\ M$$

c) $[KBr] = [(0.050\ g)/(119\ g/mol)]/(1.00\ L) = 4.20 \times 10^{-4}\ M$

$[KCl] = [(0.100\ g)/(74.55\ g/mol)]/(1.00\ L) = 1.34 \times 10^{-3}\ M$

Ionic strength $= \frac{1}{2}\ \{(0.00176\ M)(1^2) + (4.20 \times 10^{-4}\ M)(1^2) + (1.34 \times 10^{-3}\ M)(1^2)\}$

$$= 0.00176\ M$$

16. Any experimental measure of activity gives a weighted average for the activities of both negatively and positively charged ions in solution, which is represented by a term known as the mean activity coefficient ($\gamma_\pm$). The following equation shows how single ion activity coefficients for individual types of ions (γ_A and γ_B for ions A^{n+} and B^{m-}) are related to the mean activity coefficient for a strong electrolyte A_mB_n.

$$(\gamma_\pm)^{m+n} = (\gamma_A)^m \cdot (\gamma_B)^n$$

17. The mean activity coefficient for this system is described as follows.

$$(\gamma_\pm)^{1+1} = (0.85)^2 = (\gamma_{Na+})^1 \cdot (\gamma_{Cl-})^1$$

or $\quad 0.85 = \gamma_{Na+} = \gamma_{Cl-}$

This equation indicates that the individual activity coefficients for Na^+ and Cl^- are both approximately 0.85 in this case.

18. The mean activity coefficient is described as follows, where $\gamma_{K+} = \gamma_{SO_4^{2-}} = 0.75$.

$$(\gamma_\pm)^{1+2} = (0.75)^3 = (\gamma_{K+})^2 \cdot (\gamma_{SO_4^{2-}})^1$$

19. The extended Debye-Hückel equation relates ionic strength (I) and ion charge (z) to activity coefficient (γ), as given by the following relationship where A, a, and B are constants for a given type of ion and for the given solvent and temperature.

$$\log(\gamma) = \frac{-A z^2 \sqrt{I}}{1 + a B \sqrt{I}}$$

20. The form of the equation to use in this situation is shown below.

$$\log(\gamma) = \frac{-0.51 \cdot z^2 \cdot \sqrt{I}}{1 + (a\sqrt{I})/305}$$

This form must be modified if the solvent is other than water and if the temperature is other than 25 °C.

21. a) For H^+: $\log(\gamma) = -0.51(1)^2(0.005)^{1/2}/(1+(900(0.005)^{1/2}/305)) = -0.0298$ or $\gamma = 0.93$

    For NO_3^-: $\log(\gamma) = -0.51(-1)^2(0.005)^{1/2}/(1+(300(0.005)^{1/2}/305)) = -0.0337$ or $\gamma = 0.93$

    b) For K^+: $\log(\gamma) = -0.51(1)(0.020)^{1/2}/(1+ 300(0.020)^{1/2}/305)) = -0.0633$ or $\gamma = 0.86$

    For OH^-: $\log(\gamma) = -0.51(1)(0.020)^{1/2}/(1+350(0.020)^{1/2}/305)) = -0.0620$ or $\gamma = 0.87$

    c) For Ba^{2+}: $\log(\gamma) = -0.51(2^2)(0.030)^{1/2}/(1+500(0.030)^{1/2}/305)) = -0.275$ or $\gamma = 0.53$

    For Cl^-: $\log(\gamma) = -0.51(-1)^2(0.030)^{1/2}/(1+300(0.030)^{1/2}/305) = -0.0737$ or $\gamma = 0.84$

22. a) $\gamma_\pm = \{(0.93)(0.93)\}^{1/2} = 0.93$

    b) $\gamma_\pm = \{(0.86)(0.87)\}^{1/2} = 0.865$

    c) $\gamma_\pm = \{(0.53)^2(0.84)\}^{1/3} = 0.618$

23. The Debye-Hückel limiting law (DHLL) is a simplified form of the Debye-Hückel equation that is often used in work with very dilute solutions, as given below.

$$\log(\gamma) = -0.51 \cdot z^2 \cdot \sqrt{I}$$

The DHLL is only valid at quite low ionic strengths when the term $a\,(I^{1/2})/305$ is much lower than 1. The following results are obtained when using this simplified equation for Question

21.

a) For H^+ and NO_3^-: $\log(\gamma) = -0.51(1)(0.005)^{1/2} = -0.0361$ or $\gamma = 0.92$

(compared to 0.93 for H^+ and NO_3^- using the extended Debye-Hückel equation)

b) For K^+ and OH^-: $\log(\gamma) -0.51(1)(0.020)^{1/2} = -0.0721$ or $\gamma = 0.85$

(compared to 0.86 for K^+ and 0.87 for OH^- using the extended Debye-Hückel equation)

c) For Ba^{2+}: $\log \gamma = -0.51(4)(0.030)^{1/2} = -0.353$ or $\gamma = 0.44$

For Cl^-: $\log \gamma = -0.51(1)(0.030)^{1/2} = -0.883$ or $\gamma = 0.82$

(compared to 0.53 for Ba^{2+} and 0.84 for Cl^- using the extended Debye-Hückel equation)

24. The activity of a neutral compound hardly changes at all when ionic strength changes, but small changes are noted at quite high ionic strength. The activity coefficient of an aqueous neutral molecule is equal to 1.000 or slightly higher. This behavior is called the "salting out" effect and results in neutral molecules (e.g., some organic compounds) having less solubility in water at high ionic strength.

25. The salting-out effect reflects the diminished solubility of neutral molecules (e.g., some organic compounds) at a high ionic strength. A salting coefficient is a proportionality constant that is used to relate the log of the activity coefficient for a neutral chemical to the ionic strength of a solution.

26. $\log(\gamma) = 0.066(1.0) = 0.066$ or $\gamma = 1.16$

27. For the activity to be within 1% of concentration, $\gamma = 1.01$.
$\log(\gamma) = kI$, so $I = \log(1.01)/0.15 = 0.029$.

28. a) Because ion charge is squared in the equation, a larger charge (either + or -) causes a greater deviation of the activity from the concentration.

b) Size of the ion is the "a" term in the denominator of the equation so a greater value increases the denominator, thus making the log γ value smaller.

c) The square root of ionic strength is in both the numerator and denominator. Its effect is more direct in the numerator, so high ionic strength decreases log γ, except at very high ionic strength where the $a\,(I)^{1/2}/305$ becomes sizable with respect to 1.000.

29. Addition of an inert salt, such as NaCl, in a concentration much greater than that of any other charged species will cause the ionic strength to be essentially equal to the concentration of the NaCl.

30. The hydrated radius of an ion is the radius of a sphere of water molecules that surround an ion in an aqueous solution. This is the value that is reflected by the ion size parameters used in the extended Debye-Hückel equation.

31. In most analytical techniques, a calibration curve is used to relate the measured response (related to chemical activity) to the concentration or content of standards that contain known amounts of the chemical of interest. If the analysis method has been designed so that samples are treated in the same way as these standards, and therefore have similar relationships between their activity and content, this calibration curve can be used to determine the amount of the analyte in unknown samples.

32. Chemical equilibrium occurs when the rates of the forward and reverse reactions become equal. This results in the concentrations of reactants and products becoming fixed even

though both forward and reverse reactions are still occurring. Chemical equilibrium is a valuable concept to use in predicting the extent of a reaction in a chemical analysis.

33. An equilibrium constant is a ratio used to describe the relationship of the activities or concentrations of the products versus reactants for a reaction at equilibrium. This ratio can be used to help describe the expected extent of a chemical reaction at equilibrium.

34. a) $K = (a_{H_3O^+})(a_{HSO_4^-})/(a_{H_2SO_4})(a_{H_2O})$

   b) $K = (a_{Zn(NH_3)^{2+}})/(a_{Zn^{2+}})(a_{NH_3})$

   c) $K = (a_{Pb^{2+}})(a_{Cl^-})^2/(a_{PbCl_2})$

35. a) $K = [H_3O^+][HSO_4^-]/[H_2SO_4]$

   b) $K = [Zn(NH_3)^{2+}]/[Zn^{2+}][NH_3]$

   c) $K = [Pb^{2+}][Cl^-]^2$

36. $K = (3.6 \times 10^{-4})(1.0 \times 10^{-4})/(2.0 \times 10^{-4}) = 1.8 \times 10^{-4}$

37. $K = (8.4 \times 10^{-14})(8.4 \times 10^{-14}) = 7.1 \times 10^{-27}$

38. $\Delta G° = -R\,T \ln K°$ or $-\ln K° = \Delta G°/(RT)$

39. $\Delta G° = -R\,T \ln K° = -(8.314 \text{ J/mol K})(298 \text{ K})\ln(40) = -9139 \text{ J/mol} = -9{,}140 \text{ J/mol}$

40. $\Delta G° = -(8.314 \text{ J/mol K})(310 \text{ K})(\ln(2.3 \times 10^5))(1 \text{ kJ}/1000 \text{ J}) = -31.8 \text{ kJ/mol}$

41. Like an equilibrium constant, a reaction quotient is a relationship that is related to the product of the activities (or concentrations) of the products, raised to their appropriate powers, divided by the product of activities (or concentrations) of the reactants, raised to their appropriate

powers. The difference is that the activities (or concentrations) that are used in such a relationship for an equilibrium constant are those for a system that is at equilibrium, while a reaction quotient can make use of values for a system that has not yet reached equilibrium.

42. $Q = (1.3 \times 10^{-4})^2(6.3 \times 10^{-5}) = 1.1 \times 10^{-12}$. This is smaller than the equilibrium constant, so some of the solid will dissolve to form more aqueous ions.

43. a) $Q = (.58)^2/(0.81)(0.44) = 0.94$ Reaction will shift to the right.

    b) $Q = (1.35)^2/(0.078)(0.033) = 7.1 \times 10^2$ Reaction is at equilibrium, so no shift will occur.

    c) $Q = (1.50)^2/(0.034)(0.035) = 1.9 \times 10^3$ Reaction will shift to the left.

44. Table 6.5 lists questions you should consider when setting up and solving a chemical equilibrium problem.

45. A mass balance equation is an algebraic statement that the sum of the concentrations of the various forms of a substance must equal the total concentration of the substance. The "analytical concentration" of a substance is the total concentration. These are both important tools in providing a sufficient number of equations for a chemical system to allow an answer to be obtained when solving a chemical equilibrium problem.

46. $C_{\text{Phosphoric acid}} = [H_3PO_4] + [H_2PO_4^-] + [HPO_4^{2-}] + [PO_4^{3-}]$

47. $C_{\text{Copper}} = [Cu^{2+}] + [Cu(NH_3)^{2+}] + [Cu(NH_3)_2^{2+}] + [Cu(NH_3)_3^{2+}] + [Cu(NH_3)_4^{2+}]$

    $C_{\text{Ammonia}} = [NH_3] + [Cu(NH_3)^{2+}] + 2\,[Cu(NH_3)_2^{2+}] + 3\,[Cu(NH_3)_3^{2+}] + 4\,[Cu(NH_3)_4^{2+}]$

48. A charge balance equation makes use of the fact that the total charge of all ions in solution is 0. This type of equation usually takes the form of an algebraic expression that says that the total concentration of all cations (accounting for the value of their charges) will be equal to the total concentration of all anions (accounting for the value of their charges). This type of expression is another tool that can be used to provide an additional relationship for solving a chemical equilibrium problem.

49. $2[Ca^{2+}] + [Na^+] + 2[Mg^{2+}] = [Cl^-]$

50. $[H^+] = [F^-] + [OH^-]$

51. The quadratic formula (shown below) is used to find the values for x that satisfy an equation with the general form $0 = Ax^2 + Bx + C$, in which A, B, and C are constants.

$$x = \frac{-B \pm \sqrt{B^2 - 4AC}}{2A}$$

This type of equation often occurs in solving equilibrium problems, making the quadratic formula a useful tool when dealing with such problems.

52. a) Roots for x, -0.33$\underline{8}$ and -0.037$\underline{0}$

   b) Roots for x, +0.42$\underline{4}$ and -0.42$\underline{6}$

   c) Roots for x, 3.3$\underline{4}$ and 0.15$\underline{5}$

   d) First divide all terms by x, and then obtain the roots for x, which are 32.$\underline{5}$ and -0.0049$\underline{4}$.

53. a) The two roots are $-2.6\underline{1} \times 10^{-4}$ and $1.1\underline{5} \times 10^{-5}$. Because there is no such thing as a negative concentration, the correct answer is $[H^+] = 1.1\underline{5} \times 10^{-5}\ M$ (or pH = 4.94).

   b) Multiply all terms by [OH⁻] to give $[OH^-]^2 + 5.5 \times 10^{-4}[OH^-] - 1.0 \times 10^{-4} = 0$.

The two roots are -0.010$\underline{3}$ and +0.0097$\underline{3}$. The negative root makes no sense, so the hydroxide concentration is 0.0097$\underline{3}$ M (or pH = 11.99).

54. The method of successive approximations can be used to solve equations of higher order than a quadratic equation. In this method, you first guess at a value for x and solve the left side of the equation, hoping that it will equal 0. It usually does not. Next, choose another value of x, such that the sum of the terms on the left side will be closer to 0. Keep doing this until the sum of the terms on the left-hand side is close enough to 0 to satisfy the precision required in the answer. This approach can be used in a chemical equilibrium problem that involves a relationship that is more complex than what can be described by a quadratic equation or first-order relationship.

55. a) $0 = 2x^3 + 4x^2 + x + 0.5$

    Original guess: $x = +1$ Right-hand side (RHS) = 2 + 4 + 1 + 0.5 = 7.5

    Second guess: $x = -1$ RHS = -2 + 4 - 1 + 0.5 = 1.5

    Third guess: $x = -1.5$ RHS = -6.75 + 9.00 - 1.50 + 0.5 = 1.25

    Fourth guess: $x = -1.8$ RHS = -11.664 + 12.960 - 1.8 + 0.5 = -0.004

    This is probably close enough to 0 to say that $x = -1.8$.

    b) $(2.5 \times 10^{-3} + x)/x = 1.0 \times 10^{-4} x^2$

    Rearrange to give $1.0 \times 10^{-4} x^3 - x - 2.5 \times 10^{-3} = 0$.

    First guess: $x = 0$ Left-hand side (LHS) = 0 - 0 - 0.0025 = -0.0025

    Second guess: $x = 0.10$ LHS = 1.00×10^{-7} - 0.10 - 0.0025 = -0.1025

    Third guess: $x = -0.10$ LHS = -1.00×10^{-7} + 0.10 - 0.0025 = +0.0975

    Fourth guess: $x = -0.001$ LHS = 1.00×10^{-13} + 0.001 - 0.0025 = -0.0015

Fifth guess: $x = -2.5 \times 10^{-3}$ LHS $= 1.56 \times 10^{-12} + 0.0025 - 0.0025 = -1.56 \times 10^{-12}$

This is probably close enough to 0 to say that $x = -2.5 \times 10^{-3}$.

c) $9.5 x^3 = 0.25 x^2 - 1.37 x + 5.3$

Rearrange to give $9.5 x^3 - 0.25 x^2 + 1.37 x - 5.3 = 0$.

First guess: $x = 0$ RHS $= 0 - 0 + 0 - 5.3 = -5.3$

Second guess: $x = 0.50$ RHS $= 1.1875 - 0.0625 + 0.685 - 5.3 = -3.49$

Third guess: $x = 0.80$ RHS $= 4.864 - 0.1600 + 1.096 - 5.3 = 0.500$

Fourth guess: $x = 0.75$ RHS $= 4.0078 - 0.1406 + 1.0275 - 5.3 = -0.4053$

Fifth guess: $x = 0.77$ RHS $= 4.3371 - 0.1482 + 1.0549 - 5.3 = -0.0562$

Sixth guess: $x = 0.773$ RHS $= 4.3879 - 0.1494 + 1.0590 - 5.3 = -0.00242$

Final answer is $x = 0.773$.

d) $3.0 \times 10^3 x^2 = 4.1 \times 10^2 x^3 + 18.0$

Rearrange to give $410 x^3 - 3000 x^2 + 18.0 = 0$.

First guess: $x = 0$ LHS $= 0 + 0 + 18 = 18$

Second guess: $x = 1$ LHS $= 410 - 3000 + 18 = -2572$

Third guess: $x = 0.1$ LHS $= 0.41 - 30 + 18 = -11.59$

Fourth guess: $x = 0.06$ LHS $= 0.08856 - 10.8 + 18 = 7.28856$

Fifth guess: $x = 0.08$ LHS $= 0.20992 - 19.2 + 18 = -0.9901$

Sixth guess: $x = 0.078$ LHS $= 0.1946 - 18.252 + 18 = -0.0574$

Seventh guess: $x = 0.0779$ LHS $= 0.1938 - 18.2052 + 18 = -0.0114$

This is probably close enough to say that $x = 0.0779$.

56. a) $0 = [H^+]^2 + (2.5 \times 10^{-4})[H^+] - (3.0 \times 10^{-9})$

   First guess: $[H^+] = 0$ Right-hand side (RHS) = -3.0×10^{-9}

   Second guess: $[H^+] = 5 \times 10^{-5}$ RHS = $2.5 \times 10^{-9} + 1.25 \times 10^{-8} - 3.0 \times 10^{-9}$
   $$= 1.2 \times 10^{-8}$$

   Third guess: $[H^+] = 1 \times 10^{-5}$ RHS = $1.0 \times 10^{-10} + 2.5 \times 10^{-9} - 3.0 \times 10^{-9}$
   $$= -4.0 \times 10^{-10}$$

   Fourth guess: $[H^+] = 1.1 \times 10^{-5}$ RHS = $1.21 \times 10^{-10} + 2.75 \times 10^{-9} - 3.0 \times 10^{-9}$
   $$= -1.29 \times 10^{-10}$$

   Fifth guess: $[H^+] = 1.13 \times 10^{-5}$ RHS = $1.2769 \times 10^{-10} + 2.825 \times 10^{-9} - 3.0 \times 10^{-9}$
   $$= -4.731 \times 10^{-11}$$

   Sixth guess: $[H^+] = 1.15 \times 10^{-5}$ RHS = $1.3225 \times 10^{-10} + 2.875 \times 10^{-9} - 3.0 \times 10^{-9}$
   $$= 7.25 \times 10^{-12} \text{ (which is close enough to zero)}$$

   Using the quadratic formula gives $[H^+] = 1.147 \times 10^{-5}$.

   Checking for the other root gives the following results.

   First guess: $x = -1 \times 10^{-4}$ RHS = $1.0 \times 10^{-8} - 2.5 \times 10^{-8} - 3.0 \times 10^{-9}$
   $$= -1.8 \times 10^{-8}$$

   Second guess: $x = -2 \times 10^{-4}$ RHS = $4.0 \times 10^{-8} - 5.0 \times 10^{-8} - 3.0 \times 10^{-9}$
   $$= -1.3 \times 10^{-8}$$

   Third guess: $x = -4 \times 10^{-4}$ RHS = $1.6 \times 10^{-7} - 1.0 \times 10^{-7} - 3.0 \times 10^{-9}$
   $$= 5.7 \times 10^{-8}$$

   Fourth guess: $x = -3 \times 10^{-4}$ RHS = $9.0 \times 10^{-8} - 7.5 \times 10^{-8} - 3.0 \times 10^{-9}$
   $$= 1.2 \times 10^{-8}$$

Fifth guess: $x = -2.6 \times 10^{-4}$ RHS $= 6.76 \times 10^{-8} - 6.5 \times 10^{-8} - 3.0 \times 10^{-9}$

$$= -4.0 \times 10^{-10}$$

Sixth guess: $x = -2.62 \times 10^{-4}$ RHS $= 6.8644 \times 10^{-8} - 6.55 \times 10^{-8} - 3.0 \times 10^{-9}$

$$= 1.44 \times 10^{-10} \text{ (which is close enough to zero)}$$

The quadratic formula gives $x = -2.615 \times 10^{-4}$.

b) $(1 \times 10^{-4})/[OH^-] = 5.5 \times 10^{-4} + [OH^-]$

Rearrange this equation into a standard quadratic form by multiplying through by $[OH^-]$, which gives the following expression.

$[OH^-]^2 + (5.5 \times 10^{-4})[OH^-] - 1.4 \times 10^{-4} = 0$

First guess: $x = 1.0 \times 10^{-2}$ Left-hand side (LHS) $= 1.0 \times 10^{-4} + 5.5 \times 10^{-6} - 1.4 \times 10^{-4}$

$$= -3.45 \times 10^{-5}$$

Second guess: $x = 1.2 \times 10^{-2}$ LHS $= 1.44 \times 10^{-4} + 6.6 \times 10^{-6} - 1.4 \times 10^{-4}$

$$= 1.06 \times 10^{-5}$$

Third guess: $x = 1.15 \times 10^{-2}$ LHS $= 1.3225 \times 10^{-4} + 6.325 \times 10^{-6} - 1.4 \times 10^{-4}$

$$= -1.425 \times 10^{-6}$$

This result is probably close enough to zero.

Use of the quadratic formula gives $x = 0.01156$.

Because $x = [OH^-]$ and must be a positive answer, there is no need to seek the other root.

CHAPTER 7: CHEMICAL SOLUBILITY AND PRECIPITATION

1. Solubility is the maximum concentration or amount of a chemical that can be placed into a solvent to form a stable solution.

2. Ionic substances go into solution as separated ions. Most molecular compounds go into solution as intact molecules.

3. Some general factors that can affect the solubility of a chemical include the nature of the solvent, temperature, and often pH.

4. Precipitation is a process that occurs when a portion of a dissolved chemical leaves a solution to form a solid. Precipitation will take place when the amount of a chemical that is placed into a solution exceeds its thermodynamic solubility.

5. a) An unsaturated solution is a solution in which the final concentration of an added solute is given by the total amount that was added to the solution. No precipitation will occur from this type of solution.

   b) A saturated solution is a solution in which the dissolved concentration of a solute is equal to its maximum solubility at equilibrium. If the amount of chemical in a solution is in excess of what will be present in a saturated solution, precipitation will occur until the amount of chemical that is in solution is in equilibrium with the solid form of the same chemical that is also in contact with the solution.

   c) A supersaturated solution is a solution in which the concentration of a dissolved chemical is temporarily greater than its maximum solubility at equilibrium. The excess of the chemical will later precipitate until the amount of chemical that is in solution is in equilibrium with the solid form of the same chemical that is also in contact with the solution.

6. A precipitate is a solid that is formed as a dissolved chemical leaves a solution that contains more than the solubility limit of the chemical, usually as a result of two soluble reagents reacting to form an insoluble product. A crystal is a solid that is relatively pure and has a high degree of order in its structure. A colloid is a dispersion of small particles with sizes between 1 nm and 1 µm. These are all terms used to describe solid materials and/or different types of precipitates.

7. A chemical must be at least partially soluble in a sample if we are to detect or measure it in this sample. In addition, any chemicals we wish to place into a solution as reagents must be able to dissolve in a reproducible fashion and at a sufficient concentration for our purpose. There are also many analytical methods that make use of the ability of a chemical to precipitate from a solution. In a gravimetric measurement, one must be sure that essentially none of the analyte remains in solution. In a precipitation titration, the equivalence point occurs when enough titrant has been added to precipitate all the analytes, but no other interfering material has precipitated.

8. Gravimetry and precipitation titrations are two examples. In gravimetry, mass measurements of a pure precipitate are used to measure an analyte that is contained in this precipitate. In a precipitation titration, the amount of analyte is determined by measuring the volume of a reagent that is required to completely precipitate an analyte from a solution.

9. All three of these ions are precipitated with hydroxide as $Al(OH)_3$, $Fe(OH)_3$, and $Cr(OH)_3$, as occurs during the addition of a dilute ammonia solution. If concentrated NaOH is added,

however, aluminum and chromium dissolve by forming $Al(OH)_4^-$ and $Cr(OH)_4^-$, while iron hydroxide remains insoluble. The addition of H_2O_2 to the solution oxidizes the chromium to dichromate ($Cr_2O_7^{2-}$), which is precipitated as the yellow solid $BaCrO_4$ after the addition of $BaCl_2$.

10. Intermolecular forces are noncovalent, electrostatic interactions that cause separate but neighboring molecules or chemical species to attract or repel one another. The closer a solute and solvent are in terms of their types of intermolecular forces, and the extent to which the solute and solvent can interact through these forces, the greater the solubility will be for the solute in the solvent.

11. a) An ionic interaction is a type of intermolecular force in which electrostatic attraction occurs between two ions with opposite charges, or repulsion occurs between two ions with the same type of charge.

b) Hydrogen bonding is an intermolecular interaction in which a hydrogen atom is shared in a non-covalent bond between the lone electron pairs on atoms such as nitrogen, oxygen, or fluorine in molecules.

c) A dipole–dipole interaction is an intermolecular force that takes place between two chemicals that have permanent dipole moments.

d) A dispersion force is an intermolecular force that occurs when the movement of electrons in one molecule creates a temporary dipole moment, which induces another temporary but complementary dipole moment in a neighboring molecule.

12. The phrase "like dissolves like" means that polar solutes tend to dissolve in polar solvents and nonpolar solutes dissolve in nonpolar solvents. Solubility is related to the intermolecular

forces that occur between the solute and solvent.

13. A polar chemical is one in which the center of positive charge is in a different place than the center of negative charge, or a permanent dipole moment is present. A nonpolar chemical has the center of positive and negative charges at approximately the same place, giving a dipole moment that is zero or quite small. The greater the difference in the distribution of charge, the more polar the substance is said to be.

14. A dipole moment is a numerical measure of the extent and direction of charge distribution in a molecule, as given in units of a debye (D). An octanol-water partition ratio is a measure of the tendency of a solute to dissolve in a polar solvent (water) compared to a much less polar solvent (octanol), which also provides a measure of polarity.

15. a) From most polar to least: dimethylsulfoxide > acetone > chloroform > benzene; dimethylsulfoxide and acetone will mix best with water, while benzene and chloroform will mix best with octane.

    b) Acetone can undergo hydrogen bonding, dipole–dipole interactions, and interactions involving dispersion forces; benzene can undergo interactions involving dispersion forces; chloroform can undergo dipole–dipole interactions and interactions involving dispersion forces; dimethylsulfoxide can undergo dipole–dipole interactions, hydrogen bonding, and interactions involving dispersion forces

16. a) From most polar to least: acetic acid > propanol > propane > cyclohexane

    b) Acetic acid and propanol are the most soluble in water; cyclohexane and propane are the most soluble in octanol.

    c) Acetic acid can undergo hydrogen bonding, dipole–dipole interactions, and interactions involving dispersion forces; cyclohexane and propane can undergo interactions involving dispersion forces; *n*-propanol can undergo hydrogen bonding and interactions involving dispersion forces

17. a) Intermolecular forces dominate the enthalpy of mixing term, but the greater disorder that occurs during the mixing is associated with entropy, and the dissolution of a solute into a solvent results in increased entropy. The sum of these two terms gives the free energy change, which will determine the overall extent to which the solute can dissolve in the solvent at equilibrium nature.

    b) If the overall free energy change of mixing is negative, solubility will be high. This situation will occur if there are fairly strong intermolecular forces between the solvent and solute. Likewise, the system will be more random and have greater entropy if the solute and the solvent molecules are intermingled rather than in separate phases.

    c) Most chemicals will become more soluble at higher temperatures because the negative sign on the $T\Delta S_{mix}$ term will cause the free energy term to become more negative at higher temperatures.

18. Solubility equilibrium is a process in which a chemical enters a solution and reaches equilibrium between its dissolved and solid states. When a molecular solid dissolves, intermolecular attractions of solute to solute and solvent to solvent must be broken so that attractions of solute to solvent can form. When this has been completed, the solution has achieved a solubility equilibrium.

19. A solubility constant is an equilibrium constant that describes the placement of a solid chemical into a solution. This constant is directly related to solubility for a molecular solid

and can be used to calculate solubility for an ionic solid.

20. a) Naphthalene(s) ⇌ Naphthalene(benzene)

b) Vitamin C(s) ⇌ Vitamin C(aq)

c) Iodine(s) ⇌ Iodine(aq)

d) Iodine(s) ⇌ Iodine(CCl$_4$)

21. a)
$$K° = \frac{a_{Naphthalene(benzene)}}{a_{Naphthalene(s)}} = a_{Naphthalene(benzene)} \quad \text{(where } a_{Naphthalene(s)} = 1\text{)}$$

$$K = \frac{[Naphthalene]_{(benzene)}}{a_{Naphthalene(s)}} = [Naphthalene]_{(benzene)}$$

b)
$$K° = \frac{a_{Vitamin\ C(aq)}}{a_{Vitamin\ C(s)}} = a_{Vitamin\ C(aq)} \quad \text{(where } a_{Vitamin\ C(s)} = 1\text{)}$$

$$K = \frac{[Vitamin\ C]_{(aq)}}{a_{Vitamin\ C(s)}} = [Vitamin\ C]_{(aq)}$$

c)
$$K° = \frac{a_{Iodine(aq)}}{a_{Iodine(s)}} = a_{Iodine(aq)} \quad \text{(where } a_{Iodine(s)} = 1\text{)}$$

$$K = \frac{[Iodine]_{(aq)}}{a_{Iodine(s)}} = [Iodine]_{(aq)}$$

d)
$$K° = \frac{a_{Iodine(CCl_4)}}{a_{Iodine(s)}} = a_{Iodine(CCl_4)} \quad \text{(where } a_{Iodine(s)} = 1\text{)}$$

$$K = \frac{[Iodine]_{(CCl_4)}}{a_{Iodine(s)}} = [Iodine]_{(CCl_4)}$$

22. This is true for a solute with only one form in a solution because only the activity of the solid form of the solute, that appears in the denominator for K and $K°$. This will not be true for solutes that can exist in more than one form in the solution, such as acids or bases, which can form some of their conjugate acids or bases.

23. $S = (5.00\ g - 4.15\ g)/(0.250\ L) = (0.85\ g)/(0.250\ L)$

    $= 3.4\ g/L$

    $= [(3.4\ g)/(122\ g/mol)]/L = 2.79 \times 10^{-2}\ M$

24. $S = 0.0781\ g/0.250\ L = 0.312\ g/L$

    or $S = (0.312\ g/L)/(253.8\ g/mol) = 1.23 \times 10^{-3}\ M$

25. A solubility product is an equilibrium constant that describes the conditions needed to saturate a solution with a dissolved ionic solid. This constant is represented by the term K_{sp} and is the product of the activities or concentrations of the ions formed when an ionic compound is in equilibrium with the solid form of the compound. From this value and relationship, it is possible to determine the solubility of the ionic compound in the given solvent.

26. a) $K°_{sp} = (a_{Ag+})(a_{Br-})$ or $K_{sp} = [Ag^+][Br^-]$

    b) $K°_{sp} = (a_{Sr2+})(a_{F-})^2$ or $K_{sp} = [Sr^{2+}][F^-]^2$

    c) $K°_{sp} = (a_{Ca2+})^3(a_{PO43-})^2$ or $K_{sp} = [Ca^{2+}]^3[PO_4^{3-}]^2$

    d) $K°_{sp} = (a_{Mg2+})(a_{NH4+})(a_{PO43-})$ or $K_{sp} = [Mg^{2+}][NH_4^+][PO_4^{3-}]$

27. A solubility product can be used to describe the extent to which an insoluble ionic compound dissolves in a solvent of specific composition or to describe the concentration of ions

remaining in solution after a precipitation has taken place.

28. a) $S = [Zn^{2+}] = [S^{2-}] = (K_{sp})^{1/2} = (1.6 \times 10^{-24})^{1/2} = 1.3 \times 10^{-12} \, M$ for the α-form of ZnS

   or $= (2.5 \times 10^{-22})^{1/2} = 1.6 \times 10^{-11} \, M$ for the β-form

   b) $S = [Cd^{2+}][S^{2-}] = (K_{sp})^{1/2} = (8.0 \times 10^{-27})^{1/2} = 8.9 \times 10^{-14} \, M$

29. a) For AgBr: $\quad S = (5.35 \times 10^{-13})^{1/2} = 7.31 \times 10^{-7} \, M$

   For AgCl: $\quad S = (1.77 \times 10^{-10})^{1/2} = 1.33 \times 10^{-5} \, M$

   For AgI: $\quad S = (8.52 \times 10^{-17})^{1/2} = 9.23 \times 10^{-9} \, M$

   Relative solubility, from high to low: AgCl > AgBr > AgI

   b) For CaSO$_4$: $\quad S = (4.93 \times 10^{-5})^{1/2} = 0.00702 \, M$

   For calcium oxalate hydrate: $\quad S = (2.32 \times 10^{-9})^{1/2} = 4.82 \times 10^{-5} \, M$

   Relative solubility, from high to low: CaSO$_4$ > calcium oxalate hydrate

   c) For CaSO$_4$: $\quad S = (4.93 \times 10^{-5})^{1/2} = 0.00702 \, M$

   For BaSO$_4$: $\quad S = (1.08 \times 10^{-10})^{1/2} = 1.04 \times 10^{-5} \, M$

   Relative solubility, from high to low: CaSO$_4$ > BaSO$_4$

   d) For Al(OH)$_3$: $\quad S = [Al^{3+}] = (K_{sp}/27)^{1/4} = (1.3 \times 10^{-33}/27)^{1/4} = 2.6 \times 10^{-9} \, M$

   For Ca(OH)$_2$: $\quad S = [Ca^{2+}] = (K_{sp}/4)^{1/3} = (5.5 \times 10^{-6}/4)^{1/3} = 1.1 \times 10^{-2} \, M$

   Relative solubility, from high to low: Ca(OH)$_2$ > Al(OH)$_3$

   e) For Fe(OH)$_3$: $\quad S = [Fe^{3+}] = (K_{sp}/27)^{1/4} = (2.79 \times 10^{-39}/27)^{1/4} = 1.01 \times 10^{-10} \, M$

   For Fe(OH)$_2$: $\quad S = [Fe^{2+}] = (K_{sp}/4)^{1/3} = (4.7 \times 10^{-17}/4)^{1/3} = 2.3 \times 10^{-6} \, M$

   Relative solubility, from high to low: Fe(OH)$_2$ > Fe(OH)$_3$

30. Some highly soluble salts in water are KCl, NaNO$_3$, Li$_2$SO$_4$, CaBr$_2$, and MgSO$_4$.

    Some slightly soluble salts in water are AgCl, CuS, BaSO$_4$, Ca$_3$(PO$_4$)$_2$, and PbI$_2$.

31. a) "Miscible" is a term that describes two liquids that can form a stable solution when they are mixed in any proportion.

    b) "Partially miscible" is a term used to describe two liquids that give a difference in the volume for their mixture versus the total volumes of the two combined liquids.

    c) "Immiscible" is a term used to describe two liquids that do not dissolve to any appreciable extent in each another.

32. Just as with solid solutes, liquid solutes have intermolecular forces that must be overcome to dissolve in a solvent. The stronger the intermolecular interactions are that can form between the solute and solvent, the greater the amount of solute that will dissolve.

33. a) CH$_3$COOH(l) $\rightleftharpoons$ CH$_3$COOH(aq)

    The above process would also be followed to some extent by the following reaction.

    CH$_3$COOH(aq) + H$_2$O(l) $\rightleftharpoons$ CH$_3$COO$^-$(aq) + H$_3$O$^+$(aq)

    b) HOCH$_2$CH$_2$OH(l) $\rightleftharpoons$ HOCH$_2$CH$_2$OH (ethanol)

    c) Br$_2$(l) $\rightleftharpoons$ Br$_2$(CCl$_4$)

34. a) $$K° = \frac{a_{CH_3COOH(aq)}}{a_{CH_3COOH(l)}} = a_{CH_3COOH(aq)}$$

    $$K = \frac{[CH_3COOH]_{(aq)}}{a_{CH_3COOH(l)}} = [CH_3COOH](aq)$$

b)
$$K° = \frac{a_{HOCH_2CH_2OH(ethanol)}}{a_{HOCH_2CH_2OH(l)}} = a_{HOCH_2CH_2OH(ethanol)}$$

$$K = \frac{[HOCH_2CH_2OH]_{(ethanol)}}{a_{HOCH_2CH_2OH(l)}} = [HOCH_2CH_2OH]_{(ethanol)}$$

c)
$$K° = \frac{a_{Br_2(CCl_4)}}{a_{Br_2(l)}} = a_{Br_2(CCl_4)}$$

$$K = \frac{[Br_2]_{(CCl_4)}}{a_{Br_2(l)}} = [Br_2]_{(CCl_4)}$$

35. a) $[C_6H_6] = (1.79 \text{ g/L})(1 \text{ mol}/78.0 \text{ g}) = 0.0229 \, M$

b) $V = (1.79 \text{ g})(1 \text{ mL}/0.8786 \text{ g}) = 2.04 \text{ mL}$

36. $S = (1.0 \text{ mL}/0.200 \text{ L})(1.484 \text{ g/mL})(1 \text{ mol}/119.5 \text{ g}) = 0.062 \text{ mol/L}$

$S = (1.0 \text{ mL}/0.200 \text{ L})(1.484 \text{ g/mL})(10^3 \text{ mg/g}) = 7.4 \times 10^3 \text{ mg/L (or ppm)}$

$S = (1.0 \text{ mL}/0.200 \text{ L})(1.484 \text{ g/mL})(10^6 \text{ µg/g}) = 7.4 \times 10^6 \text{ µg/L (or ppb)}$

37. $S = (42 \text{ g/L})(1 \text{ mol}/88.0 \text{ g}) = 0.477 \, M$

$V = (1 \text{ L}/0.477 \text{ mol})(0.50 \text{ mol}) = 1.05 \text{ L}$

38. An emulsion is a dispersion of tiny droplets of a liquid in another liquid in which it is nearly insoluble.

39. Acetic acid is a weak acid that can dissociate in water to form acetate, thus increasing the apparent solubility of acetic acid in water. A similar situation occurs for hexanoic acid (caproic acid, $C_5H_{11}COOH$), which has a solubility of 4 grams per liter in pure water; this

acid becomes much more soluble if a strong base is added, converting the hexanoic acid into $C_5H_{11}CO_2^-$.

40. Henry's law, as given below, states that the solubility of a gas is proportional to the partial pressure of that gas in contact with the surface of the liquid.

$$C_{solute} = K_H P_{solute}$$

Each gas has its own proportionality constant (K_H) in this relationship.

41. If we were dealing with pure oxygen gas, the following result would be obtained.

$$C_{solute} = K_H P_{solute} = (1.26 \times 10^{-3} \text{ mol/L} \cdot \text{bar})(600 \text{ torr})(1 \text{ bar}/750.06 \text{ torr})$$

$$= 1.01 \times 10^{-3} M$$

If we were dealing instead with air, which has a partial pressure for O_2 of 0.21, the following result would be obtained.

$$C_{solute} = K_H P_{solute} = (1.26 \times 10^{-3} \text{ mol/L} \cdot \text{bar})(600 \text{ torr} \times 0.21 \text{ } O_2)(1 \text{ bar}/750.06 \text{ torr})$$

$$= 2.12 \times 10^{-4} M$$

42. Because we are dealing with air, we need to consider the partial pressures for each of the given gases. These values can be obtained from a source such as the *CRC Handbook of Chemistry and Physics* and are as follows in air at 1 atm: O_2, 0.21; N_2, 0.78; H_2, 0.010; and CO_2, 0.00034. We can then calculate the following solubilities by using this information.

Solubility of O_2 = $(1.26 \times 10^{-3}$ mol/L $\cdot$ bar$)(0.21 \cdot 1$ atm$)(1.013$ bar/1 atm$) = 2.7 \times 10^{-4} M$

Solubility of N_2 = $(6.40 \times 10^{-4}$ mol/L $\cdot$ bar$)(0.78 \cdot 1$ atm$)(1.013$ bar/1 atm$) = 5.1 \times 10^{-4} M$

Solubility of H_2 = $(7.80 \times 10^{-4}$ mol/L $\cdot$ bar$)(0.010 \cdot 1$ atm$)(1.013$ bar/1 atm$) = 7.9 \times 10^{-6} M$

Solubility of CO_2 = $(3.34 \times 10^{-2}$ mol/L $\cdot$ bar$)(0.00034 \cdot 1$ atm$)(1.013$ bar/1 atm$) = 1.2 \times 10^{-5} M$

43. Carbon dioxide (CO_2) in water can react with the water to form H_2CO_3, which partially dissociates into HCO_3^- and H_3O^+. Any base that will react with the hydronium ion will cause greater amounts of CO_2 to dissolve.

44. You can add a known mass of solute to a known quantity of solvent, allow equilibrium to be attained, and then filter off and weigh the solute that did not dissolve. The difference in mass can then be used to find the solubility of the solute. This approach works best for solutes that are reasonably soluble. If the solute is not very soluble, a second approach is to add the solute to solvent, allow equilibrium to be attained, and measure the concentration of the dissolved solute by whatever means is suitable for its analysis. (Note: Other approaches are also given in this chapter.)

    To examine a mixture of two liquids, a clear solution with no visible layers or cloudiness would represent a miscible mixture. The presence of a distinct boundary between the liquid phases in the mixture would indicate that a saturated solution is present, and a change in volume for this mixture versus the total volume of the combined liquids would indicate that these liquids are partially miscible. If any cloudiness is present, then an emulsion has formed.

45. The *CRC Handbook of Chemistry and Physics* and *Lange's Handbook of Chemistry* contain lists on the general solubility of many organic and inorganic chemicals in water, acidic or basic solutions, and organic solvents. The International Union of Pure and Applied Chemistry (IUPAC) also has a large series of books devoted to the solubilities of particular classes of substances. In addition, information on solubility is sometimes provided for a chemical in its materials safety data sheet.

46. Add a small amount of solute to the solvent in a test tube, shake well, and observe whether the solute dissolved.

47. a) Mix known amounts of solute and solvent such that there is excess solute. After equilibrium is attained, filter, dry, and weigh the excess solute. This works best if a weighable amount of solute dissolves.

    b) Mix known amounts of solute and solvent such that there is excess solute. After equilibrium is attained, filter off the excess solute, and then evaporate away the solvent and weigh the solute that remained after solvent was evaporated.

    c) Mix an excess amount of solute in solvent. After equilibrium is attained, measure the concentration of dissolved solute by whatever means is suitable. This works best when the solubility if quite low.

48. The first step in precipitation is the formation of small particles (or "nuclei") of the precipitating chemical. These nuclei act as centers onto which more of the desired chemical can add to form even larger particles. The slow addition of more molecules or ions to these nuclei gives rise to a competing process known as crystal growth. This, in turn, can produce a pure precipitate or even a crystal of the precipitating chemical.

49. a) Nucleation is the first step during precipitation, during which small particles (or "nuclei") of the precipitating chemical are formed.

    b) Crystal growth is a process that occurs during the addition of more ions or molecules to the nuclei to form larger particles to lead to precipitation.

50. If a solution is merely saturated, no increased amount of solid will form over time. If the

solution is supersaturated, precipitation will occur as this solution approaches a more thermodynamically stable situation, during which the dissolved solute is in equilibrium with the precipitate.

51. Rapid precipitation will lead to the formation of a large number of nuclei and small solid particles that may be difficult to filter. Slow precipitation allows more time for crystal growth to compete with nucleation and tends to produce larger precipitate particles. Rapid precipitation typically occurs from a solution that is very supersaturated. Slow precipitation will occur if the solution is only slightly supersaturated.

52. a) Occlusion is a condition in which tiny quantities of the solvent or molecular impurities are trapped inside a precipitate.

    b) Inclusion is a condition in which ionic impurities occupy some lattice sites in a precipitate and are trapped in this material.

    c) Adsorption is a condition in which impurities stick to the surface of a precipitate.

53. Coprecipitation is a collective term that refers to the presence of impurities that precipitate along with the desired substance. This effect can lead to a systematic error when using precipitation for quantitative analysis.

54. A gathering agent is used to intentionally cause coprecipitation to bring down a material that is difficult to precipitate by itself.

55. Postprecipitation involves the slow formation of one substance after the main precipitation has occurred. In coprecipitation, the desired and undesired substance precipitate simultaneously, but in postprecipitation the two substances precipitate sequentially.

56. The first precipitate was the Al_2O_3 formed from $Al(OH)_3$ contaminated with Fe_2O_3 from the impurity which came out of the solution as $Fe(OH)_3$. After reprecipitation, most of the iron was not allowed to precipitate, so the smaller mass represents the iron that was present the first time but not the second time. The mass of Fe_2O_3 was 0.2543 – 0.2487 or 0.0056 g. This number is not useful in estimating the amount of iron because it surely is not the total amount of iron.

57. $Ni(dmg)_2$ precipitates are known to form small crystals that are extremely difficult and slow to filter. The second, more patient, student had a precipitate that was made up of much larger and more easily filtered crystals.

58. Reprecipitation is a procedure in which the initial precipitate containing coprecipitated impurities is redissolved, thereby releasing the impurities, and precipitated again. The resulting precipitate contains far less impurity material than the initial precipitate. A disadvantage is that invariably some of the desired material is lost in the process.

59. You can use the concentrations (or activities) of the ions in the corresponding ion product expression to assess if the ion product is greater than or less than the K_{sp}. If the calculated ion product is greater than the K_{sp} value, precipitation will occur. If it is less than the K_{sp} value, precipitation will not occur.

60. a) Moles of added Pb^{2+} = 0.050 L (0.025 mol Pb^{2+}/L) = 0.00125 mol Pb^{2+}

    Maximum conc. Pb^{2+} in final solution = 0.00125 mol Pb^{2+}/0.150 L = 0.00833 M

    Moles of added SO_4^{2-} = 0.100 L (0.010 mol SO_4^{2-}/L) = 0.00100 mol SO_4^{2-}

    Maximum conc. SO_4^{2-} in final solution = 0.00100 mol SO_4^{2-}/0.150 L = 0.00667 M

$Q = [Pb^{2+}][SO_4^{2-}] = (0.00833\ M)(0.00677\ M) = 5.64 \times 10^{-5} = 5.6 \times 10^{-5}$

$Q = 5.6 \times 10^{-5} > K_{sp} = 2.53 \times 10^{-8}$ (precipitation will occur; sulfate is the limiting reagent)

$K_{sp} = [Pb^{2+}][SO_4^{2-}]$ (at equilibrium)

$2.53 \times 10^{-8} = (0.00833 - x)(0.00667 - x)$ or $0 = x^2 - 0.01500\ x + 5.553 \times 10^{-5}$

Using the quadratic formula gives $x = 0.00665\ M$ (other root, $0.00835\ M$).

Mass of $PbSO_4$ formed $= (0.00665\ \text{mol/L})(0.150\ L)(303.26\ \text{g/mol}) = 0.30\ g$

b) Moles of added $Pb^{2+} = 0.050\ L\ (0.0080\ \text{mol}\ Pb^{2+}/L) = 0.000400\ \text{mol}\ Pb^{2+}$

Maximum conc. Pb^{2+} in final solution $= 0.000400\ \text{mol}\ Pb^{2+}/0.250\ L = 0.00160\ M$

Moles of added $Cl^- = 0.200\ L\ (0.0050\ \text{mol}\ Cl^-/L) = 0.00100\ \text{mol}\ Cl^-$

Maximum conc. Cl^- in final solution $= 0.00100\ \text{mol}\ Cl^-/0.250\ L = 0.00400\ M$

$Q = [Pb^{2+}][Cl^-]^2 = (0.00160\ M)(0.00400\ M)^2 = 2.56 \times 10^{-8} = 2.6 \times 10^{-8}$

$Q = 2.6 \times 10^{-8} < K_{sp} = 1.7 \times 10^{-5}$ (precipitation will not occur)

61. a) $[Ag^+] = (K_{sp})^{1/2} = (8.52 \times 10^{-17})^{1/2} = 9.23 \times 10^{-9}\ M$

b) $[Pb^{2+}] = (K_{sp}/4)^{1/3} = (1.70 \times 10^{-5}/4)^{1/3} = 1.62 \times 10^{-2}\ M$

c) $[OH^-] = (2\ K_{sp})^{1/3} = (2 \cdot 5.5 \times 10^{-6})^{1/3} = 2.2 \times 10^{-2}\ M$

62. It is possible to use a solubility product to calculate the concentration of an ionic analyte that will remain in solution after precipitate. This concentration can be used to find the moles or mass of analyte in solution, which can then be compared with the total amount of analyte that was originally present in order to determine the amount or fraction that was actually precipitated.

63. If precipitation were 100.000% complete, the mass of $PbCl_2$ would be as follows.

$$\text{Mass } PbCl_2 = (0.050 \text{ mol/L})(0.050 \text{ L})(278.1 \text{ g/mol}) = 0.69\underline{5} \text{ g}$$

The Pb^{2+} and Cl^- ions in solution are produced in 1:2 ratio from $PbCl_2$, so the concentration remaining in solution after precipitation would be as given below:

$$\text{Solubility} = [Pb^{2+}] = (K_{sp}/4)^{1/3} = \{(1.70 \times 10^{-5})/4\}^{1/3} = 0.0162 \text{ } M \text{ in 100 mL of water}$$

This corresponds to a mass of $PbCl_2 = (0.0162 \text{ mol/L})(0.100 \text{ L})(278.1 \text{ g/mol}) = 0.450\underline{5}$ g.

Therefore, the mass precipitated = $0.69\underline{5}$ g – $0.450\underline{5}$ g = $0.244\underline{5}$ g = 0.24 g

64. a) $[Fe^{3+}] = K_{sp}/[OH^-]^3 = (2.79 \times 10^{-39})/(1.0 \times 10^{-9})^3 = 2.8 \times 10^{-12}$ M

This is an extremely small fraction of all the iron, so the mass of precipitate is as follows.

Mass of precipitate = $(1.50 \times 10^{-3} \text{ mol/L})(0.500 \text{ L})(106.87 \text{ g/mol}) = 0.0802$ g

b) Mass Fe in solution = $(2.8 \times 10^{-12} \text{ mol/L})(0.500 \text{ L})(55.845 \text{ g/mol}) = 7.8 \times 10^{-11}$ g

c) Fraction not precipitated = $(2.8 \times 10^{-12} M)/(1.50 \times 10^{-3} M) = 1.1 \times 10^{-9}$ or 1.1 ppb

65. a) Total amount of chloride = $(0.200 \text{ mol/L})(0.0010 \text{ L}) = 2.0 \times 10^{-4}$ mol

Total amount of silver = $(0.0250 \text{ mol/L})(0.0010 \text{ L}) = 2.5 \times 10^{-5}$ mol

Silver is the limiting reagent in this case. The amount of undissolved silver ion will be quite small compared to the total silver in this case, so the mass of AgCl that forms will be approximately equal to the following.

$$\text{Mass of AgCl} = (2.5 \times 10^{-5} \text{ mol})(143.3 \text{ g/mol}) = 3.6 \times 10^{-3} \text{ g}$$

b) Mass of chloride remaining = $(2.0 \times 10^{-4} \text{ mol} - 2.5 \times 10^{-5} \text{ mol})(35.45 \text{ g/mol}) = 6.2 \times 10^{-3}$ g

c) % Cl^- precipitated = $100 (2.5 \times 10^{-5} \text{ mol})/(2.0 \times 10^{-4} \text{ mol}) = 12.5\%$

66. The solubility of many substances depends, to a great extent, on pH and, to a much smaller extent, on ionic strength. Many substances can be prevented from precipitating by adding a coordinating ligand to mask the metal ion from the precipitating agent.

67. The common ion effect is a specific example of La Châtelier's principle. Adding a higher concentration of one of the ions of an insoluble salt causes greater precipitation of the salt.

68. $[Ag^+] = K_{sp}^{1/2} = (5.35 \times 10^{-13})^{1/2} = 7.3 \times 10^{-7}$ M in pure water

   $[Ag^+] = K_{sp}/[Br^-] = (5.35 \times 10^{-13})/(0.010) = 5.4 \times 10^{-11}$ M in 0.010 M KBr

69. a) Amount of $Cd(IO_3)_2 = (K_{sp}/4)^{1/3} = (2.5 \times 10^{-8}/4)^{1/3} = 1.8 \times 10^{-3}$ mol in a liter

   b) Amount of $Cd(IO_3)_2 = K_{sp}/[IO_3^-]^2 = (2.5 \times 10^{-8})/(0.01)^2 = 2.5 \times 10^{-4}$ mol in a liter

70. a) $[CuI] = (K_{sp})^{1/2} = (1.27 \times 10^{-12})^{1/2} = 1.13 \times 10^{-6}$ M

   b) $[CuI] = K_{sp}/(0.05) = (1.27 \times 10^{-12})/(0.05) = 2.54 \times 10^{-11}$ M

71. a) $K_{sp,AgCl} = [Ag^+][Cl^-] = 1.77 \times 10^{-10}$ and $K_{sp,AgBr} = [Ag^+][Br^-] = 5.35 \times 10^{-13}$

   There is only one silver ion concentration, so the following must also be true.

   $$K_{sp,AgCl}/[Cl^-] = K_{sp,AgBr}/[Br^-]$$

   Rearrange this equation to solve for $[Cl^-]/[Br^-] = (1.77 \times 10^{-10})/(5.35 \times 10^{-13}) = 330$.

   Because equal moles of silver and of halide are mixed, $[Ag^+] = [Cl^-] + [Br^-]$.

   $[Ag^+] = [Cl^-] + [Cl^-]/330 = [Cl^-]\{1 + 1/330\} = [Cl^-](1.00303)$

   $1.77 \times 10^{-10} = [Cl^-]\{1.003 [Cl^-]\} = 1.003 [Cl^-]^2$

   $[Cl^-] = \{1.77 \times 10^{-10}/1.003\}^{1/2} = 1.33 \times 10^{-5}$ M

   b) $[Ag^+] = 330 [Br^-] + [Br^-] = 331 [Br^-]$

   $5.35 \times 10^{-13} = 331 [Br^-]^2$

$[Br^-] = \{(5.35 \times 10^{-13})/331\}^{1/2} = 4.02 \times 10^{-8}$ M

c) $[Cl^-] = K_{sp}/[Ag^+] = (1.77 \times 10^{-10})/(0.0050) = 3.5 \times 10^{-8}$ M

d) $[Br^-] = K_{sp}/[Ag^+] = (5.35 \times 10^{-13})/(0.0050) = 1.1 \times 10^{-10}$ M

72. If the anion is the conjugate base of a weak acid, adding a strong acid will protonate the anion and increase the solubility of the ionic solid. An example is calcium fluoride. Adding HCl will protonate the fluoride to make HF, thereby lowering the actual concentration of F⁻ and increasing the solubility of CaF_2.

    If a ligand such as ammonia or EDTA is added to the solution of a metal ion that forms a strong complex with that ligand, the solubility will be increased because most of the soluble metal ion will be in the form of the complex ion. An example of this occurs when ammonia is added to a silver ion solution in the presence of chloride ion. Adding ammonia will cause AgCl to dissolve even in the presence of excess chloride.

73. a) $[OH^-] = (K_{sp}/[Fe^{3+}])^{1/3} = \{(2.79 \times 10^{-39})/(1.0 \times 10^{-5})\}^{1/3} = 6.5 \times 10^{-12}$ M

    pOH = -log(6.5 × 10⁻¹²) = 11.19 or pH = 14.00 − 11.19 = 2.81

    b) $[OH^-] = 10^{-6.5} = 3.16 \times 10^{-7}$ M

    $[Fe^{3+}] = (2.79 \times 10^{-39})/(3.16 \times 10^{-7})^3 = 8.8 \times 10^{-20}$ M

74. $K_{sp} = [Mg^{2+}][OH^-]^2 = \{[OH^-]/2\}[OH^-]^2$

    $[OH^-]^3 = 2(K_{sp})$ $[OH^-] = \{2(5.61 \times 10^{-12})\}^{1/3} = 2.24 \times 10^{-4}$ M

    pOH = -log(2.24 × 10⁻⁴) = 3.65

    pH = 14.00 − 3.65 = 10.35

79. a) Solubility product: $K_{sp} = [Ba^{2+}][SO_4^{2-}] = 1.08 \times 10^{-10}$

   Acid-dissociation constant: $K_{a2} = [H^+][SO_4^{2-}]/[HSO_4^-] = 1.03 \times 10^{-2}$

   Mass balance: $[Ba^{2+}] = [HSO_4^-] + [SO_4^{2-}]$

   $[Ba^{2+}] = [H^+][SO_4^{2-}]/K_{a2} + [SO_4^{2-}] = [SO_4^{2-}]\{1 + (1.0 \times 10^{-2})/(1.03 \times 10^{-2})\} = [SO_4^{2-}](2)$

   $[Ba^{2+}]\{[Ba^{2+}]/2\} = 1.08 \times 10^{-10}$

   $[Ba^{2+}] = \{2.16 \times 10^{-10}\}^{1/2} = 1.5 \times 10^{-5}\ M$

   b) Assume a body weight of 70 kg (154 lb).

   Toxicity limit = $\{(0.0192\ g/kg)(70\ kg)/(244.3\ g/mol)\}/(1\ L) = 5.5 \times 10^{-3}\ M$

   Therefore, the barium ions that are produced from $BaSO_4$ are well below the toxicity limit.

80. a) Solubility product: $K_{sp} = [Ca^{2+}][F^-]^2 = 5.3 \times 10^{-9}$

   Base ionization constant: $K_b = [HF][OH^-]/[F^-] = 1.5 \times 10^{-11}$

   Mass balance: $2[Ca^{2+}] = [HF] + [F^-]$

   pH = 2.00, so $[OH^-] = 1.0 \times 10^{-12}\ M$

   $2[Ca^{2+}] = K_b[F^-]/[OH^-] + [F^-] = [F^-]\{K_b/[OH^-] + 1\} = [F^-]\{(5.3 \times 10^{-9}//(1.0 \times 10^{-12})\}$

   $[F^-] = 2[Ca^{2+}]/(5.3 \times 10^3) = [Ca^{2+}](3.77 \times 10^{-4})$

   $[Ca^{2+}]\{[Ca^{2+}](3.77 \times 10^{-4})\}^2$

   $[Ca^{2+}]^3 = 1.42 \times 10^{-7}$

   $[Ca^{2+}] = 5.2 \times 10^{-3}\ M$

   This is a much greater concentration than would be present if the solution were not acidic.

   b) At pH 2.00, the fraction of fluoride as F^- can be calculated as shown below.

   $\alpha_{F^-} = K_w/\{K_w + K \cdot 10^{-pH}\} = (1.01 \times 10^{-14})/\{1.01 \times 10^{-14} + (1.5 \times 10^{-11})(1.0 \times 10^{-2})\}$

   $= 0.0625$

If NaF is the only important source of fluoride, then the following is true.

$[F^-] = 0.0625(0.010) = 6.25 \times 10^{-4}\,M$

$[Ca^{2+}] = K_{sp}/[F^-]^2 = (5.3 \times 10^{-9})/(6.25 \times 10^{-4})^2 = 0.0136\,M$

However, we must consider the total fluoride from both NaF and CaF$_2$, as shown below.

$[F^-] = 0.0625\{(0.010) + 2\,[Ca^{2+}]\} = 6.25 \times 10^{-4} + 0.125\,[Ca^{2+}]$

$[Ca^{2+}] = K_{sp}/[F^-]^2 = (5.3 \times 10^{-9})/(6.25 \times 10^{-4} + 0.125\,[Ca^{2+}])^2$

$\qquad = (5.3 \times 10^{-9})/(3.91 \times 10^{-7} + 1.56 \times 10^{-4}\,[Ca^{2+}] + 0.0156\,[Ca^{2+}]^2)$

$[Ca^{2+}]\,(3.91 \times 10^{-7} + 1.56 \times 10^{-4}\,[Ca^{2+}] + 0.0156\,[Ca^{2+}]^2) = 5.3 \times 10^{-9}$

$0.0156\,[Ca^{2+}]^3 + 1.56 \times 10^{-4}\,[Ca^{2+}]^2 + 3.91 \times 10^{-7}\,[Ca^{2+}] - 5.3 \times 10^{-9} = 0$

$[Ca^{2+}] = 4.1 \times 10^{-3}\,M$

CHAPTER 8: ACID–BASE REACTIONS

1. In the Arrhenius model, an acid is a substance that, when added to water, causes hydrogen ions to be formed. A base is a substance that, when added to water, causes hydroxide ions to be formed. Typical Arrhenius acids are HCl, H_2SO_4, and CH_3COOH. Typical Arrhenius bases are NaOH, $Ca(OH)_2$, and NH_3 (NH_4OH).

2. In the Brønsted–Lowry model, an acid is a substance that reacts with water to form hydronium ions and a base is a substance that reacts to accept a hydrogen ion from an acid. Typical Brønsted acids are HCl, H_2SO_4, and CH_3COOH. Typical Brønsted bases are OH^-, NH_3, F^-, and CH_3COO^-. Water acts both as a Brønsted base and acid.

    Conjugate acids and conjugate bases are compounds that form from bases and acids, respectively, and differ from these parent species only by the presence or absence of a hydrogen ion.

3. Chemist 1 is using the Brønsted definition and Chemist 2 is using the Arrhenius model. Arrhenius thought of bases as having a hydroxide ion in their structure that could be lost by ionization. Brønsted recognized that the atoms making up the hydroxide ion can originate as part of a water molecule that donates an H^+ to the base and leaves behind its conjugate base, a hydroxide ion.

4. Acids and bases can be used as titrants in acid–base titrations. Both can also be used to adjust and control the pH of a solution. This pH control can be used to adjust other types of reactions that make use of acids or bases. Examples include the use of pH control and

variation in methods such as precipitation, electrochemical techniques, and separation methods.

5. A strong acid is an acid that essentially undergoes complete dissociation in a solvent, such as the reaction of HCl in water to form hydronium ions and Cl⁻. A weak acid is an acid that has only partial dissociation in a solvent to form its conjugate base. HF and acetic acid (CH_3COOH) are typical weak acids.

6. An acid dissociation constant is an equilibrium constant describing the ionization (or "dissociation") of an acid when dissolved in a given solvent. The value of an acid dissociation constant decreases as we go from a strong acid to a weak acid.

7. a) $K_a^\circ = (a_{H_3O^+})(a_{OCN^-})/(a_{HOCN})(a_{H_2O})$

   b) $K_a^\circ = (a_{H_3O^+})(a_{C_2H_5COO^-})/(a_{C_2H_5COOH})(a_{H_2O})$

   c) $K_a^\circ = (a_{S^{2-}})(a_{H_3O^+})/(a_{HS^-})(a_{H_2O})$

   d) $K_a^\circ = (a_{CH_3COOH_2^+})(a_{CH_3COO^-})/(a_{CH_3COOH})^2$

8. a) $K_a = [OCN^-][H_3O^+]/[HOCN]$

   b) $K_a = [C_2H_5COO^-][H_3O^+]/[C_2H_5COOH]$

   c) $K_a = [S^{2-}][H_3O^+]/[SH^-]$

   d) $K_a = [CH_3COOH_2^+][CH_3COO^-]/(a_{CH_3COOH})^2 = [CH_3COOH_2^+][CH_3COO^-]$

9. a) $H_3PO_4 > H_2PO_4^- > HPO_4^{2-}$

   b) $HClO > HBrO > HIO$

   c) $H_3PO_2 > H_3PO_3 > H_3PO_4$

In (a), all three species have P in the same oxidation state (+5), but different numbers of hydrogen atoms are bonded to oxygen. The acid with the largest number of such H atoms is strongest. The other two are partially neutralized with base.

In (b), the three acids have the same structure, but they have different halogen atoms as the central atom. The acid containing the smallest halogen, with the highest electronegativity, is the strongest acid.

In (c), these three acids have different numbers of oxygens bonded to phosphorous, which means that the P atoms are in different oxidation states. The acid with the lowest oxidation state is the strongest acid for the loss of the first H^+. Even though each has three H atoms, only the ones bonded to oxygen are acidic. Thus, phosphoric is triprotic, phosphorous is diprotic, and hypophosphorous is monoprotic.

10. a) $CH_2ClCOOH \sim CH_2BrCOOH > CH_3COOH$

   b) 4-nitrophenol ~ 2 nitrophenol > 3 nitrophenol >> phenol

   c) formic > acetic ~ butanoic ~ propanoic

   In (a), the presence of the highly electronegative halogen on the alpha carbon makes loss of the acidic hydrogen easier.

   In (b), nitro groups in the 2- or 4-position (ortho or para to the OH) have a similar effect in making the acids stronger than the nonsubstituted phenol. A nitro in the 3-(meta) position has less of an effect, but the resulting acid is still much stronger than phenol.

   In (c), formic acid has only the carbon atom of the carboxylic acid. Formic acid is considerably stronger than the other three acids shown, which are quite similar in strength.

11. A strong base is a base that totally ionizes to form hydroxide ions in water (or creates a species that then accepts protons from the solvent). Examples of strong bases in water are NaOH, KOH, LiOH, and $(CH_3)_4NOH$. A weak base in water is a base that only partly ionizes or accepts a proton from the solvent. An example of a weak base is ammonia.

12. A base ionization constant is a special equilibrium constant that is used to describe the strength of a base and its ability to accept a hydrogen ion from an acid. The value of this constant increases with the strength of a base.

13. a) $K_b = (a_{RNH_3+})(a_{OH-})/(a_{RNH_2})(a_{H_2O})$

 b) $K_b = (a_{HClO})(a_{OH-})/(a_{ClO-})(a_{H_2O})$

 c) $K_b = (a_{RCOOH})(a_{OH-})/(a_{RCOO-})(a_{H_2O})$

 d) $K_b = (a_{CH_3OH_2+})(a_{CH_3O})/(a_{CH_3OH})^2$

14. a) $K_b = [RNH_3^+][OH^-]/[RNH_2]$

 b) $K_b = [HClO][OH^-]/[ClO^-]$

 c) $K_b = [RCOOH][OH^-]/[RCOO^-]$

 d) $K_b = [CH_3OH_2^+][CH_3O^-]/(a_{CH_3OH})^2 = [CH_3OH_2^+][CH_3O^-]$

15. a) Methylamine is slightly stronger than ethylamine, which is considerably stronger than ammonia.

 b) $PO_4^{3-} > HPO_4^{2-} > H_2PO_4^-$

 c) Piperidine >> pyridine > 2,2'-bipyridine

 In (a), the alkyl groups make the nitrogen a stronger base than ammonia.

In (b), the strongest base in this set is the phosphate ion. The other two bases in this list are partially neutralized by acid.

In (c), the nonaromatic compound piperidine is much more basic than either of the two aromatic heterocyclic nitrogen compounds that are shown.

16. An amphiprotic substance is a chemical that can either donate or accept a hydrogen ion. Water can accept a hydrogen ion to form hydronium ion or can lose a hydrogen ion to form hydroxide ion.

17. Autoprotolysis is the act of one molecule donating a hydrogen ion to another molecule of the same compound.

   Autoprotolysis of water: $H_2O + H_2O \rightleftarrows H_3O^+ + OH^-$

   Autoprotolysis of methanol: $CH_3OH + CH_3OH \rightleftarrows CH_3OH_2^+ + CH_3O^-$

18. An "autoprotolysis constant" is an equilibrium constant that is used to describe an autoprotolysis reaction. The thermodynamic and concentration-dependent autoprotolysis constants for water are given below.

   $K_w^o = (a_{H_3O+})(a_{OH-})/(a_{H_2O})^2$ $K_w = [H_3O^+][OH^-]$

19. a) $[OH^-] = K_w/[H^+]$
   $= (1.01 \times 10^{-14})/(2.7 \times 10^{-4}) = 3.7 \times 10^{-11}\ M$

   b) $[H^+] = K_w/[OH^-]$
   $= (1.01 \times 10^{-14})/(5.10 \times 10^{-5}) = 1.98 \times 10^{-10}\ M$

   c) $[H^+] = (K_w)^{1/2} = (2.92 \times 10^{-15})^{1/2} = 5.40 \times 10^{-8}\ M$

20. a) $[H^+] = (0.365 \text{ g}/1000 \text{ L})(1 \text{ mol}/63 \text{ g})(1 \text{ mol } H^+/\text{mol } HNO_3) = 5.8 \times 10^{-6} \text{ M}$

   b) $[H^+] = (36.5 \text{ g}/1000)(1 \text{ mol}/63 \text{ g})(1 \text{ mol } H^+/\text{mol } HNO_3 = 5.8 \times 10^{-4} \text{ M}$

   Trout could survive in the first type of environment, but not in the second.

21. The "leveling effect" is a consequence of strong acids reacting quantitatively to form hydronium ions in water. No acid stronger than a hydronium ion can exist in water. The same thing is true for bases. No base stronger than a hydroxide ion can exist in water. A similar effect occurs in other solvents that can also act as acids or bases.

22. $HClO_4$ and HCl are both strong acids in water. However, in concentrated (glacial) acetic acid, $HClO_4$ is still a strong acid and HCl is not. A basic solvent, such as pyridine or ethylenediamine, can be used to discriminate among the strengths of strong bases.

23. The relationship between an acid and its conjugate base in water is described for a dilute solution by the relationship $K_w = K_a \cdot K_b$, where K_w is the autoprotolysis constant for water, K_a is the dissociation constant for an acid and K_b is the base ionization constant for its conjugate base. If a different solvent is present, a different autoprotolysis constant will be needed in place of K_w.

24. a) $HCOO^-$ $K_b = K_w/K_a = (1.01 \times 10^{-14})/(1.77 \times 10^{-4}) = 5.71 \times 10^{-11}$

   b) $C_6H_5O^-$ $K_b = (1.01 \times 10^{-14})/(1.28 \times 10^{-10}) = 7.89 \times 10^{-5}$

   c) $HCrO_4^-$ $K_b = (1.01 \times 10^{-14})/(1.8 \times 10^{-1}) = 5.6 \times 10^{-14}$

   d) CrO_4^{2-} $K_b = (1.01 \times 10^{-14})/(3.20 \times 10^{-7}) = 3.16 \times 10^{-8}$

25. a) NH_4^+ $K_a = K_w/K_b = (1.01 \times 10^{-14})/(1.80 \times 10^{-5}) = 5.67 \times 10^{-10}$

   b) $C_5H_6N^+$ $K_a = (1.01 \times 10^{-14})/(1.6 \times 10^{-9}) = 6.3 \times 10^{-6}$

   c) HPO_4^{2-} $K_a = 4.22 \times 10^{-13}$ from Table 8.1

   d) $H_2PO_4^-$ $K_a = 6.34 \times 10^{-8}$ from Table 8.1

26. This equation assumes that solvent is water and that the activity of water is one. This second assumption means that the acid–base solution should be relatively dilute for the expression to be strictly valid.

27. Notational definition: $pH = -\log(a_{H+}) \approx -\log([H^+])$

   This definition and equation are valuable when you are carrying out chemical calculations that involve hydrogen ions, but this same relationship is not a practical definition for pH because it describes a pH based on the activity of a single ion (H^+). This is a problem because this activity cannot be measured independently from the counter ions (for instance, OH^-) that must also be present to maintain charge balance in the system. A practical consequence of this fact is that any pH measurement must be performed by a method that has been calibrated with reference solutions having accepted pH values.

28. At pH 6.8, $[H^+] = 10^{-6.8} = 1.\underline{6} \times 10^{-7} M$

   At pH 5.2, $[H^+] = 10^{-5.2} = 6.\underline{3} \times 10^{-6} M$

29. At pH 7.2: $a_{H+} = 10^{-7.2} = 6.3 \times 10^{-8}$ (upper limit of normal range in blood)

   At pH 7.6: $a_{H+} = 10^{-7.6} = 2.5 \times 10^{-8}$ (lower limit of normal range in blood)

   At pH 7.4: $a_{H+} = 10^{-7.4} = 4.0 \times 10^{-8}$ (most common value in blood)

30. a) pH = 5.20 b) pK_b = 7.084
     c) pOH = 14.00 − 2.667 = 11.332 d) pCl = 3.82

31. a) $[H^+] = 5 \times 10^{-10}\,M$

     b) $K_a = 6.0 \times 10^{-3}$

     c) $[OH^-] = 1.6 \times 10^{-3}\,M$

     d) $[Ca^{2+}] = 2 \times 10^{-6}\,M$

32. This relationship arises by taking the negative log of each side of the equation $K_w = [H^+][OH^-]$, which describes the autoprotolysis constant for water.

33. The value of pK_w only equals 14.000 at 24 °C but is quite close to that value at 25 °C (13.997). At other temperatures, it has considerably different values. At 20 °C, the sum of pH and pOH in water is 14.17 and at 10 °C the sum of these terms is 14.54.

34. Any solution having a greater concentration of hydrogen ion than hydroxide ion is regarded as acidic. At 25 °C that means that any pH less than 7.0 is acidic and any pH higher than 7.0 is basic. A solution having a pH equal to 7.0 is neutral.

35. a) An increase in acid concentration lowers the pH.

     b) An increase in base concentration raises the pH.

     c) Changing the solvent from water to acetic acid will lower the pH.

36. The values of K_a, K_b, and K_w are temperature dependent, so the solution pH will change with temperature due to the change in temperature of these values. Also, for a strong acid or base, a

temperature change will change the solution density, which results in a change in the molarity (but not the molality) and of the solution. This effect will also change the pH. Finally, a variation in temperature can alter activity coefficients, such as for H^+, which will further affect the pH.

37. A monoprotic acid is an acid that can donate only one hydrogen ion to a base. Likewise, a monoprotic base is a base that can accept only one hydrogen ion. Examples are HCl and NaOH.

38. a) pH = -log(0.030) = 1.52

    b) pH = 14.00 + log (0.0060) = 11.78

    c) pH = -log({0.50 g/L}/{63.0 g/mol}) = 2.10

    d) pH = 14.00 + log({0.2 g/L}/{40.0 g/mol}) = 11.7

    Each of these calculations assumes that complete dissociation is present and that the activity coefficients for the resulting species are all 1.00.

39. pH = 14.00 − log(0.0600) = 12.778

    [HCl] = [NaOH](25 mL)/(20 mL) = 0.075 M

40. [HCl] = (12.0 M){(10.00 mL)/(250.00 mL)} = 0.480 M

    pH = -log(0.480) = 0.319

41. Moles of HCl = (0.100 mol/L)(0.02500 L) = 2.50×10^{-3} mol

    Moles of NaOH = (0.0500 mol/L)(0.01000 L) = 5.00×10^{-4} mol

    After reaction [H^+] = (2.00×10^{-3} mol/0.035 L) = 0.0571 M or   pH = 1.243

42. At very high concentrations, the activity coefficients will differ dramatically from 1.00. At very low concentrations, you must include the dissociation of water as a source of hydrogen or hydroxide ions when estimating the pH.

43. Each of these solutions are too dilute to use the approximation that concentration of the solute is the same as the concentration of H^+ or OH^-. Therefore $[OH^-]$ in Parts (a) and (d) can be solved instead by using a more complete expression, such as $0 = [OH^-]^2 - C[OH^-] - K_w$. In the same manner, $[H^+]$ in Parts (b) and (c) must be solved using $0 = [H^+]^2 - C[H^+] - K_w$. (Note: This same equation can also be used in Parts (a) and (d) because $[OH^-]$ can be related to $[H^+]$ through $[OH^-] = K_w/[H^+]$.)

   a) $0 = [OH^-]^2 - C[OH^-] - K_w$ $[OH^-] = 1.16 \times 10^{-7} M$ pOH = 6.94   pH = 7.06

   b) $0 = [H^+]^2 - C[H^+] - K_w$ $[H^+] = 6.26 \times 10^{-7} M$ pH = 6.20

   c) $0 = [H^+]^2 - C[H^+] - K_w$ $C = \{(1.0 \times 10^{-6} \text{ g/L})/(80.9 \text{ g/mol})\} = 1.2\underline{4} \times 10^{-8} M$

           $[H^+] = 1.06\underline{9} \times 10^{-7} M$ pH = 6.97

   d) $0 = [OH^-]^2 - C[OH^-] - K_w$ $C = \{(0.2 \times 10^{-6} \text{ g/L})/(40.0 \text{ g/mol})\} = 5.0 \times 10^{-9} M$

           $[OH^-] = 1.0\underline{3} \times 10^{-7} M$ pOH = 6.99   pH = 7.01

44. a) $[HCl] = (5.00 \times 10^{-4})(1.00/1000.00) = 5.00 \times 10^{-7} M$

   $[H^+]^2 - C[H^+] - K_w = 0$

   $[H^+]^2 - (5.00 \times 10^{-7})[H^+] - 1.00 \times 10^{-14} = 0$

   $[H^+] = 5.19 \times 10^{-7} M$

   pH = 6.28

   b) $[H^+]^2 - (5.00 \times 10^{-7})[H^+] - 6.81 \times 10^{-15} = 0$

$[H^+] = 5.13 \times 10^{-7}\ M$

pH = 6.29

45. Ionic strength of a strong acid or base causes the activity coefficient values to differ from 1.0 and, therefore, the concentration is quite different from the activity. The same holds true for weak acids and bases, but for these chemicals there is the additional complication that the acidity or basicity constants also depend on ionic strength. These features become most important at high values of ionic strength (i.e., values generally greater than 0.01 M).

46. Strong acids and strong bases undergo essentially complete ionization. In water, a strong acid or strong base is an acid or base that essentially has a complete reaction with water to form hydrogen ion or hydroxide ions. In this case, the initial concentration of the strong acid or strong base can be used to easily obtain the concentration of H^+ or OH^- and pH. Weak acids or weak bases ionize only to a small extent, as governed by the relative strength of the acid or base. In this situation, more complex equations are needed to account for the incomplete ionization and to allow pH to be determined from the amount of acid or base that is present in the solution.

47. To show the difference between using a simplified equation and the appropriate quadratic equation, these answers will be calculated both ways.

	Simplified approach	Quadratic approach
a)	$[H^+] = (K_a C)^{1/2}$	$[H^+]^2 + K_a [H^+] - K_a C = 0$
	$[H^+] = \{(6.28 \times 10^{-5})(0.0500\ M)\}^{1/2}$	$[H^+]^2 + (6.28 \times 10^{-5})[H^+] - (3.14 \times 10^{-6}) = 0$
	$[H^+] = 1.77 \times 10^{-3}\ M$	$[H^+] = 1.74 \times 10^{-3}\ M$

pH = 2.75	pH = 2.76

b) $[H^+] = \{(1.77 \times 10^{-4})(1.75 \times 10^{-4})\}^{1/2}$ $[H^+]^2 + 1.77 \times 10^{-4} [H^+] - 3.10 \times 10^{-8} = 0$

$[H^+] = 1.76 \times 10^{-4} \, M$ $[H^+] = 1.09 \times 10^{-4} \, M$

pH = 3.75   pH = 3.96

c) $C = (20 \text{ g/L})/(167.0 \text{ g/mol}) = 0.12 \, M$

$[H^+] = \{(0.12)(6.6 \times 10^{-3})\}^{1/2}$ $[H^+]^2 + (6.6 \times 10^{-3})[H^+] - 7.9 \times 10^{-4} = 0$

$[H^+] = 2.8 \times 10^{-2} \, M$ $[H^+] = 2.5 \times 10^{-2} \, M$

pH = 1.55   pH = 1.60

d) $C = (0.150 \text{ g/L})(1 \text{ mol}/94.5 \text{ g}) = 1.59 \times 10^{-3} \, M$

$[H^+] = \{(1.59 \times 10^{-3})(1.40 \times 10^{-3})\}^{1/2}$ $[H^+]^2 + (1.40 \times 10^{-3})[H^+] - 2.22 \times 10^{-6} = 0$

$[H^+] = 1.49 \times 10^{-3} \, M$ $[H^+] = 9.5 \times 10^{-4} \, M$

pH = 2.83   pH = 3.02

The ratio of C/K_a for these four answers are (a) 796, (b) 0.97, (c) 18, and (d) 1.17. Only the Part (a) has a value of C/K_a that approaches 1000 and only this answer gives values for pH that are in good agreement by the two calculation methods. The ionic strength for all four solutions is low enough that any activity corrections would be minimal.

48. To show the difference between using a simplified equation and the appropriate quadratic equation, these answers will be calculated both ways. (Note: The quadratic equations used in this case are related to those given in the chapter through the fact that $[OH^-] = K_w/[H^+]$ and $K_a K_b = K_w$.)

a) Simplified approach:

$[OH^-] = \{(0.0040 \, M)(1.80 \times 10^{-5})\}^{1/2} = 2.68 \times 10^{-4} \, M$

pOH = 3.57   pH = 10.43

Quadratic approach:

$$[OH^-]^2 + K_b[OH^-] - K_b C = 0 \qquad [OH^-]^2 + (1.80 \times 10^{-5}) - 7.20 \times 10^{-8} = 0$$

$[OH^-] = 2.59 \times 10^{-4}\ M$ pOH = 3.59   pH = 10.41

b) Simplified approach:

$$[OH^-] = \{(6.50 \times 10^{-5}\ M)(1.6 \times 10^{-9})\}^{1/2} = 3.22 \times 10^{-7}\ M$$

pOH = 6.49   pH = 7.51

Quadratic approach:

$$[OH^-]^2 + (1.6 \times 10^{-9})[OH^-] - 1.04 \times 10^{-13} = 0$$

$[OH^-] = 3.22 \times 10^{-7}\ M$

pOH = 6.49   pH = 7.51

c) Simplified approach:

$$[OH^-] = \{(7.25 \times 10^{-4}\ M)(1.0 \times 10^{-14})/(3.3 \times 10^{-4})\}^{1/2} = 1.4\underline{9} \times 10^{-7}\ M$$

pOH = 6.83   pH = 7.17

Quadratic approach:

$$[OH^-]^2 + (3.03 \times 10^{-11})[H^+] - 2.197 \times 10^{-14} = 0$$

$[OH^-] = 1.49 \times 10^{-7}\ M$ pOH = 6.83   pH = 7.17

d) Simplified approach:

$$[OH^-] = \{(9.5 \times 10^{-3})(4.47 \times 10^{-4})\}^{1/2} = 2.06 \times 10^{-3}\ M$$

pOH = 2.69   pH = 11.31

Quadratic approach:

$$[OH^-]^2 + (4.47 \times 10^{-4})[OH^-] - 4.25 \times 10^{-6} = 0$$

$[OH^-] = 1.84 \times 10^{-3} \, M$ pOH = 2.73 pH = 11.27

The values of C/K_b in the above problems are (a) 222, (b) 3.8×10^4, (c) 2.4×10^7, and (d) 20. The answer to (d) shows the least agreement between the two types of calculations.

49. Using the simplified approach:

$$[H^+] = \{(0.80 \, M)(6.8 \times 10^{-4})\}^{1/2} = 2.33 \times 10^{-2} \, M \qquad pH = 1.63$$

50. a) Simplified approach:

$$[H^+] = \{(1.28 \times 10^{-10})(2.5 \times 10^{-3})\}^{1/2} = 5.66 \times 10^{-7} \, M \qquad pH = 6.25$$

Quadratic approach:

$$[H^+]^2 + (1.28 \times 10^{-10})[H^+] - 3.20 \times 10^{-13} = 0$$

$[H^+] = 5.66 \times 10^{-7} \, M$ pH = 6.25

b) Simplified approach:

$$[H^+] = \{(2.95 \times 10^{-8})(8.1 \times 10^{-4})\}^{1/2} = 4.89 \times 10^{-6} \, M \qquad pH = 5.31$$

Quadratic approach:

$$[H^+]^2 + (2.95 \times 10^{-8})[H^+] - 2.38\underline{9} \times 10^{-11} = 0$$

$[H^+] = 4.87 \times 10^{-6} \, M$ pH = 5.31

c) Simplified approach (using K_a from *CRC Handbook of Chemistry and Physics*):

$$C = (1.0 \times 10^{-3} \, g/L)/(99.17 \, g/mol) = 1.01 \times 10^{-5} \, M$$

$$K_b = (1.01 \times 10^{-14})/(2.19 \times 10^{-11}) = 4.62 \times 10^{-4}$$

$$[OH^-] = \{(1.01 \times 10^{-5})(4.62 \times 10^{-4})\}^{1/2} = 6.83 \times 10^{-5} \, M$$

pOH = 4.16   pH = 9.83

Quadratic approach:

$[OH^-]^2 + (4.62 \times 10^{-4})[OH^-] - 4.67 \times 10^{-9} = 0$

$[OH^-] = 9.98 \times 10^{-6}\,M$ pOH = 5.00   pH = 9.00

d) Simpified approach:

$C = (0.200\text{ g/L})/(59.1\text{ g/mol}) = 3.38 \times 10^{-3}\,M$

$[H^+] = \{(3.38 \times 10^{-3})(8.0 \times 10^{-2})\}^{1/2} = 1.64 \times 10^{-2}\,M$ pH = 1.78

Quadratic approach:

$[H^+]^2 + (8.0 \times 10^{-2})[H^+] - 2.70 \times 10^{-4} = 0$

$[H^+] = 3.25 \times 10^{-3}\,M$ pH = 2.49

The value of C/K for the four solutions are (a) 2.5×10^7, (b) 2.7×10^4, (c) 0.022, and (d) 0.042.

This explains why there is good agreement for Parts (a) and (b), but not for (c) and (d).

51. Using the simplified expression:

$K_a = 1.35 \times 10^{-4}$ $[H^+] = 10^{-6.5} = 3.16 \times 10^{-7}\,M$

$[\text{lactic acid}] = [H^+]^2/K_a = (3.16 \times 10^{-7})^2/(1.35 \times 10^{-4}) = 7.39 \times 10^{-10}\,M$

Using the quadratic expression:

$[H^+]^2 + K_a[H^+] - K_a C = 0$ rearranges to $C = \{[H^+]^2 + K_a[H^+]\}/K_a$

$C = \{(3.16 \times 10^{-7})^2 + (3.16 \times 10^{-7})(1.35 \times 10^{-4})\}/(1.35 \times 10^{-4}) = 3.17 \times 10^{-7}\,M$

Of course, the lactic acid in muscle cells is not this dilute at this pH because it is partially neutralized, and its matrix is buffered by other components.

52. a) $C = (0.010 \text{ g}/0.10000 \text{ L})(1 \text{ mol}/176 \text{ g}) = 5.68 \times 10^{-4}$ M and $K_a = 7.94 \times 10^{-5}$

$$[H^+]^2 + (7.94 \times 10^{-5})[H^+] - 4.51 \times 10^{-8} = 0$$

$[H^+] = 1.76 \times 10^{-4}$ M \qquad pH = 3.75

b) $C = (0.010 \text{ g}/0.10000 \text{ L})(1 \text{ mol}/175 \text{ g}) = 5.71 \times 10^{-4}$ M and $K_b = 1.26 \times 10^{-10}$

$$[OH^-] = \{(5.71 \times 10^{-4})(1.26 \times 10^{-10})\}^{1/2} = 2.68 \times 10^{-7} \ M$$

pOH = 6.57

pH = 14.00 − 6.57 = 7.43

53. Quadratic method:

$$[OH^-]^2 + [OH^-] K_b - K_b C = 0$$

$$[OH^-]^2 + [OH^-](1.80 \times 10^{-5}) - (1.80 \times 10^{-5})(9.0 \times 10^{-6}) = 0$$

$[OH^-] = 1.03 \times 10^{-9}$ M \qquad pOH = 8.99 \qquad pH = 5.01

Clearly this result is incorrect as it represents an acidic solution.

The dissociation of water must be considered for this dilute ammonia solution to give a correct answer. This leads to the following cubic equation that can be solved by successive approximations or some comparable method.

For 9 μM:

$$[OH^-]^3 + K_b [OH^-]^2 - (K_w + C K_b) - K_w K_b = 0$$

$$[OH^-]^3 + (1.80 \times 10^{-5})[OH^-]^2 - (1.01 \times 10^{-14} + (9.0 \times 10^{-6})(1.80 \times 10^{-5})[OH^-]$$

$$- (1.01 \times 10^{-14})(1.80 \times 10^{-5}) = 0$$

$[OH^-] = 6.59 \times 10^{-6}$ M \qquad pOH = 5.18 \qquad pH = 8.82

For 30 μM:

$$[OH^-] = 1.59 \times 10^{-5} \, M \qquad pOH = 4.79\underline{8} \qquad pH = 9.20$$

54. An increase in ionic strength has the effect of raising the effective value of K_a of an acid or K_b of a base. In both cases, a greater extent of ionization occurs. Strong acids and bases are fully ionized so these cannot ionize to any greater extent.

55. This solution is quite dilute ($C = 4.1 \times 10^{-8} \, M$), so the hydrogen ion formed from the autoprotolysis of water must be considered. This requires the solution of a cubic equation that can be solved by successive approximations or some comparable method.

$$[H^+]^3 + K_a[H^+]^2 - (K_w + C\,K_a)[H^+] - K_w K_a = 0$$

$$[H^+]^3 + (6.28 \times 10^{-5})[H^+]^2 - \{1.01 \times 10^{-14} + (4.1 \times 10^{-8})(6.28 \times 10^{-5})\}[H^+]$$
$$- (1.01 \times 10^{-14})(6.28 \times 10^{-5}) = 0$$

$$[H^+]^3 + (6.28 \times 10^{-5})[H^+]^2 - (1.01 \times 10^{-14} + (4.1 \times 10^{-8})(6.28 \times 10^{-5})[H^+]$$
$$- (1.01 \times 10^{-14})(6.28 \times 10^{-5}) = 0$$

$$[H^+]^3 + (6.28 \times 10^{-5})[H^+]^2 - (2.585 \times 10^{-12})[H^+] - (6.28 \times 10^{-19}) = 0$$

$$[H^+] = 1.22 \times 10^{-7} \, M \qquad pH = 6.91$$

56. a) $pH = pK_a + \log\{[A^-]/[HA]\} = 4.757 + \log\{(2.5 \times 10^{-4})/(1.0 \times 10^{-3})\}$

$= 4.757 + (-0.60) = 4.15\underline{4} = 4.15$

b) $pH = 4.202 + \log(5.0 \times 10^{-4}/4.6 \times 10^{-4})$

$= 4.202 + 0.03\underline{6} = 4.23\underline{8} = 4.24$

c) $pH = 9.250 + \log(6.9 \times 10^{-5}/1.5 \times 10^{-4})$

$= 9.250 + (-0.33\underline{7}) = 8.90\underline{9} = 8.91$

57. a) $C = (0.125 \text{ g}/139 \text{ g/mol})/(0.100 \text{ L}) = 9.0 \times 10^{-3} M$ $K_a = 6.8 \times 10^{-8}$

$[H^+] = \{(9.0 \times 10^{-3})(6.8 \times 10^{-8})\}^{1/2} = 2.47 \times 10^{-5} M$ or   pH = 4.61

b) Moles of acid = 9.0×10^{-4} mol

After 25 mL: Moles of NaOH = $(0.0100 M)(0.02500 \text{ L}) = 2.5 \times 10^{-4}$ mol

After 50 mL: Moles of NaOH = $(0.0100 M)(0.05000 \text{ L}) = 5.0 \times 10^{-4}$ mol

In these two cases, the number of moles of conjugate base formed is the same as the moles of NaOH added.

After 100 mL: Moles of NaOH = $(0.0100 M)((0.10000 \text{ L}) = 10.0 \times 10^{-4}$ mol

Because only 9.0×10^{-4} mol of acid were present, the acid becomes the limiting reagent and only 9.0×10^{-4} mol of the conjugate base of nitrophenol will form (i.e., 1.0×10^{-4} mol of hydroxide will remain).

c) After 25 mL: $pH = pK_a + \log\{[A^-]/[HA]\} = 7.17 + \log\{(2.5 \times 10^{-4})/(6.5 \times 10^{-4})\} = 6.76$

After 50 mL: $pH = 7.17 + \log\{(5.0 \times 10^{-4})/(4.0 \times 10^{-4})\} = 7.17 + 0.097 = 7.26$

After 100 mL: $[OH^-] = (1.0 \times 10^{-4} \text{ mol}/0.200 \text{ L}) = 5.0 \times 10^{-4} M$

pOH = 3.30   pH = 10.70

58. a) $C = (0.2500 \text{ g})/(94.1 \text{ g/mol})/(0.2500 \text{ L}) = 0.01063 M$

$pK_b = 14.00 - 9.37 = 4.63$ or $K_b = 2.3\underline{44} \times 10^{-5}$

Simplified approach:

$[OH^-] = \{(0.01064 M)(2.3\underline{44} \times 10^{-5})\}^{1/2} = 4.9\underline{9} \times 10^{-4} M$

pOH = 3.30   pH = 10.70

Quadratic method:

$[OH^-]^2 + [OH^-] K_b - K_b C = 0$

$[OH^-]^2 + [OH^-](2.344 \times 10^{-5}) - (2.344 \times 10^{-5})(0.01064) = 0$

$[OH^-] = 4.88 \times 10^{-4} M$ pOH = 3.31   pH = 10.69

b) moles of base = (0.2500 g)/(94.1 g/mol) = $2.65\underline{7} \times 10^{-3}$ mol

After 50 mL of acid: moles of HCl = (0.01500 M)(0.05000 L) = 7.50×10^{-4} mol

After 100 mL of acid: moles of HCl = (0.01500 M)(0.1000 L) = 1.500×10^{-3} mol

In both cases the number of moles of the conjugate acid of 4-aminopyridine equals the number of moles of HCl added.

c) After adding 50 mL:

$pH = (pK_w - pK_b) - \log([HB^+]/[B])$

$= (14.00 - 4.63) - \log(7.5 \times 10^{-4}/(2.659 - 0.75) \times 10^{-3})$

$= (14.00 - 4.63) - (-0.406)$

$= 9.78$

After adding 100 mL:

$pH = (14.00 - 4.63) - \log[(1.500 \times 10^{-3})/\{(2.659 - 1.50) \times 10^{-3})\}]$

$= (14.00 - 4.63) - 0.113$

$= 9.26$

59. A buffer solution is a solution that contains a mixture of a weak acid and its conjugate base. This type of mixture resists a change in pH with the addition of an acid or a base. Two examples of buffers are acetate and phosphate buffers.

60. The Henderson–Hasselbalch equation has the form shown below. This equation assumes that equilibrium is already present in the solution and that the effect of water dissociation can be safely ignored.

$$pH = pK_a + \log\left(\frac{[A^-]}{[HA]}\right)$$

61. $[H_3PO_4] + [H_2PO_4^-] = 0.050\ M$

$K_{a1} = [H_2PO_4^-][H^+]/[H_3PO_4] = 7.11 \times 10^{-3}$

$[H_2PO_4^-]/[H_3PO_4] = (7.11 \times 10^{-3})/(1.00 \times 10^{-3}) = 7.11$

$[H_3PO_4] + (7.11)[H_3PO_4] = 0.050\ M$

$[H_3PO_4] = (0.050\ M)/(8.11) = 6.16 \times 10^{-3}\ M$

$[H_2PO_4^-] = (0.050\ M) - (6.16 \times 10^{-3}\ M) = 4.38 \times 10^{-2}\ M$

62. $pH = pK_a + \log\{(1.9 \times 10^{-5})/(2.3 \times 10^{-6})\}$

$= 9.89 + \log(8.26)$

$= 9.89 + 0.917 = 10.80\underline{7} = 10.81$

63. $pH = pK_a + \log\{[Ac^-]/[HAc]\} = 4.757 + \log\{7.5 \times 10^{-5}/5.0 \times 10^{-5}\}$

$= 4.757 + 0.17\underline{6} = 4.9\underline{33} = 4.93$

64. a) pH 5.0: use acetic acid ($pK_a = 4.757$)

b) pH 6.0: use MES ($pK_a = 6.27$, as listed in Table 8.9)

c) pH 7.0: use ACES ($pK_a = 6.85$, as listed in Table 8.9)

d) pH 8.0: use TRICINE ($pK_a = 8.14$, as listed in Table 8.9)

65. a) $\log\{[Na_2HPO_4]/[NaH_2PO_4]\} = pH - pKa$

For pH 7.0: $\log\{[Na_2HPO_4]/[NaH_2PO_4]\} = 7.00 - 7.198 = -0.198$

$[Na_2HPO_4]/[NaH_2PO_4] = 0.634$

For pH 6.5: $\log\{[Na_2HPO_4]/[NaH_2PO_4]\} = 6.50 - 7.198 = -0.698$

$[Na_2HPO_4]/[NaH_2PO_4] = 0.200$

For pH = 7.5: $\log\{[Na_2HPO_4]/[NaH_2PO_4]\} = 7.50 - 7.198 = 0.302$

$[Na_2HPO_4]/[NaH_2PO_4] = 2.00$

b) Total concentration of the phosphate species will be 0.20 M. The approximate fractions for HPO_4^{2-} and $H_2PO_4^-$ in each buffer can be found by using the ratios for $[Na_2HPO_4]/[NaH_2PO_4]$ that were calculated in Part (a). At pH 7.0, these fractions are found as shown below by assuming that HPO_4^{2-} and $H_2PO_4^-$ are the only significant phosphate species at the given pH values (i.e., $\alpha_{HPO_4^{2-}} + \alpha_{H_2PO_4^-} = 1.00$).

$\alpha_{HPO_4^{2-}}/\alpha_{H_2PO_4^-} = \alpha_{HPO_4^{2-}}/(1 - \alpha_{HPO_4^{2-}}) = [Na_2HPO_4]/[NaH_2PO_4]$

or $\alpha_{HPO_4^{2-}} = ([Na_2HPO_4]/[NaH_2PO_4])/(1 + [Na_2HPO_4]/[NaH_2PO_4])$

For pH 7.0 ($[Na_2HPO_4]/[NaH_2PO_4] = 0.634$):

$\alpha_{HPO_4^{2-}} = (0.634)/(1 + 0.634) = 0.39$

$\alpha_{H_2PO_4^-} = 1.00 - \alpha_{H_2PO_4^-} = 1.00 - 0.39 = 0.61$

For pH 6.5 ($[Na_2HPO_4]/[NaH_2PO_4] = 0.200$):

$\alpha_{HPO_4^{2-}} = (0.200)/(1 + 0.200) = 0.17$

$\alpha_{H_2PO_4^-} = 1.00 - \alpha_{H_2PO_4^-} = 1.00 - 0.17 = 0.83$

For pH = 7.5 ($[Na_2HPO_4]/[NaH_2PO_4] = 2.00$):

$\alpha_{HPO_4^{2-}} = (2.00)/(1 + 2.00) = 0.67$

$$\alpha_{H2PO4-} = 1.00 - \alpha_{H2PO4-} = 1.00 - 0.67 = 0.33$$

c) For pH 7.2 $\log\{[Na_2HPO_4]/[NaH_2PO_4]\} = 7.2 - 7.198 = 0.00\underline{2}$ (essentially zero)

$$[Na_2HPO_4]/[NaH_2PO_4] = 1.0$$

Therefore, equal volumes (500 mL each) of the two solutions are needed.   Each solution will require (0.500 L)(0.20 mol/L) = 0.100 moles of solute.

For $Na_2HPO_4 \cdot 2H_2O$:   (178 g/mol)(0.100 mol) = 17.8 g

For $NaH_2PO_4 \cdot 2H_2O$:   (156 g/mol)(0.100 mol) = 15.6 g

66. a) $C = (10.68/61.83)/0.2500 = 0.6908\ M$

Simplified approach:

$$[H^+] = (K_a C)^{1/2} = \{(5.79 \times 10^{-10})(0.6908)\}^{1/2} = 2.00 \times 10^{-5}\ M \qquad pH = 4.70$$

Quadratic approach:

$$[H^+]^2 + [H^+] K_a - K_a C = 0$$

$$[H^+]^2 + [H^+](5.79 \times 10^{-10}) - (5.79 \times 10^{-10})(0.6908) = 0$$

$$[H^+] = 2.00 \times 10^{-5}\ M \qquad\qquad pH = 4.70$$

b)   Amount borate = moles NaOH = (0.0500 L)(0.500 M) = 0.0250 mol

Amount boric acid = (0.100 L)(0.6908 mol/L) − moles NaOH

$$= 0.06908\ mol - 0.02500\ mol = 0.04408\ mol$$

$$pH = pK_a + \log\{[borate]/[boric\ acid]\}$$

$$= pK_a + \log\{mol\ borate/mol\ boric\ acid\}$$

(Note: This second equation is true because borate and boric acid are in the same solution volume, so their volumes cancel in the ratio of [borate]/[boric acid].)

$$pH = 9.237 + \log[(0.02500\ mol)/(0.04408\ mol)]$$

$= 9.237 + \log(0.5672)$

$= 9.237 - 0.246 = 8.991$

67. All buffers resist a change in pH but can be overwhelmed by addition of too much acid or base. A concentrated buffer will require a larger amount of acid or base to change its pH than will a dilute buffer. The "buffer capacity" is defined as the moles of strong acid or strong base that must be added per liter of buffer to produce a pH change of 1.0 units in the buffer. This value is useful in determining how effective a buffer will be in resisting a change in pH when a small amount of acid or base is added to such a solution.

68. At pH 5.00, $\log\{[Py]/[HPy^+]\} = 5.00 - 5.25 = -0.25$, so $[Py]/[HPy^+] = 0.562$

    $[Py] + [HPy^+] = 0.050\ M$

    $\qquad = (0.562)[HPy^+] + [HPy^+]$

    $\qquad = (1.562)[HPy^+]$

    By combining the above information, we get $[HPy^+] = 0.320\ M$ and $[Py] = 0.0180\ M$

    $6.00 = 5.25 + \log\{(0.0320 + x)/(0.0180 - x)\}$

    $0.75 = \log\{(0.0320 + x)/(0.0180 - x)\}$

    Taking the anti-logarithm of both sides gives the following result.

    $5.623 = (0.0320 + x)/(0.0180 - x)$

    $5.623(0.0180 - x) = (0.0320 + x)$

    $0.1012 - 5.623\ x = 0.0320 + x$

    $0.0692 = 6.623\ x$

    $x = 0.01045$ mol   (i.e., the moles of NaOH that will raise the pH by 1.0 units)

69. Malic acid is a diprotic acid having $pK_{a1} = 3.40$ and $pK_{a2} = 5.11$.

At pH 3.40, all three species are present to some extent.

$$\alpha_{H_2A} = 1/\{1 + 0.980 + 0.019\} = 1/(1.999) = 0.500$$

$$\alpha_{HA^-} = (0.980)/(1.999) = 0.490$$

$$\alpha_{A^{2-}} = (0.019)/(1.999) = 0.0095$$

If an acid is added:

$$2.40 = 3.40 + \log\{(0.10 - x)/(0.10 + x)\}$$

$$-1.00 = \log\{(0.10 - x)/(0.10 + x)\}$$

$$0.10 = (0.10 - x)/(0.10 + x)$$

$$0.10(0.10 + x) = (0.10 - x)$$

$$0.010 + 0.10\,x = 0.10 - x$$

$$1.1\,x = 0.09$$

$x = 0.082$ mol HCl needed to lower the pH by 1.0 units

If a base is added:

(Note: Malic acid will begin to form the dianion to a significant extent, which makes the question of buffer capacity trickier when a base is added.)

At pH 4.40, the fractions of the three forms become as follows.

$$\alpha_{H_2A} = 1/\{1 + 9.799 + 1.919\} = 1/(12.718) = 0.0786$$

$$\alpha_{HA^-} = 9.799/(12.718) = 0.7704$$

$$\alpha_{A^{2-}} = 1.919/(12.718) = 0.1509$$

The analytical concentration of all species is 0.20 M.

The result of adding NaOH to reach pH 4.40 is a decrease of $(0.421)(0.20)$ mol H_2A =

0.0842 mol, an increase of (0.280)(0.20) = 0.056 mol HA⁻, and (0.141)(0.20) = 0.0282 mol A²⁻

Therefore, the buffer capacity is 0.0842 mol NaOH per liter.

70. The buffer index is a derivative describing the pH change for a small, given amount of added acid or base, whereas the buffer capacity is the amount of acid or base that must be added to produce a change of one pH unit in the buffer solution.

71. The curve reaches a maximum at the pK_a of the acid making up the buffer. The curve would show a maximum at pH = 6.99 for imidazole, the pH that is equal to the pK_{a1} of imidazole.

72. A polyprotic acid is an acid that can donate more than one hydrogen ion. A polyprotic base is a base that can accept more than one hydrogen ion. An examples of a polyprotic acid is phosphoric acid (H_3PO_4). An example of a polyprotic base is ethylenediamine ($H_2NCH_2CH_2NH_2$).

73. A "fraction of species equation" is an expression that makes it possible to describe and calculate the fraction of the total amount of a substance that is present in a particular form. This type of equation is useful when describing the solution of a polyprotic acid or polyprotic base by making it possible to determine the fraction of all possible acid–base species over a range of pH values or at a particular pH.

74. a) $C_{H_2S} = [H_2S] + [HS^-] + [S^{2-}]$

   Fraction of $H_2S = [H_2S]/\{[H_2S] + [HS^-] + [S^{2-}]\}$

   $= [H_2S]/\{[H_2S] + [H_2S]K_{a1}/[H^+] + [H_2S]K_{a1}K_{a2}/[H^+]^2\}$

   $= [H^+]^2/\{[H^+]^2 + K_{a1}[H^+] + K_{a1}K_{a2}\}$

Fraction of $HS^- = K_{a1}[H^+]/\{[H^+]^2 + K_{a1}[H^+] + K_{a1}K_{a2}\}$

Fraction of $S^{2-} = K_{a1}K_{a2}/\{[H^+]^2 + K_{a1}[H^+] + K_{a1}K_{a2}\}$

b) $C_{succinic} = [H_2C_4H_4O_4] + [HC_4H_4O_4^-] + [C_4H_4O_4^{2-}]$

Fraction of $H_2C_4H_4O_4 = [H_2C_4H_4O_4]/\{[H_2C_4H_4O_4] + [HC_4H_4O_4^-] + [C_4H_4O_4^{2-}]\}$

$$= [H^+]^2/\{[H^+]^2 + K_{a1}[H^+] + K_{a1}K_{a2}\}$$

Fraction of $HC_4H_4O_4^- = K_{a1}[H^+]/\{[H^+]^2 + K_{a1}[H^+] + K_{a1}K_{a2}\}$

Fraction of $C_4H_4O_4^{2-} = K_{a1}K_{a2}/\{[H^+]^2 + K_{a1}[H^+] + K_{a1}K_{a2}\}$

c) $C_{citric} = [H_3C_5H_5O_7] + [H_2C_5H_5O_7^-] + [HC_5H_5O_7^{2-}] + [C_5H_5O_7^{3-}]$

Fraction $H_3C_5H_5O_7 = [H_3C_5H_5O_7]/\{[H_3C_5H_5O_7] + [H_2C_5H_5O_7^-] + [HC_5H_5O_7^{2-}] + [C_5H_5O_7^{3-}]\}$

$$= [H^+]^3/\{[H^+]^3 + K_{a1}[H^+]^2 + K_{a1}K_{a2}[H^+] + K_{a1}K_{a2}K_{a3}\}$$

Fraction of $H_2C_5H_5O_7^- = K_{a1}[H^+]^2/\{[H^+]^3 + K_{a1}[H^+]^2 + K_{a1}K_{a2}[H^+] + K_{a1}K_{a2}K_{a3}\}$

Fraction of $HC_5H_5O_7^{2-} = K_{a1}K_{a2}[H^+]/\{[H^+]^3 + K_{a1}[H^+]^2 + K_{a1}K_{a2}[H^+] + K_{a1}K_{a2}K_{a3}\}$

Fraction of $C_5H_5O_7^{3-} = K_{a1}K_{a2}K_{a3}/\{[H^+]^3 + K_{a1}[H^+]^2 + K_{a1}K_{a2}[H^+] + K_{a1}K_{a2}K_{a3}\}$

d) $C_{lysine} = [H_3C_6H_{14}O_2N_2] + [H_2C_6H_{14}O_2N_2^-] + [HC_6H_{14}O_2N_2^{2-}] + [C_6H_{14}O_2N_2^{3-}]$

Fraction $H_3C_6H_{14}O_2N_2 =$

$[H_3C_6H_{14}O_2N_2]/\{[H_3C_6H_{14}O_2N_2] + [H_2C_6H_{14}O_2N_2^-] + [HC_6H_{14}O_2N_2^{2-}] + [C_6H_{14}O_2N_2^{3-}]\}$

$= [H^+]^3/\{[H^+]^3 + K_{a1}[H^+]^2 + K_{a1}K_{a2}[H^+] + K_{a1}K_{a2}K_{a3}\}$

Fraction $H_2C_6H_{14}O_2N_2^- = K_{a1}[H^+]^2/\{[H^+]^3 + K_{a1}[H^+]^2 + K_{a1}K_{a2}[H^+] + K_{a1}K_{a2}K_{a3}\}$

Fraction $HC_6H_{14}O_2N_2^{2-} = K_{a1}K_{a2}[H^+]/\{[H^+]^3 + K_{a1}[H^+]^2 + K_{a1}K_{a2}[H^+] + K_{a1}K_{a2}K_{a3}\}$

Fraction $C_6H_{14}O_2N_2^{3-} = K_{a1}K_{a2}K_{a3}/\{[H^+]^3 + K_{a1}[H^+]^2 + K_{a1}K_{a2}[H^+] + K_{a1}K_{a2}K_{a3}\}$

75. Piperazine is a diprotic base having $K_{a1} = 2.76 \times 10^{-6}$ and $K_{a2} = 1.48 \times 10^{-10}$.

Fraction of $H_2Pip^{2+} = [H^+]^2/\{[H^+] + K_{a1}[H^+] + K_{a1}K_{a2}\}$

$= (1.0 \times 10^{-6})^2/\{(1.0 \times 10^{-6})^2 + (2.76 \times 10^{-6})(1.0 \times 10^{-6}) +$

$(2.76 \times 10^{-6})(1.48 \times 10^{-10})\}$

$= 0.27$

Fraction of $HPip^+ = ([H^+]K_{a1})/\{[H^+] + K_{a1}[H^+] + K_{a1}K_{a2}\}$

$= (1.0 \times 10^{-6})(2.76 \times 10^{-6})/\{(1.0 \times 10^{-6})^2 + (2.76 \times 10^{-6})(1.0 \times 10^{-6}) +$

$(2.76 \times 10^{-6})(1.48 \times 10^{-10})\}$

$= 0.73$

Fraction of $Pip = (K_{a1}K_{a2})/\{[H^+] + K_{a1}[H^+] + K_{a1}K_{a2}\}$

$= (2.76 \times 10^{-6})(1.48 \times 10^{-10})/\{(1.0 \times 10^{-6})^2 + (2.76 \times 10^{-6})(1.0 \times 10^{-6})$

$+ (2.76 \times 10^{-6})(1.48 \times 10^{-10})\}$

$= 0.00011$

76. Tryptophan is a diprotic species with $K_{a1} = 4.2\underline{7} \times 10^{-3}$ and $K_{a2} = 4.6\underline{8} \times 10^{-10}$.

At pH 7.4, $[H^+]$ is equal to 3.98×10^{-8}.

Fraction of $H_2Trp^{2+} = 1/\{1 + K_{a1}/[H^+] + K_{a1}K_{a2}/[H^+]^2\}$

$= 1/\{1 + (4.2\underline{7} \times 10^{-3})/(3.98 \times 10^{-8}) + (2.0\underline{1} \times 10^{-12})/(1.58 \times 10^{-15})\}$

$= 1/\{1 + 1.0\underline{8} \times 10^5 + 1.2\underline{7} \times 10^3\}$

$= 1/\{1.0\underline{93} \times 10^5\} = 9.1 \times 10^{-6}$

In a similar fashion:

Fraction of $HTrp^+ = (1.08 \times 10^5)/(1.0\underline{93} \times 10^5) = 0.99$

Fraction of Trp = $(1.27 \times 10^3)/(1.093 \times 10^5) = 0.011$

77. a) Fraction of $NH_3 = K_a/([H^+] + K_a)$ (using K_a for the ammonium ion)

$$= (5.62 \times 10^{-10})/\{1.0 \times 10^{-11} + 5.62 \times 10^{-10}\}$$

$$= 0.983 = 0.98$$

b) $[NH_3] = 0.983 \,(0.025\ M) = 0.0246\ M$

c) Fraction of $NH_3 = K_a/([H^+] + K_a)$

$$= (5.62 \times 10^{-10})/\{1.0 \times 10^{-7} + 5.62 \times 10^{-10}\}$$

$$= 0.00559 = 0.0056$$

Thus, there is much more free ammonia at pH 11 than at pH 7.

78. a) HCl is a strong acid so it essentially is 100% ionized to H^+ and Cl^- at pH 2.00.

b) HNO_3 is a strong acid, so it is 100% ionized to H^+ and NO_3^- at pH 2.00.

c) $\alpha_{Ac^-} = 1/\{1 + [H^+]/K_a\} = 1/\{1 + 1.0 \times 10^{-2}/1.75 \times 10^{-5}\} = 1/\{1 + 571\} = 0.0017$

d) Phosphoric acid is triprotic, but only the first ionization needs to be considered at pH 2.00.

$\alpha_{H_2PO_4^-} = 1/\{1 + [H^+]/K_{a1}\} = 1/\{1 + 1.0 \times 10^{-2}/7.11 \times 10^{-3}\} = 1/\{1 + 1.406\} = 0.415$

79. The first situation we might encounter is a solution where we begin with only the most acidic form of the polyprotic agent. Thus, it is usually safe in this case to consider only formation of the first conjugate base from the acid and use this reaction alone to estimate the hydrogen ion concentration of a solution. This approach essentially treats the polyprotic acid as a weak monoprotic acid.

A second, similar situation occurs when we prepare a solution that initially has only the most basic form of a polyprotic acid–base system. We can determine the pH of this solution by treating the dissolved form as a weak monoprotic base.

A third situation is when we prepare a solution that contains a mixture of two closely related acid–base forms. If we are dealing with a relatively concentrated solution, allowing us to ignore the effects of water dissociation, we can find the pH of these mixtures using the same approach as used for mixtures of monoprotic acids and their conjugate bases.

A fourth situation is when we prepare a solution containing only the conjugate base of a polyprotic weak acid. Generally, the pH equals $(pK_{a1} + pK_{a2})/2$ in this situation.

80. An amphiprotic compound is a compound that can either accept or donate a hydrogen ion. Water, the bicarbonate ion (HCO_3^-), and bisulfate (HSO_4^-) are examples.

81. a) $[H^+] = \{(2.50 \times 10^{-3})(7.11 \times 10^{-3})\}^{1/2} = 4.22 \times 10^{-3}$ M or pH = 2.37

This assumes that only the first ionization is important which is a good assumption and also assumes that the fraction of ionization is small, which is not a good assumption. A more accurate pH would be calculated using the quadratic equation, which gives pH = 2.71.

b) $[OH^-] = (K_{b1} C)^{1/2} = \{(0.0237)(4.3 \times 10^{-4})\}^{1/2} = 3.19 \times 10^{-3}$ M or pOH = 2.50, giving pH = 11.50. This assumes that only the first basicity is important, which is a good assumption. It also assumes that the fraction of ionization is small, which is not a good assumption. A more accurate pH would be calculated using the quadratic formula, which gives pH 10.63 and pOH = 3.37.

c) pH = $(pK_{a1} + pK_{a2})/2 = (2.148 + 7.198)/2 = 4.67$

d) pH = $(pK_{a2} + pK_{a3})/2 = (7.198 + 12.375)/2 = 9.78$

Parts (c) and (d) both assume that only two species are present at significant concentrations.

82. Chromic acid is diprotic with $K_{a1} = 0.18$ and $K_{a2} = 3.1 \times 10^{-7}$.

a) Chromic acid is almost a strong acid, so a good first approximation is $[H^+] = 5.0 \times 10^{-5}\ M$, or pH = 4.30.

b) $[OH^-] = \{(5 \times 10^{-5})(1.01 \times 10^{-14})/(3.1 \times 10^{-7})\}^{1/2} = 1.28 \times 10^{-6}\ M$

pOH = 5.90   pH = 8.10

c) pH = pK_{a1} + log($HCrO_4^-$)/(H_2CrO_4)

= 0.74 + log (1.00) = 0.74

83. Equation 8.51 assumes that only a small fraction of the intermediate species either gains or loses a proton, making it possible to assume $[HA^-] = C_{HA}$.

Equation 8.52 assumes that the concentration of the intermediate species is much greater than K_{a1} and that the product of the concentration and the second acidity constant is much greater than K_w. The result is that Equation 8.52 is reasonably accurate for a fairly concentrated solution and if the acid is reasonably weak, but not so weak that its values of K_a approaches K_w.

84. A zwitterion is a chemical that possesses a net charge of zero but contains groups with an equal number of negative and positive charges. Amino acids have a carboxylic acid functional group (–COOH) and an amine group (–NH$_2$), as well as other possible charged groups, which can create a zwitterion when there is an equal number of positively and negatively charged groups in this structure.

85. The isoelectric point (pI) is the pH at which the average charge is zero for a chemical that can form a zwitterion. Knowing the pI value for an amino acid, peptide, protein or other type of

zwitterion can be extremely useful in identifying and isolating these chemicals. The isoelectric point of a zwitterion is also important in determining the solubility of such a compound.

88. a) pI = (2.16 + 8.73)/2 = 5.44

   b) pI = (2.32 + 9.76)/2 = 6.29

   c) pI = (2.18 + 9.08)/2 = 5.63

   d) pI = (1.95 + 10.64)/2 = 5.32

89. pI = (2.33 + 9.74)/2 = 6.04

90. a) pI = (2.16 + 4.30)/2 = 3.23

   b) pI = (5.97 + 9.28)/2 = 7.63

   c) pI = (1.7 + 8.36)/2 = 5.03

   d) pI = (2.41 + 8.67)/2 = 5.54

93. It was shown earlier in Chapter 4 that the relationship between the precision of a measured pH value and the resulting hydrogen ion concentration is as follows.

$$(s_{a_{H+}}/a_{H+}) = 2.303\, s_{pH}$$

This equation indicates that we need to select a particular pH (or value of a_{H+}) at which we will carry out this calcuation. As an example, let's see what result is obtained at pH 4.00 (or $a_{H+} = 1.0 \times 10^{-6}\,M$).

For $s_{pH} = 0.01$:

$$s_{a_{H+}} = 2.303 \cdot (a_{H+} \cdot s_{pH})$$
$$= 2.303 \cdot (1.0 \times 10^{-4}\,M) \cdot (0.01)$$
$$= 0.02\underline{3} \times 10^{-4}\,M \quad (2.\underline{3}\%)$$

For $s_{pH} = 0.02$:
$$s_{a_{H+}} = 2.303 \cdot (a_{H+} \cdot s_{pH})$$
$$= 2.303 \cdot (1.0 \times 10^{-4} M) \cdot (0.02)$$
$$= 0.04\underline{6} \times 10^{-4} M \quad (4.\underline{6}\%)$$

For $s_{pH} = 0.05$:
$$s_{a_{H+}} = 2.303 \cdot (a_{H+} \cdot s_{pH})$$
$$= 2.303 \cdot (1.0 \times 10^{-4} M) \cdot (0.005)$$
$$= 0.01\underline{2} \times 10^{-4} M \quad (1.\underline{2}\%)$$

It is also possible to address this problem by using the following approximate approach. Suppose the pH is truly 4.00 and the random errors in measurement gives 4.01 as the reading. This again represents a difference in [H$^+$] of $1.0\underline{0} \times 10^{-4} M - 9.7\underline{7} \times 10^{-5} M = 2.\underline{3} \times 10^{-6} M$ or $100 \cdot (2.27 \times 10^{-6} M)/(1.00 \times 10^{-4} M) = 2.3\%$. Similar calculations using a measured pH of 4.02 or 4.005 give a relative difference of 4.$\underline{5}$% or 1.$\underline{1}$%.

CHAPTER 9: COMPLEX FORMATION

1. Complex formation is a reaction in which there is reversible binding between two or more distinct chemical species. Examples are the formation of copper ammonia complexes and calcium forming an EDTA complex.

2. EDTA is added to mayonnaise to bind Fe^{3+}, a process which prevents this metal ion from taking part in the oxidation of fats. This is reversible binding process and, as such, represents complex formation process.

3. One way that complex formation can be utilized is to measure the amount of an analyte. In the case of EDTA and metal ions such as Fe^{3+}, this approach often involves the use of a titration. A second way that complex formation can be used in chemical analysis is to control the effective amount of an analyte that is available for other reactions. This effect is illustrated by the addition of EDTA to mayonnaise for binding of Fe^{3+}, a process which prevents this metal ion from taking part in the oxidation of fats. A third way that complex formation can be used in analytical methods is as a tool for the separation of chemicals. For instance, complexing agents can be attached to a solid support for binding and separating their target compounds from other substances in a sample. This approach is found in some types of extractions and in liquid chromatography.

4. Alfred Werner was the first to understand the nature of complex compounds. He showed that the ligand wasn't changed during the reaction, except that it bonded to the metal ion by what we now call coordinate covalent bonds.

5. a) A coordinate bond is the type of bond that forms when one chemical shares a pair of electrons with a metal ion.

b) The chemical that shares its electron pair with the metal ion is known as the ligand.

c) The product of the reaction of a ligand with a metal ion is a metal-ligand complex, or metal-coordination complex.

6. A Lewis acid is a chemical that can accept a pair of electrons from another substance, and a Lewis base is a chemical that can donate a pair of electrons to another substance. These are broader definitions than those used in the Brønsted–Lowry model of acids and bases, in which an acid is a proton donor and a base is a proton acceptor.

7. The Cu^{2+} ion accepts pairs of nonbonded electrons from ammonia molecules. This makes Cu^{2+} a Lewis acid in this reaction and ammonia a Lewis base.

8. a) Mg^{2+} is the Lewis acid and OH^- is the Lewis base.

b) Ag^+ is the Lewis acid and Cl^- is the Lewis base.

c) Fe^{3+} is the Lewis acid and $EDTA^{4-}$ is the Lewis base.

d) CH_3COOH is the Brønsted acid and NH_3 is the Brønsted base. These are also the Lewis acid and base, respectively, because the H^+ from acetic acid accepts an electron pair from ammonia to form ammonium.

9. A nonbonded electron pair from water can be donated to form a coordinate bond to most metal ions, which makes water act as a ligand for these metal ions.

10. The Brønsted–Lowry definition of an acid or a base is less broad than that of the Lewis definitions. The Brønsted–Lowry definitions specifically involve the transfer of an H^+,

which also involves the donation of an electron pair by the species that accepts this proton. However, some Lewis acid–base reactions can occur in a system where there are no hydrogen atoms at all, but an electron pair is still being donated from one species to another.

11. A monodentate ligand is a ligand that can donate only a single pair of electrons. Examples include fluoride ion, water, and the cyanide ion (CN^-).

12. Reaction 1 shows what really happens as water molecules are substituted by ammonia. Reaction 2 is a simplified representation and emphasizes what most people regard as the important changes that are occurring in this reaction.

13. The addition of Cl^- to a silver ion solution will initially cause precipitation of AgCl, but continued addition will cause the AgCl to dissolve to form $AgCl_2^-$. Likewise, the addition of hydroxide ions to an aluminum ion solution causes precipitation of $Al(OH)_3$, but further addition of hydroxide ions causes this solid to dissolve to form $Al(OH)_4^-$.

14. A formation constant is the equilibrium constant for a particular step for the formation of a complex between a metal ion and ligand. Metal ions can generally accept more than one monodentate ligand, so there are several sequential formation constants that can be used to describe each step in this process.

15. a) $K_f = [BaOH^+]/[Ba^{2+}][OH^-]$ b) $K_f = [Cu(NH_3)_2^{2+}]/[Cu^{2+}][NH_3]^2$
    c) $K_f = [Ni(CN)_4^{2-}]/[Ni^{2+}][CN^-]^4$ d) $K_f = [FeF_6^{3-}]/[Fe^{3+}][F^-]^6$

16. a) $K°_f = (a_{BaOH+})/\{(a_{Ba2+})(a_{OH-})\}$

    $K°_f = K_f (\gamma_{BaOH+})/\{(\gamma_{Ba2+})(\gamma_{OH-})\}$

b) $K°_f = (a_{Cu(NH3)22+})/\{(a_{Cu2+})(a_{NH3})^2\}$

$K°_f = K_f (\gamma_{Cu(NH3)22+})/\{(\gamma_{Cu2+})(\gamma_{NH3})^2\}$

c) $K°_f = (a_{Ni(CN)42-})/\{(a_{Ni2+})(a_{CN-})^4\}$

$K°_f = K_f (\gamma_{Ni(CN)42-})/\{(\gamma_{Ni2+})(\gamma_{CN-})^4\}$

d) $K°_f = (a_{FeF63-})/\{(a_{Fe3+})(a_{F-})^6\}$

$K°_f = K_f (\gamma_{FeF63-})/\{(\gamma_{Fe3+})(\gamma_{F-})^6\}$

17. As a metal ion adds a ligand, the effective positive charge decreases, so there is less of an attraction for subsequent pairs of electrons. Additionally, there are fewer locations available for additional ligands to bind.

18. These metal ions have room for six ligands in an octahedral arrangement. The sequential formation constants are somewhat similar, so a second ammonia will bond to some of the metal ions before all of the metal ions have bound their first ligand.

19. An overall formation constant is the product of all the stepwise formation constants.

20. $\beta_4 = K_{f1} K_{f2} K_{f3} K_{f4}$

    $= (1.02 \times 10^6)(1.3 \times 10^5)(3.4 \times 10^4)(1.9 \times 10^2)$

    $= 8.6 \times 10^{17}$

21. $K_{f1} = 10^{1.77} = 59$

    $K_{f2} = 10^{(2.6-1.77)} = 6.7$

    $K_{f3} = 10^{(3.0-2.6)} = 2.5$

    $K_{f4} = 10^{(2.3-3.0)} = 0.2$

22. $C_{Ni^{2+}} = (0.250\ M)(1/101) = 2.48 \times 10^{-3}\ M$ $C_{NH_3} = (0.058\ M)(100/101) = 0.057\underline{4}\ M$

This is a 23-fold excess of ammonia, so nearly all of the nickel will be complexed by ammonia. Several species of nickel-ammonia complexes will actually be present when the ammonia concentration is 0.057 M.

The fraction existing as Ni^{2+} can be calculated as follows.

$$\alpha_{Ni^{2+}} = \frac{1}{1 + (5.2\underline{5} \times 10^2)(0.057\underline{4}) + (7.5\underline{9} \times 10^4)(0.057\underline{4})^2 + (3.4\underline{7} \times 10^6)(0.057\underline{4})^3 + (4.6\underline{8} \times 10^7)(0.057\underline{4})^4 + (2.1\underline{4} \times 10^8)(0.057\underline{4})^5 + (2.0\underline{0} \times 10^8)(0.057\underline{4})^6}$$

$= 1/(1 + 30.1 + 250 + 656 + 508 + 133 + 7.1) = 1/1585 = 6.3 \times 10^{-4}$

Therefore, $[Ni^{2+}] = (2.48 \times 10^{-3}\ M)(6.3 \times 10^{-4}) = 1.6 \times 10^{-6}\ M$

23. a) $C_{Hg} = [Hg^{2+}] + [HgCl^+] + [HgCl_2] + [HgCl_3^-] + [HgCl_4^{2+}]$

b) $\alpha_{Hg^{2+}} = 1/(1 + \beta_1[Cl^-] + \beta_2[Cl^-]^2 + \beta_3[Cl^-]^3 + \beta_4[Cl^-]^4)$

c) $\alpha_{Hg^{2+}} = 1/\{1 + 5.01 \times 10^6(0.05) + 1.58 \times 10^{13}(0.05)^2 + 1.26 \times 10^{14}(0.05)^3$

   $+ 1.26 \times 10^{15}(0.05)^4\}$

   $= 1.58 \times 10^{-11}$

24. a) $C_{Cu(II)} = [Cu^{2+}] + [Cu(NH_3)^{2+}] + [Cu(NH_3)_2^{2+}] + [Cu(NH_3)_3^{2+}] + [Cu(NH_3)_4^{2+}]$
    $+ [Cu(NH_3)_5^{2+}] + [Cu(NH_3)_6^{2+}]$

b) $\alpha_{Cu^{2+}} = 1/(1 + \beta_1[NH_3] + \beta_2[NH_3]^2 + \beta_3[NH_3]^3 + \beta_4[NH_3]^4 + \beta_5[NH_3]^5 + \beta_6[NH_3]^6)$

$\alpha_{Cu(NH_3)_4^{2+}} = \beta_4[NH_3]^4/(1 + \beta_1[NH_3] + \beta_2[NH_3]^2 + \beta_3[NH_3]^3 + \beta_4[NH_3]^4$
$+ \beta_5[NH_3]^5 + \beta_6[NH_3]^6)$

c) $\alpha_{Cu^{2+}} = 1/\{1 + (1.35 \times 10^4)(0.10) + (4.07 \times 10^7)(0.10)^2 + (3.02 \times 10^{10})(0.10)^3 +$

   $(3.89 \times 10^{12})(0.10)^4 + (3.47 \times 10^{12})(0.10)^5 + (3.47 \times 10^{10})(0.10)^6\}$

$$= 1/(1 + 1.35 \times 10^3 + 4.07 \times 10^5 + 3.02 \times 10^7 + 3.89 \times 10^8 + 3.47 \times 10^7 + 3.47 \times 10^4)$$

$$= 1/(4.54 \times 10^8) = 2.2 \times 10^{-9}$$

$\alpha_{Cu(NH3)4} = (3.89 \times 10^8)/(4.54 \times 10^8) = 0.857$

25. $\alpha_{Zn2+} = [Zn^{2+}]/\{[Zn^{2+}] + [Zn(SCN)^+] + [Zn(SCN)_2]\}$

$\quad = [Zn^{2+}]/\{[Zn^{2+}] + K_{f1}[Zn^{2+}][SCN^-] + K_{f1}K_{f2}[Zn^{2+}][SCN^-]^2\}$

$\quad = 1/\{1 + K_{f1}[SCN^-] + K_{f1}K_{f2}[SCN^-]^2\}$

$\quad = 1/\{1 + \beta_1[SCN^-] + \beta_2[SCN^-]^2\}$

$\alpha_{Zn(SCN)+} = K_{f1}[SCN^-]/\{1 + K_{f1}[SCN^-] + K_{f1}K_{f2}[SCN^-]^2\}$

$\quad = \beta_1[SCN^-]/\{1 + \beta_1[SCN^-] + \beta_2[SCN^-]^2\}$

$\alpha_{Zn(SCN)2} = K_{f1}K_{f2}[SCN^-]^2/\{1 + K_{f1}[SCN^-] + K_{f1}K_{f2}[SCN^-]^2\}$

$\quad = \beta_2[SCN^-]^2/\{1 + \beta_1[SCN^-] + \beta_2[SCN^-]^2\}$

26. $\alpha_{Sn2+} = 1/\{1 + K_{f1}[F^-] + K_{f1}K_{f2}[F^-]^2 + K_{f1}K_{f2}K_{f3}[F^-]^3\}$

$\quad = 1/\{1 + \beta_1[F^-] + \beta_2[F^-]^2 + \beta_3[F^-]^3\}$

$\alpha_{SnF+} = K_{f1}[F^-]/\{1 + K_{f1}[F^-] + K_{f1}K_{f2}[F^-]^2 + K_{f1}K_{f2}K_{f3}[F^-]^3\}$

$\quad = \beta_1[F^-]/\{1 + \beta_1[F^-] + \beta_2[F^-]^2 + \beta_3[F^-]^3\}$

$\alpha_{SnF2} = K_{f1}K_{f2}[F^-]^2/\{1 + K_{f1}[F^-] + K_{f1}K_{f2}[F^-]^2 + K_{f1}K_{f2}K_{f3}[F^-]^3\}$

$\quad = \beta_2[F^-]^2/\{1 + \beta_1[F^-] + \beta_2[F^-]^2 + \beta_3[F^-]^3\}$

$\alpha_{SnF3-} = K_{f1}K_{f2}K_{f3}[F^-]^3/\{1 + K_{f1}[F^-] + K_{f1}K_{f2}[F^-]^2 + K_{f1}K_{f2}K_{f3}[F^-]^3\}$

$\quad = \beta_3[F^-]^3/\{1 + \beta_1[F^-] + \beta_2[F^-]^2 + \beta_3[F^-]^3\}$

27. A chelating agent is a ligand that can donate more than one pair of electrons located on different atoms to the same metal ion. A monodentate ligand can only donate a single pair of electrons.

28. Ethylenediamine ($H_2NCH_2CH_2NH_2$) has a lone pair of nonbonded electrons on each nitrogen atom. These two pairs of nonbonded electrons can each bond to the same metal ion simultaneously forming a five-membered ring of atoms.

29. A series of five-membered rings are produced when EDTA forms a complex with a metal ion. These rings are created by the interaction of electron pairs on the nitrogen and oxygen atoms on EDTA with the metal ion.

30. a) A bidentate ligand is a ligand that has two pairs of electrons that can coordinate to a metal ion.

    b) A tridentate ligand is a ligand that has three pairs of electrons that can coordinate to a metal ion.

    c) A tetradentate ligand is a ligand that has four pairs of electrons that coordinate to a metal ion.

    d) A polydentate ligand is a ligand with several (greater than one) pairs of electrons that can coordinate to a metal ion.

31. A chelate is a complex formed when a metal ion is coordinated to a multidentate ligand. The structure that is produced is a ring of five to six atoms.

32. The two nitrogen atoms of ethylenediamine (en) are approximately the same basicity as ammonia, based on their K_b values. $Co(en)_3$ is more stable than $Co(NH_3)_3$ for two reasons.

The first reason is that it has more nitrogens bonded to the metal ion. The second reason is that once the first nitrogen has bonded, the second one is so nearby that it has a much greater chance to bond as opposed to if it were an independent molecule that were free to diffuse away from the metal ion.

33. The chelate effect is the tendency of chelating agents to give more stable complexes with metal ions and provide larger overall formation constants than monodentate ligands. This effect can be explained by the greater positive change in entropy that is associated with forming a chelate compared to a complex with several monodentate ligands.

34. The chelate effect is useful because it creates much larger equilibrium constants for complex formation reactions involving multidentate ligands. This effect also simplifies the fraction of species calculations because there are fewer species to consider.

35. EDTA is an abbreviated name for the chemical ethylenediaminetetraacetic acid. This molecule was designed and synthesized to form highly stable complexes with metal ions. It is potentially a hexadentate ligand that can form several five-membered chelate rings with a metal ion.

36. $Pb^{2+} + EDTA^{4-} \rightleftharpoons Pb(EDTA)^{2-}$ $K_f = 1.0 \times 10^{18}$

    Both nitrogen atoms of EDTA and its four carboxylate oxygens can bond simultaneously to a large metal ion such as Pb^{2+}.

37. a) $Ca^{2+} + EDTA^{4-} \rightleftharpoons CaEDTA^{2-}$ b) $K_f = 4.5 \times 10^{10}$
    $Mg^{2+} + EDTA^{4-} \rightleftharpoons MgEDTA^{2-}$ $K_f = 6.2 \times 10^{8}$

    c) mol Ca^{2+} = (0.5764 g)/(100.1 g/mol) = 5.759×10^{-3} mol

Volume of EDTA = $(5.76 \times 10^{-3}$ mol$)/(0.0132$ mol/L$) = 0.436$ L or 436 mL

38. Mass = $(0.0200$ mol/L$)(0.500$ L$)(372.24$ g/mol$) = 3.7224$ g

    Molarity = $(7.50$ g$)/(372.24$ g/mol$)/(0.750$ L$) = 0.0269$ M

39. Both the amino groups and the carboxylic acid groups on EDTA behave as Brønsted–Lowry acids and bases. These are the same regions that bind to metal ions.

40. a)
$$\alpha_{EDTA^{4-}} = \frac{K_{a1} K_{a2} K_{a3} K_{a4} K_{a5} K_{a6}}{[H^+]^6 + K_{a1}[H^+]^5 + K_{a1}K_{a2}[H^+]^4 + K_{a1}K_{a2}K_{a3}[H^+]^3 + K_{a1}K_{a2}K_{a3}K_{a4}[H^+]^2 + K_{a1}K_{a2}K_{a3}K_{a4}K_{a5}[H^+] + K_{a1}K_{a2}K_{a3}K_{a4}K_{a5}K_{a6}}$$

$= (3.12 \times 10^{-23})/\{[H^+]^6 + (1.0 \times 10^0)[H^+]^5 + (3.20 \times 10^{-2})[H^+]^4 + (3.26 \times 10^{-4})[H^+]^3$
$+ (6.98 \times 10^{-7})[H^+]^2 + (4.83 \times 10^{-13})[H^+] + (3.12 \times 10^{-23})\}$

where $[H^+] = 10^{-7.50} = 3.16 \times 10^{-8}$ M

$\alpha_{EDTA^{4-}} = 1.9\underline{5} \times 10^{-3} = 2.0 \times 10^{-3}$

b) One can tell either by examining Figure 9.7 or using calculations like the preceding fraction of species calculation that the dominant species at pH 7.5 is HEDTA^{3-}.

41. a)
$$\alpha_{EDTA^{4-}} = \frac{K_{a1} K_{a2} K_{a3} K_{a4} K_{a5} K_{a6}}{[H^+]^6 + K_{a1}[H^+]^5 + K_{a1}K_{a2}[H^+]^4 + K_{a1}K_{a2}K_{a3}[H^+]^3 + K_{a1}K_{a2}K_{a3}K_{a4}[H^+]^2 + K_{a1}K_{a2}K_{a3}K_{a4}K_{a5}[H^+] + K_{a1}K_{a2}K_{a3}K_{a4}K_{a5}K_{a6}}$$

$= (3.12 \times 10^{-23})/\{[H^+]^6 + (1.0 \times 10^0)[H^+]^5 + (3.20 \times 10^{-2})[H^+]^4 + (3.26 \times 10^{-4})[H^+]^3$
$+ (6.98 \times 10^{-7})[H^+]^2 + (4.83 \times 10^{-13})[H^+] + (3.12 \times 10^{-23})\}$

where $[H^+] = 10^{-9.50} = 3.16 \times 10^{-10}$ M $\qquad \alpha_{EDTA^{4-}} = 1.7\underline{0} \times 10^{-1} = 1.7 \times 10^{-1}$

b) One can tell either by examining Figure 9.7 or using calculations like the preceding

b) One can tell either by examining Figure 9.7 or using calculations like the preceding fraction of species calculation that the dominant species at pH 9.5 is HEDTA^{3-}.

42. According to Table 9.8, the fraction of EDTA that will be present in the EDTA^{4-} form at 25 °C and pH 5.0 is $4.1\underline{0} \times 10^{-7}$, while the fraction of the same form will be $6.0\underline{6} \times 10^{-2}$ at pH 9.0. The stability constants of metal-EDTA complexes are written in terms of the tetraanion, so the binding of any metal will be much less at pH 5.0 than at pH 9.0 because less EDTA is present in a form that is suitable for binding at the lower pH.

43. This effect happens because the fraction of EDTA present as the tetraanion is much smaller at pH 6.0 than at pH 10.0.

44. A conditional formation constant is an equilibrium constant that describes the complex formation under a given set of reaction conditions. This type of equilibrium constant is used to examine and predict the effects of various conditions on complex formation, especially pH.

45. A conditional formation constant can be used to correct for the acid–base properties of EDTA by making this conditional formation constant equal to the true formation constant for the metal-EDTA complex multiplied by the fraction of the EDTA that is present at the given pH in the tetraanion form.

46. a) $K'_f = K_f\, \alpha_{EDTA4-} = (6.02 \times 10^{18})(4.2\underline{4} \times 10^{-9}) = 2.5\underline{5} \times 10^{10} = 2.6 \times 10^{10}$

 b) $K'_f = K_f\, \alpha_{EDTA4-} = (2.51 \times 10^{18})(6.0\underline{6} \times 10^{-2}) = 1.5\underline{2} \times 10^{17} = 1.5 \times 10^{17}$

 c) $K'_f = K_f\, \alpha_{EDTA4-} = (3.16 \times 10^{16})(6.3\underline{3} \times 10^{-3}) = 2.0\underline{0} \times 10^{14} = 2.0 \times 10^{14}$

 d) $K'_f = K_f\, \alpha_{EDTA4-} = (6.16 \times 10^{8})(2.9\underline{5} \times 10^{-11}) = 1.8\underline{2} \times 10^{-2} = 1.8 \times 10^{-2}$

47. The fraction of EDTA that is present as the tetraanion at the pH values used in this problem can be found through fraction of species calculations to be $1.9\underline{5} \times 10^{-3}$ (pH 7.5), $1.0\underline{8} \times 10^{-8}$ (pH 4.2), $1.2\underline{4} \times 10^{-5}$ (pH 5.8), or $1.4\underline{0} \times 10^{-4}$ (pH 6.5).

   a) $K'_f = K_f \alpha_{EDTA4-} = (2.29 \times 10^{15})(1.9\underline{5} \times 10^{-3}) = 4.4\underline{6} \times 10^{12} = 4.5 \times 10^{12}$

   b) $K'_f = K_f \alpha_{EDTA4-} = (1.26 \times 10^{23})(1.0\underline{8} \times 10^{-8}) = 1.3\underline{6} \times 10^{15} = 1.4 \times 10^{15}$

   c) $K'_f = K_f \alpha_{EDTA4-} = (1.58 \times 10^{23})(1.2\underline{4} \times 10^{-5}) = 1.9\underline{6} \times 10^{18} = 2.0 \times 10^{18}$

   d) $K'_f = K_f \alpha_{EDTA4-} = (3.16 \times 10^{16})(1.4\underline{0} \times 10^{-4}) = 4.4\underline{2} \times 10^{12} = 4.4 \times 10^{12}$

48. Conditional formation constants can be used to describe the effect of any side reaction on complex formation. This is particularly easy to do for ligands like EDTA that also have acid–base properties. For instance, a ligand like ammonia must be nonprotonated at the site of its electron pair before it can donate this pair of electrons to a metal ion. As a result, a conditional formation constant that includes the effect of pH on the amount of this ligand that is present in the nonprotonated form will decrease at lower pH values. Ammonia has a pK_b of 4.74, which means that below pH 9.26 more than half of the ammonia is protonated. This ligand will have its highest fraction available for metal complexation above this pH.

49. The fraction of nonprotonated ammonia and conditional formation constants at the given pH values are as follows.

    At pH 5: $\alpha_{NH_3} = 1/\{1 + 1.0 \times 10^{-5}/5.49 \times 10^{-10}\} = 5.4\underline{9} \times 10^{-5}$

    $K'_f = (5.4\underline{9} \times 10^{-5})(525) = 0.029$

    At pH 10: $\alpha_{NH_3} = 1/\{1 + 1.0 \times 10^{-10}/5.49 \times 10^{-10}\} = 0.84\underline{6}$

    $K'_f = (0.84\underline{6})(525) = 444$

    As expected, the value at pH 10 is much closer to the true formation constant of 525.

50. a) $K'_{f1} = K_f \alpha_{NH_3}$ $\beta'_2 = \beta_f (\alpha_{NH_3})^2$ $\beta'_3 = \beta_3 (\alpha_{NH_3})^3$ $\beta'_4 = \beta_4 (\alpha_{NH_3})^4$

b) At pH 4.0: $\alpha_{NH_3} = 1/\{1 + [H^+]/K_a\} = 5.5 \times 10^{-6}$

At pH 6.0, 8.0, and 10.0: $\alpha_{NH_3} = 6.5 \times 10^{-4}$, 5.2×10^{-2}, and 8.5×10^{-1}, respectively.

c) The overall formation constants for the $Co(NH_3)^{2+}$ complexes are $K_{f1} = 97.7$, $\beta_2 = 3.16 \times 10^3$, $\beta_3 = 2.9 \times 10^4$, and $\beta_4 = 1.17 \times 10^5$. These values plus those from Part a) result in the following conditional formation constants.

Conditional Formation Constant (K'_f or β') at a Given pH

Complex	pH = 4.0	pH = 6.0	pH 8.0	pH = 10.0
$Co(NH_3)$	$(97.7)(5.5 \times 10^{-6})$ $= 5.4 \times 10^{-4}$	$(97.7)(6.5 \times 10^{-4})$ $= 0.063$	$(97.7)(5.2 \times 10^{-2})$ $= 5.1$	$(97.7)(0.85)$ $= 82.4$
$Co(NH_3)_2$	$(3.16 \times 10^3)(5.5 \times 10^{-6})^2$ $= 9.5 \times 10^{-8}$	$(3.16 \times 10^3)(6.5 \times 10^{-4})^2$ $= 1.33 \times 10^{-3}$	$(3.16 \times 10^3)(5.2 \times 10^{-2})^2$ $= 8.54$	$(3.16 \times 10^3)(0.85)^2$ $= 2.28 \times 10^3$
$Co(NH_3)_3$	$(2.9 \times 10^4)(5.5 \times 10^{-6})^3$ $= 4.8 \times 10^{-12}$	$(2.9 \times 10^4)(6.5 \times 10^{-4})^3$ $= 7.9 \times 10^{-6}$	$(2.9 \times 10^4)(5.2 \times 10^{-2})^3$ $= 4.1$	$(2.9 \times 10^4)(0.85)^3$ $= 1.8 \times 10^4$
$Co(NH_3)_4$	$(1.17 \times 10^5)(5.5 \times 10^{-6})^4$ $= 1.1 \times 10^{-16}$	$(1.17 \times 10^5)(6.5 \times 10^{-4})^4$ $= 2.1 \times 10^{-8}$	$(1.17 \times 10^5)(5.2 \times 10^{-2})^4$ $= 0.86$	$(1.17 \times 10^5)(0.85)^4$ $= 6.1 \times 10^4$

51. At a very high pH, the soluble form of $CaOH^+$ is produced. This decreases the conditional formation constant just as the protonation of the EDTA decreases the value of this constant. This effect is smaller than that of the protonation of EDTA, and so is often ignored.

52. $$A + L \rightleftharpoons A\text{-}L \qquad K_f = \frac{[A\text{-}L]}{[A][L]}$$

53. Possible forces are hydrogen bonding, dipole–dipole interactions, dispersion forces, and ionic interactions.

54. An association constant is an equilibrium constant for the formation of a complex of any sort, not just metal-ligand complexes. This is another term for a formation constant.

55. A dissociation constant is the equilibrium constant for the dissociation of a complex. It is numerically the inverse of the association constant.

56. $K_A = 1/(6.3 \times 10^{-10}\, M) = 1.6 \times 10^9\, M^{-1}$

57. For the analyte: $K_D = 1/(2.7 \times 10^{10}\, M^{-1}) = 3.7 \times 10^{-11}\, M$

    For the contaminant: $K_D = 1/(4.0 \times 10^6\, M^{-1}) = 2.5 \times 10^{-7}\, M$

    The analyte, with the larger association constant and smaller dissociation constant, binds more tightly than the contaminant to the antibody.

58. An antibody is a protein produced by the body's immune system that has the ability to specifically bind to a foreign agent. An antibody has the same ability to form reversible complexes with analytes through two identical binding sites that are located at the upper ends of its structure.

59. An antibody can form a reversible complex with many types of analytes. However, an antibody differs from binding agents discussed earlier in this chapter in that metal-ligand interactions are not usually involved in this process. Instead, processes such as hydrogen bonding, dipole–dipole interactions, ionic interactions, and dispersion interactions lead to the formation of this complex.

60. An immunoassay is an analytical method that uses an antibody as a reagent. Two common ways in which antibodies are used are in the competitive-binding immunoassay and the sandwich immunoassay.

61. A cyclodextrin is a cyclic compound formed by several glucose molecules polymerizing into a ring compound rather than the more common linear, acyclic compounds. The most common number of glucose molecules are six, seven, and eight. The resulting molecule has the shape of a truncated cone, with the interior quite hydrophobic and the exterior hydrophilic.

62. A nonpolar molecule can insert itself into the interior of the cyclodextrin. The hydrophilic nature of the exterior of the cyclodextrin allows it and its inclusion partner to dissolve in water. The size of the formation constant for this complex will depend on a number of factors, including the fit of the compound into the cyclodextrin cavity and its ability to form hydrogen bonds or other interactions with the alcohol groups on the cyclodextrin.

68. a) $K = K_{f,FeEDTA}/K_{f,CaEDTA} = 10^{25.1}/10^{10.65} = 10^{14.45} = 2.8 \times 10^{14}$

b) If we assume K is approximately equal to $K°$, then the following relationship can be used to find the change in standard free energy.

$\Delta G° = -R\,T \ln(K) = -(8.314 \text{ J/mol K})(298 \text{ K})(33.26)(10^{-3} \text{ kJ/J})$

$\quad\quad = -82.4 \text{ kJ/mol}$

75. $k = k_{-H_2O}\,K_{os} = (5.8 \times 10^{-7})(0.16) = 9.3 \times 10^{-8} \text{ s}^{-1}$

CHAPTER 10: OXIDATION–REDUCTION REACTIONS

1. Oxidation is the loss of electrons by a chemical substance. Reduction is the gain of electrons by a chemical substance. These occur simultaneously because any electron lost by one substance will be picked up by another.

2. An oxidation–reduction reaction is a reaction in which one substance loses one or more electrons and is oxidized, while another substance gains electrons and is reduced. These reactions are also sometimes called "electrochemical reactions" because they involve the transfer of electrons between chemicals.

3. Corrosion of a metal involves the formation of an oxide or other compound of a metal through an oxidation process. Some other chemical is also reduced as part of this process. This makes corrosion a type of oxidation–reduction reaction.

4. One important application of these reactions is a redox titration, which is a method that can be used to measure the concentration of an analyte that can undergo either oxidation or reduction in the presence of a particular reagent. Oxidation–reduction reactions are also often used to pretreat samples prior to a chemical analysis. In the method of coulometry, the amount of electrons that are transferred during an oxidation–reduction reaction is used to determine the moles of a chemical that is being reduced or oxidized. In voltammetry, the change in current of a chemical system is examined as changes are made in the electrical potential applied to this system.

5. An electron is gained by each iron(III) ion to form iron(II), which is a reduction process. At the same time, the nitrogen in NH_2OH is oxidized from an oxidation state of -1 to +1 in

N₂O. (Note: Another possible product is N$_2$, in which case the nitrogen has an oxidation state of zero.)

6. A redox couple is a pair of chemical species in which at least one atom changes its oxidation number by loss or gain of an electron as it forms the other species in the redox couple. This term is analogous to concept of a conjugate acid–base pair in acid–base chemistry. Every oxidation–reduction reaction involves two redox couples. One member of each redox couple is oxidized or reduced on one side of the overall reaction, and the other member is reduced or oxidized on the other side. The overall result is a balanced reaction.

7.
a) Zn/Zn^{2+} and Ag^+/Ag

b) Pb/Pb^{2+} and PbO_2/Pb^{2+}

c) I_2/I^- and $S_2O_3^{2-}/S_4O_6^{2-}$

8.
a) Ru^{3+}/Ru^{2+}

b) $Ni(OH)_3/Ni(OH)_2$

c) $HOCl/Cl_2$

d) H_2S/S

9. Both oxidation–reduction and acid–base reactions involve the transfer of a simple entity from one reactant to another. In the case of acid-base reactions, the transferred species is a hydrogen ion, and in oxidation–reduction reactions, the transferred species is an electron.

10. An oxidation number is the charge that an element in a chemical would have if this element existed as a solitary ion, while still possessing the same number of electrons that it has in the chemical. Table 10.1 summarizes some rules that can be used to determine the

oxidation number for each element in a molecule or ion.

11.
a) Cl, 0
b) Au, 0
c) Ca, +2; O, -2

d) K, +1; S, -6; O, -2
e) O, -2; Fe, 8/3
f) H, +1; O, -1

g) F, -1; Xe, +4
h) N, -3; H, +1; Cr, +6; O, -2

i) H, +1; O, -2; C, -2

12.
a) Cl, -1
b) Ca, +2
c) H, +1; O, -2

d) Mn, +7; O, -2
e) N, -3; H, +1
f) O, -2; I, +5

g) Cr, +6; O, -2
h) H, +1; O, -2; P, +5
i) O, -2; As, +5

13. In an oxidation–reduction reaction, the oxidation number of at least one element changes. This change in oxidation number can be used, in turn, to identify a reaction as being an oxidation–reduction reaction.

14. a) Fe is reduced from +3 to 0; Al is oxidized from 0 to +3; O is always at an oxidaton number of -2.

b) Co is reduced from +3 to +2; Sn is oxidized from 0 to +2; O is at an oxidation number of -2 and H is +1.

c) One O is reduced from -1 to -2, and the other oxygen is oxidized from -1 to 0; H is at +1.

d) F is reduced from 0 to -1; Br is oxidized from -1 to 0.

15. a) No

b) Yes, Cl is reduced from 0 to -1 and I is oxidized from -1 to 0.

c) No

d) Yes, Zn is oxidized from 0 to +2 and Cu is reduced from +2 to 0.

16. A half-reaction is a chemical "reaction" that is written to show electrons among either the products or reactants. An oxidation half-reaction has electrons as products (having been lost by reactant) and a reduction half-reaction has electrons as reactants (to be gained by reactant).

17. They are artificial because free electrons are not free in solution as implied by the half-reaction. Rather, they are transferred directly from the species being oxidized to the species being reduced. Also, oxidation and reduction reactions do not occur separately, but always occur as an oxidation–reduction reaction pair. There must always be both a species that is being oxidized and one that is being reduced during such a process.

18. Fe^{2+} is oxidized according to the half-reaction: $Fe^{2+} \rightleftarrows Fe^{3+} + 1\ e^-$.

    Cr(VI) is reduced according to the half-reaction: $Cr_2O_7^{2-} + 14\ H^+ + 6\ e^- \rightleftarrows 2\ Cr^{3+} + 7\ H_2O$

19. Only (b) and (d) in Problem 15 are examples of oxidation–reduction reactions. The two half-reactions that are involved in these reactions are given below.

    b) $Cl_2 + 2\ e^- \rightleftarrows 2\ Cl^-$ (reduction half-reaction)

    $2\ I^- \rightleftarrows I_2 + 2\ e^-$ (oxidation half-reaction)

    d) $Zn(s) \rightleftarrows Zn^{2+} + 2\ e^-$ (oxidation half-reaction)

    $Cu^{2+} + 2\ e^- \rightleftarrows Cu(s)$ (reduction half-reaction)

20. a) Sn^{2+} is oxidized and Hg(II) is reduced. Two electrons are transferred.

$Sn^{2+} \rightleftarrows Sn^{4+} + 2\ e^-$ and $2\ HgCl_2 + 2\ e^- \rightleftarrows Hg_2Cl_2 + 2\ Cl^-$

b) I_2 is reduced and S(II) is oxidized. Two electrons are transferred.

$I_2 + 2\ e^- \rightleftarrows 2\ I^-$ and $2\ S_2O_3^{2-} \rightleftarrows S_4O_6^{2-} + 2\ e^-$

c) Fe^{2+} is oxidized and O(-I) is reduced. Two electrons are transferred.

$Fe^{2+} \rightleftarrows Fe^{3+} + 1\ e^-$ and $H_2O_2 + 2\ H^+ + 2\ e^- \rightleftarrows 2\ H_2O$

d) Br^- is oxidized and Br(V) is reduced. As written, five electrons are transferred.

$2\ Br^- \rightleftarrows Br_2 + 2\ e^-$ and $BrO_3^- + 6\ H^+ + 5\ e^- \rightleftarrows \frac{1}{2}\ Br_2 + 3\ H_2O$

21. a) $K^\circ = (a_{Sn4+})(a_{Hg2Cl2})(a_{Cl-})^2 / \{(a_{Sn2+})(a_{HgCl2})^2\}$

b) $K^\circ = (a_{I-})^2 (a_{S4O6 2-}) / \{(a_{I2})(a_{S2O3 2-})^2\}$

c) $K^\circ = (a_{H2O})^2 (a_{Fe3+})^2 / \{(a_{H2O2})(a_{Fe2+})^2 (a_{H+})^2\}$

d) $K^\circ = (a_{Br2})^3 (a_{H2O})^3 / \{(a_{Br-})^5 (a_{BrO3-})(a_{H+})^6\}$

22. a) $K = ([Sn^{4+}][Cl^-]^2)/([Sn^{2+}][HgCl_2]^2)$

b) $K = ([I^-]^2[S_4O_6^{2-}])/([I_2][S_2O_3^{2-}]^2)$

c) $K = [Fe^{3+}]^2/([H_2O_2][Fe^{2+}]^2[H^+]^2)$

d) $K = [Br_2]^3/([Br^-]^5[BrO_3^-][H^+]^6)$

23. a) $Q = ([Sn^{4+}][Cl^-]^2)/([Sn^{2+}][HgCl_2]^2)$

b) $Q = ([I^-]^2[S_4O_6^{2-}])/([I_2][S_2O_3^{2-}]^2)$

c) $Q = [Fe^{3+}]^2/([H_2O_2][Fe^{2+}]^2[H^+]^2)$

d) $Q = [Br_2]^3/([Br^-]^5[BrO_3^-][H^+]^6)$

Although these relationships have the same form as those for K, they differ in that these equations use concentrations at some time when the reaction is not at equilibrium. In contrast to this, the concentrations that are used in an expression for K must be those that are present at equilibrium.

24. Red is the limiting reagent. When it is at an equilibrium value of 0.00450 M, the concentration has decreased by (1.0 M − 0.00450 M) = 0.98550 M. Therefore, the equilibrium concentration of Ox is {1.0 M − (1.0 M − 0.98550 M)/2} = 0.50725 M. The concentration of Red' is 1.0 M − 0.50725 M = 0.49275 M and concentration of Ox' is 0.98550 M.

$$K = [\text{Red'}][\text{Ox'}]^2/[\text{Red}]^2[\text{Ox}]$$

$$= (0.49275)(0.99550)^2/\{(0.00450)^2(0.50725)\}$$

$$= 4.75 \times 10^5 = 4.8 \times 10^5$$

25. $K = ([Fe^{3+}]^5[Mn^{2+}])/([Fe^{2+}]^5[MnO_4^-][H^+]^8) = 1.4 \times 10^{62}$

At pH 2.50: $[H^+] = 3.16 \times 10^{-3} \, M$

At equilibrium: $[Fe^{3+}] = 0.0500 \, M$, $[Mn^{2+}] = 0.0100 \, M$, $[Fe^{2+}] = 5 \, [MnO_4^-]$

$K = \{(0.0500)^5(0.0100)\}/\{(5 \, [MnO_4^-])^5[MnO_4^-](3.16 \times 10^{-3})^8\} = 1.4 \times 10^{62}$

$K = (3.125 \times 10^{-9})/\{3125 \, [MnO_4^-]^6 (9.9 \times 10^{-21})\} = 1.4 \times 10^{62}$

$[MnO_4^-]^6 = (3.125 \times 10^{-9})/\{(3125)(9.9 \times 10^{-21})(1.4 \times 10^{62})\} = 7.2 \times 10^{-55}$

$[MnO_4^-] = 9.5 \times 10^{-10} \, M$

26. The standard electrode potential is the potential expected for a half-reaction under standard conditions when compared to a standard hydrogen electrode. The conditions that are used in determining the value of a standard electrode potential include a temperature of 25 °C, a concentration of exactly 1 M for all dissolved chemicals, and a pressure of 1 bar for all gases that are reactants or products in the half-reaction.

27. The reference half-reaction that is used in determining a standard electrode potential is that of the standard hydrogen electrode, as shown below.

$$2 H^+ + 2 e^- \rightleftarrows H_2$$

The potential of this electrode is assigned a value of exactly 0.0000 V. If some other electrode has a potential greater than 0.0000 V, it will be reduced by hydrogen. If some other electrode has a potential lower than 0.0000 V, it will be oxidized by hydrogen ion.

28. The ability to undergo reduction is also the ranking of their strength as oxidizing agents: $O_3 > Cl_2 > Cr^{3+} > Zn^{2+} > Na^+$.

29. The ability to undergo oxidation is the same as the ranking of the strength of reducing agents: $Na > Zn > Cr > Cl^- > O_2$.

30. $E°_{Net} = E°_{Reduction} - E°_{Oxidation}$

31. a) $E°_{Net} = (-0.23 \text{ V}) - (-0.40 \text{ V}) = +0.17 \text{ V}$

    b) $E°_{Net} = (-0.76 \text{ V}) - (+0.34 \text{ V}) = -1.10 \text{ V}$

    c) $E°_{Net} = (-0.14 \text{ V}) - (-0.41 \text{ V}) = +0.27 \text{ V}$

    The order of decreasing tendency for these reactions to occur is c > a > b.

32. $Ag^+ + e^- \rightleftarrows Ag$ and $Bi \rightleftarrows Bi^{3+} + 3 e^-$

    $E°_{Net} = (+0.80 \text{ V}) - (+0.31 \text{ V}) = 0.49 \text{ V}$

33. $\Delta G° = -n F E°_{Net}$

34. Faraday's constant is the number of coulombs that is equal to the charge on one mole of electrons. The value of Faraday's constant is 96,485 C/mol electrons.

35. $E°_{Net} = (+0.73 \text{ V}) - (-0.76 \text{ V}) = +1.49 \text{ V}$

$\Delta G° = -(2 \text{ mol electrons})(96{,}485 \text{ C/mol electrons})(+1.49 \text{ V})(1 \text{ J/V·C})$

$= -2.88 \times 10^5 \text{ J (or } -288 \text{ kJ)}$

36. $E°_{Net} = (+1.40 \text{ V}) - (+.53 \text{ V}) = +0.87 \text{ V}$

$\Delta G° = -(2 \text{ mol electrons})(96{,}485 \text{ C/mol electrons})(0.87 \text{ V})(1 \text{ J/V·C})$

$= -1.68 \times 10^5 \text{ J (or } -168 \text{ kJ)}$

37. $$E°_{Net} = \frac{(0.05916 \text{ V})}{n} \log(K°) \quad \text{or} \quad K° = 10^{(n\, E°_{Net})/(0.05916 \text{ V})}$$

$$\Delta G° = -n F E°_{Net} \quad \text{or} \quad E°_{Net} = -\frac{\Delta G°}{nF}$$

A large positive value of $E°_{Net}$ gives a large value for $K°$ and a negative value for $\Delta G°$, which implies that a favorable spontaneous reaction is present. A large negative value of $E°_{Net}$ results in a small value for $K°$ and a large positive value of $\Delta G°$, which indicates that a nonspontaneous reaction is present.

38. The reaction in a Daniell cell is $Cu^{2+} + Zn \rightleftarrows Cu + Zn^{2+}$. The net change in standard potential for this reaction and corresponding equilibrium constant is as follows.

$E°_{Net} = (+0.34 \text{ V}) - (0.76 \text{ V}) = +1.10 \text{ V}$

$\text{Log } K = n E°_{Net}/0.05916 = 2 \cdot (+1.10 \text{ V})/(0.05916 \text{ V}) = 37.1\underline{9}$

$K = 1.5 \times 10^{37}$

39. $E°_{Net} = \{(0.05916 \text{ V})/n\} \log(K°)$

$= \{(0.05916 \text{ V})/1\} \log(100)$

$= 0.118 \text{ V}$

40. a) $Zn + Cu^{2+} \rightleftharpoons Cu + Zn^{2+}$

$E°_{Net} = (+0.34 \text{ V}) - (-0.76 \text{ V}) = +1.10 \text{ V}$

$\log K = 2 \cdot (1.10 \text{ V})/(0.05916 \text{ V}) = 37.1\underline{9}$

$K = 1.5\underline{4} \times 10^{37} = 1.5 \times 10^{37}$

b) mol Zn = $(1.00 \text{ g})/(65.39 \text{ g/mol}) = 1.53 \times 10^{-2}$ mol

mol Cu^{2+} = $(0.050 \text{ mol/L})(0.050 \text{ L}) = 2.50 \times 10^{-3}$ mol

These results indicate that Cu^{2+} is the limiting reagent.

At equilibrium, $[Zn^{2+}] = 0.050 \ M$ and $[Cu^{2+}] = (0.050 \ M)/(1.5\underline{4} \times 10^{37}) = 3.2 \times 10^{-39} \ M$.

This value of $[Cu^{2+}]$ is extremely small and may be regarded as essentially being equal to zero.

The mass of copper metal at equilibrium will be $(2.50 \times 10^{-3} \text{ mol})(63.5 \text{ g/mol}) = 0.159$ g.

The mass of zinc metal at equilibrium will be $(1.53 \times 10^{-2} \text{ mol} - 2.50 \times 10^{-3} \text{ mol})(65.39$ g/mol$) = 0.837$ g.

41. a) $Ce^{4+} + Fe^{2+} \rightleftharpoons Ce^{3+} + Fe^{3+}$

In 1.0 M HCl:

$E°_{Net} = (+1.47 \text{ V}) - (+0.73 \text{ V}) = 0.74 \text{ V}$

$\log K = (1)(0.74 \text{ V})/(0.05916 \text{ V}) = 12.5\underline{1}$ or $K = 3.\underline{2} \times 10^{12} = 3 \times 10^{12}$

b) With this large a value for K, $[Fe^{3+}] = [Ce^{3+}] = (0.0800 \text{ mol})/(2.00 \text{ L}) = 0.0400 \ M$.

$[Fe^{2+}] = [Ce^{4+}] = \{(0.0400 \ M)^2/3.\underline{2} \times 10^{12}\}^{1/2} = 2.2 \times 10^{-8} \ M$.

42. An electrochemical cell is a device in which the oxidation and reduction processes of an oxidation–reduction reaction occur in different locations, with electrons flowing from one location to the other through an external circuit. This type of device can be used to either

control or study oxidation–reduction reactions by measuring or controlling factors such as the current and potential that are involved in such processes.

43. A Daniell cell involves the following two half reactions:

$Cu^{2+} + 2\ e^- \rightleftharpoons Cu(s)$ and $Zn(s) \rightleftharpoons Zn^{2+} + 2\ e^-$

These two half-reactions and their corresponding half-cells are separated by a salt bridge in a Daniell cell.

44. a) An electrode is a conducting material at which one of the half-reactions in an electrochemical cell is taking place.

b) The anode is the electrode at which oxidation occurs in an electrochemical cell.

c) The cathode is the electrode at which reduction occurs in an electrochemical cell.

d) A salt bridge is a device that separates, but electrically connects, the solutions surrounding the anode and cathode in an electrochemical cell. This device allows current to flow between two electrodes, while preventing the mixing of their electrolytes.

e) An electrolyte is the solution of ions that surrounds an electrode in an electrochemical cell.

45. A galvanic cell is an electrochemical cell in which an oxidation–reduction reaction occurs spontaneously, resulting in a flow of electrons. An electrolytic cell is an electrochemical cell in which an external power source is used to apply an electric current and cause a particular oxidation–reduction reaction to occur.

46. An oxidation–reduction reaction that is spontaneous can be used as a galvanic cell and, if sufficient voltage is applied in opposition to the galvanic current, can be made into an electrolytic cell. An example is the reaction of hydrogen with oxygen, which is the basis of a

fuel cell when it is carried out, so that the electrons flow through an electrical circuit. Because this reaction is spontaneous, it is a galvanic cell that produces a voltage and current. Application of sufficient voltage in the opposite direction causes water to be electrolyzed into its elements, oxygen and hydrogen.

47. A standard cell potential is the potential that develops between an anode and a cathode when all of the components in an electrochemical cell are in their standard states. This is the potential observed if all species are at an activity of 1.000 at standard temperature and pressure. This value can be calculated by using the difference in the standard electrode potentials for the cathode and anode.

48. a) $E°_{Cell} = (+1.72 \text{ V}) - (-0.76 \text{ V}) = +2.48 \text{ V}$

    b) $E°_{Cell} = (+1.51 \text{ V}) - (+0.77 \text{ V}) = +0.74 \text{ V}$

    c) $E°_{Cell} = (+1.36 \text{ V}) - (+1.00 \text{ V}) = +0.36 \text{ V}$

    d) $E°_{Cell} = (-2.00 \text{ V}) - (-2.36 \text{ V}) = +0.36 \text{ V}$

49. $U^{3+} + 3 e^- \rightleftarrows U$ and $V^{3+} + 1 e^- \rightleftarrows V^{2+}$

    $3 V^{3+} + U \rightleftarrows U^{3+} + 3 V^{2+}$

    $E°_{Cell} = (-0.26 \text{ V}) - (-1.8 \text{ V}) = 1.5\underline{4} \text{ V} = 1.5 \text{ V}$

50. There are several reasons that the actual measured potential of a cell may be different from the calculated standard cell potential. For instance, we may not be working under standard conditions and the activities of our reactants and products may not be equal to one. The potential may also be affected by the fact that we have placed additional components in our system when constructing a device that separates the two half-reactions. A common example of this occurs when the presence of a salt bridge leads to

the formation of a junction potential at each boundary between the electrolyte and salt bridge. There will also be some resistance to the flow of ions through the electrolyte between the electrodes and salt bridge when current is passing through the cell, creating a change in potential due to an effect known as "IR drop".

51. There is no way to measure the potential of only one electrode. Thus, a comparison must always be made between the potentials for two electrodes or half-reactions.

52. A standard hydrogen electrode consists of a platinum wire coated with platinum black in the presence of 1.00 M HCl and a bubbling stream of hydrogen gas at a pressure of 1.00 bar. This is the electrode against which all other electrode potentials are measured and compared.

53. Because the standard potential of the standard hydrogen electrode (SHE) is exactly 0.0000 V, the value of the measured cell potential versus the SHE will be equal to the electrode potential of the system being compared to the SHE.

$E°_{cell} = E°_{Cathode} - E°_{SHE}$

$= E°_{Cathode} - (0.000\ V) = E°_{Cathode}$

54. $E°_{cell} = E_{Calomel} - E°_{SHE} = (+0.242\ V) - (0.000\ V) = 0.242\ V$

55. a) Cathode: $Cu^{2+} + 2\ e^- \rightleftarrows Cu(s)$

   Anode: $H_2 \rightleftarrows 2\ H^+ + 2\ e^-$

   b) $E°_{cell} = (+0.34\ V) - (0.000\ V) = +0.34\ V$

56. The Nernst equation is an expression used for a reversible half-reaction to relate the reduction potential under nonstandard conditions (E) to the activities of the reactants and products and half-reaction's standard reduction potential ($E°$). This equation makes it

possible to calculate electrode potentials and cell potentials of such a system under nonstandard conditions.

57. $E_{Cell} = E° - \{(RT)/(nF)\} \ln\{a_{Red}/a_{Ox}\}$

At 25 °C: $E_{Cell} = E° - \{(0.05916 \text{ V})/n\} \log\{a_{Red}/a_{Ox}\}$

$= E° - (0.05916/n) \log\{a_{Red}/a_{Ox}\}$

58. a) The only differences between this equation and the one given in this chapter are the use of addition instead of subtraction for the logarithmic term and the use of the inverse ratio in this term compared to that given earlier in this chapter. Using the inverse ratio in the logarithm term creates a negative sign in front of the logarithm, making these two equations equivalent.

b) The use of a temperature of 25 °C (298.15 K) and the fact that $2.303 \log(x) = \ln(x)$ allows this equation to be converted into the following form.

$$E = E° + \frac{0.05916 \text{ V}}{n} \log\left(\frac{a_{Ox}}{a_{Red}}\right)$$

59. a) $E = E° - \{RT/(1F)\} \ln\{a_{Cu+}/a_{Cu2+}\}$

b) $E = E° - \{RT/(2F)\} \ln\{1/a_{Hg2+}\}$

c) $E = E° - \{RT/(1F)\} \ln\{1/a_{Tl+}\}$

d) $E = E° - \{RT/(2F)\} \ln\{a_{Sn2+}/a_{Sn4+}\}$

60. a) $E = E° - (0.05916 \text{ V}) \log\{a_{Cu+}/a_{Cu2+}\}$

b) $E = E° - (0.02958 \text{ V}) \log\{1/a_{Hg2+}\}$

c) $E = E° - (0.05916 \text{ V}) \log\{1/a_{Tl+}\}$

d) $E = E° - (0.02958 \text{ V}) \log\{a_{Sn2+}/a_{Sn4+}\}$

61. $E = -0.41 - (0.05916/4) \log\{1/[Fe^{2+}]^2\}$

$E = -0.41 - (0.05916/2) \log\{1/[Fe^{2+}]\}$

It is true that $\log(x) = \frac{1}{2} \log(x^2)$, so the above two equations give the same overall result.

62. a) $E = 0.222 - \{RT/F\} \ln(a_{Cl^-})$

b) $E = -0.83 - \{RT/(2F)\} \ln\{(a_{H_2})(a_{OH^-})^2\}$

c) $E = +0.94 - \{RT/(2F)\} \ln\{(a_{HNO_2})/[(a_{NO_3^-})(a_{H^+})^3]\}$

d) $E = 1.63 - \{RT/(2F)\} \ln\{(a_{Cl_2})/[(a_{HOCl})^2(a_{H^+})^2]\}$

63. a) $E = 0.222 - 0.05916 \log\{a_{Cl^-}\}$

b) $E = -0.83 - 0.02958 \log\{(a_{H_2})(a_{OH^-})^2\}$

c) $E = +0.94 - 0.02958 \log\{(a_{HNO_2})/[(a_{NO_3^-})(a_{H^+})^3]\}$

d) $E = +1.63 - 0.02958 \log\{(a_{Cl_2})/[(a_{HOCl})^2(a_{H^+})^2]\}$

64. $E = -0.28 - (0.05916/2) \log\{(a_{H_3PO_3})/[(a_{H_3PO_4})(a_{H^+})^2]\}$

$E = +0.58 - (0.05916/2) \log\{(a_{H_3AsO_3})/[(a_{H_3AsO_4})(a_{H^+})^2]\}$

65. The net cell potential is the difference in the two electrode potentials. Therefore, write the Nernst expression for the cathode reaction and subtract from it the Nernst expression for the anode reaction.

66. $E_{Cell} = \{E°_{Cu} - 0.02598 \log(1/a_{Cu^{2+}})\} - \{E°_{Zn} - 0.02958 \log(1/a_{Zn^{2+}})\}$

$= (E°_{Cu} - E°_{Zn}) - 0.02958 \log\{[Zn^{2+}]/[Cu^{2+}]\}$

$= \{(+0.34 \text{ V}) - (-0.76 \text{ V})\} - (0.02958 \log\{0.340/0.135\})$

$= (+1.10 \text{ V}) - 0.02958 \log\{2.518\}$

$= (+1.10 \text{ V}) - 0.02958 \{0.4011\}$

$= (+1.10 \text{ V}) - (0.012 \text{ V}) = +1.09 \text{ V}$

67. a) $E_{Cathode} = 1.09 - (0.02958) \log\{[Br^-]^2/[Br_2]\}$

$E_{Anode} = 0.58 - (0.02958) \log\{[H_3AsO_3]/[H_3AsO_4][H^+]^2\}$

b) $E_{Cell} = \{1.09 - (0.02958) \log (0.244)^2/(0.00500)\}$

$- \{0.58 - (0.02958) \log\{(0.128)/(0.00367)(1.0 \times 10^{-2})^2\}$

$= \{1.09 - (0.02958) \log(11.907)\} - \{0.58 - (0.02958) \log(3.488 \times 10^5)$

$= \{1.09 - 0.032\} - \{0.58 - 0.164\} = 1.05\underline{8} - 0.41\underline{6} = 0.64\underline{2} = 0.64$ V

68. $3 Ag^+ + Bi_{(s)} \rightleftarrows Bi^{3+} + 3 Ag_{(s)}$

$E_{Ag} = (+0.80\ V) - (0.05916/3) \log\{1/[Ag^+]^3\}$

$E_{Bi} = (+0.31\ V) - (0.05916/3 \log\{1/[Bi^{3+}]\})$

$E_{Cell} = (0.80\ V - 0.31\ V) - 0.01972 \log([Bi^{3+}]/[Ag^+]^3)$

$= 0.49\ V - 0.01972 \log\{(0.0100)/(.0100)^3\}$

$= 0.49\ V - 0.01972(4.00) = 0.49\ V - 0.079\ V = 0.41\ V$

69. a) $Ni^{2+} + 2 e^- \rightleftarrows Ni(s)$ $Co^{2+} + 2 e^- \rightleftarrows Co(s)$

b) $E°$ for the nickel half-reaction is -0.23 V (vs. SHE) and $E°$ for the cobalt half-reaction is -0.28 V (vs. SHE); because the nickel $E°$ value is more positive, we might expect that nickel would be the cathode and cobalt would be the anode.

c) $E_{Cell} = E_{Ni} - E_{Co} = \{(-0.23\ V) - (-0.28\ V)\} - 0.02958 \log(0.050/0.00040)$

$= (+0.05\ V) - 0.02958 \log(125) = 0.05\ V - 0.062\ V = -0.01\ V$

d) The negative sign on the answer to part c) tells us that we have incorrectly chosen the nickel electrode as the cathode. Actually, the cell potential is +0.01 V, with cobalt being the cathode and nickel being the anode.

70. a) The two half-reactions, both written as reduction processes, are as follows.

$Ag^+ + 1\,e^- \rightleftarrows Ag(s)$ $\qquad\qquad$ $Cu^{2+} + 2\,e^- \rightleftarrows Cu(s)$

b) $E_{Cell} = (0.80 - 0.34) - 0.02958 \log(0.00500)/(0.235)^2$

$= (0.46) - 0.02958\,(-1.043) = 0.46 + 0.031 = 0.49\underline{1} = 0.49$ V

c) Copper is oxidized, making the copper metal electrode the anode. Silver ions are reduced, making the silver metal electrode the cathode.

71. The Cu^{2+} will be reduced to $Cu(s)$ at the cathode and $Cu(s)$ will be oxidized to Cu^{2+} at the cathode. The reaction will continue until the two copper ion activities become equal, at which time the cell potential will be 0.0000 V.

$E_{Cell} = \{E°_{Cu} - 0.02958 \log(1/0.875)\} - \{E°_{Cu} - 0.02958 \log(1/0.236)\}$

$= -0.02958 \log(0.236/0.875) = 0.0168$ V

72. The Nernst equation is ideally written in terms of activities. The activity of an ionic substance depends on the ionic strength, but the concentration does not. If the ionic strength is high, errors due to this difference in activity versus concentration will tend to be larger (the exception is when the molar concentration of all species is 1.00, which also corresponds to an activity of 1.00 for each species).

73. $3\,Ag^+ + Bi_{(s)} \rightleftarrows Bi^{3+} + 3\,Ag_{(s)}$

$$(E_{Net} - E_{Net,Conc}) = -\frac{RT}{nF} \ln \frac{(\gamma_{Bi3+})}{(\gamma_{Ag+})^3}$$

The ionic strength in this case is $I = 0.07$ M. We obtain the following activity coefficients if we assume we can use the extended Debye–Hückel equation to estimate the activity coefficients. (Note: The given ionic strength is a little higher than we would normally use with this equation; the actual activity coefficients are probably a little higher than what we estimate here.)

Using a = 900 pm for Bi^{3+}: $\gamma_{Bi3+} = 0.20\underline{8}$

$\gamma_{Ag+} = 0.77\underline{5}$

$(E_{Net} - E_{Net,Conc}) = -[(8.314 \text{ J/mol·K})(298 \text{ K})/\{3\cdot(96{,}485 \text{ C/mol})\}] \ln\{(0.20\underline{8})/(0.77\underline{5})^3\}$

$= (-0.00856) \ln(0.44\underline{7})$

$= +0.0069 \text{ V}$ (This error is probably not significant.)

74. $2 \text{ Ag}^+ + \text{Cu(s)} \rightleftarrows 2 \text{ Ag(s)} + \text{Cu}^{2+}$

$$(E_{Net} - E_{Net,Conc}) = -\frac{RT}{nF} \ln \frac{(\gamma_{Cu2+})}{(\gamma_{Ag+})^2}$$

Ionic strength in this case is controlled by $AgNO_3$, giving $I = 0.235$ M. If we assume we can use the extended Debye–Hückel equation to estimate the activity coefficients, we obtain the following. (Note: The given ionic strength is higher than we would normally use with this equation; the actual activity coefficients are probably a little higher than what we estimate here.)

$\gamma_{Cu2+} = 0.31\underline{2}$ $\gamma_{Ag+} = 0.66\underline{5}$

$(E_{Net} - E_{Net,Conc}) = [-(8.314 \text{ J/mol·K})(298 \text{ K})/\{2\cdot(96{,}485 \text{ C/mol})\}] \ln\{(0.31\underline{2})/(0.66\underline{5})^2\}$

$= 0.0045 \text{ V}$ (This error is probably not significant.)

75. a) $E_{Cell} = \{(+1.51 \text{ V}) - (+0.77 \text{ V})\} - (0.05916/5) \log \{[Mn^{2+}][Fe^{3+}]^5/[MnO_4^-][Fe^{2+}]^5[H^+]^8\}$

$= 0.74 - 0.01183 \log\{(0.0439)(0.00764)^5/(0.0764)(0.00235)^5(0.10)^8\}$

$= 0.74 - 0.01183 (10.319) = 0.74 \text{ V} - 0.122 \text{ V} = +0.62 \text{ V}$

b) $\log \gamma_1 = -0.51(1)^2(0.10)^{1/2}/\{1 + (0.10)^{1/2}) = -0.122$ $\gamma_1 = 0.755$

$\log \gamma_2 = -0.51(2)^2(0.10)^{1/2}/\{1 + (0.10)^{1/2}) = -0.490$ $\gamma_2 = 0.323$

$$\log \gamma_3 = -0.51(3)^2(0.10)^{1/2}/\{1 + (0.10)^{1/2}\} = -1.102 \qquad \gamma_3 = 0.0789$$

The combined activity coefficients have the following value.

$$(0.323)(0.0789)^5/(0.755)(0.323)^5(0.755)^8 = 0.00352$$

$\log(0.00352) = -2.45$, so this value would be multiplied by $(0.05916/5)$, giving a 0.03 V error.

76. The reactions that are directly affected by a change in pH are (a), (c), and (d).

77. If the pH increases by one unit, the activity of OH^- will increase by 10-fold. This change will result in the logarithmic term of the equation changing by 0.011 V per pH unit.

78.
$$E_{Cr_2O_7^{2-}/Cr^{3+}} = E°_{Cr_2O_7^{2-}/Cr^{3+}} - \frac{0.05916\ V}{6} \log\left[\frac{(a_{Cr^{3+}})^2}{(a_{Cr_2O_7^{2-}})(a_{H^+})^{14}}\right]$$

If both Cr^{3+} and $Cr_2O_7^{2-}$ are at standard activity, then the following holds.

$$E = 1.36 - 0.009860 \log(1/(10^{-7})^{14}) = 1.36 - 0.009860 \log(1.0 \times 10^{98})$$

$$= 1.36 - 0.009860(98) = 1.36 - 0.966 = 0.39\ V$$

$$E_{Cr_2O_7^{2-}/Cr^{3+}} = E°_{Cr_2O_7^{2-}/Cr^{3+}} - \frac{0.05916\ V}{6} \log\left[\frac{(a_{Cr^{3+}})^2(a_{OH^-})^{14}}{(a_{Cr_2O_7^{2-}})}\right]$$

$$= 1.36 - 0.009860 \log(1.0 \times 10^{-7})^{14}$$

$$= 1.36 - 0.009860 \log(1.0 \times 10^{98})$$

$$= 1.36 - 0.966 = 0.39\ V$$

The two methods give the same answer. There is the problem, however, in that $Cr(OH)_3$ would precipitate out of solution if 1.0 M Cr^{3+} were in solution at pH 7.0 because the K_{sp} for $Cr(OH)_3$ is approximately 1×10^{-30}.

79. At pH 2.00:

$E_{Cell} = \{E°_{Mn} - 0.05916/5 \log([Mn^{2+}]/[MnO_4^-][H^+]^8)\} - \{E°_{Sn} - 0.05916/2 \log([Sn^{2+}]/[Sn^{4+}])\}$

$= \{1.51 - 0.01183 \log\{(0.023)/(0.0015)(1.0 \times 10^{-2})^8\}\} - \{0.14 - 0.02958 \log(0.014)/(0.025)$

$= \{1.51 - 0.01183 \log(1.53 \times 10^{17})\} - \{0.14 - 0.02958 \log(0.56)\}$

$= \{1.51 - 0.01183(17.185)\} - \{0.14 - 0.02958(-0.252)\} = \{1.51 - 0.203\} - \{0.14\text{ V} + 0.007\text{ V}\}$

$= 1.30\underline{7}\text{ V} - 0.14\underline{7}\text{ V} = 1.16\text{ V}$

At pH 4.00:

$E_{Cell} = \{E°_{Mn} - 0.05916/5 \log([Mn^{2+}]/[MnO_4^-][H^+]^8)\} - \{E°_{Sn} - 0.05916/2 \log([Sn^{2+}]/[Sn^{4+}])\}$

$= \{1.51 - 0.01183 \log\{(0.023)/(0.0015)(1.0 \times 10^{-4})^8\}\} - \{0.14 - 0.02958 \log(0.014)/(0.025)$

$= \{1.51 - 0.01183 \log(1.53 \times 10^{33})\} - \{0.14 - 0.02958 \log(0.56)\}$

$= \{1.51 - 0.01183(33.185)\} - \{0.14 - 0.02958(-0.252)\} = \{1.51 - 0.393\} - \{0.14 + 0.007\}$

$= 1.11\underline{7}\text{ V} - 0.14\underline{7}\text{ V} = 0.97\text{ V}$

80. The equilibrium constants of the side reaction products can be included appropriately in the log term of the Nernst equation. This makes it possible to include the effects of a side reaction as part of the Nernst equation.

81. a) $AgCl + 1\text{ e}^- \rightleftarrows Ag + Cl^-$ and $2\text{ H}^+ + 2\text{ e}^- \rightleftarrows H_2$

$E_{AgCl} = E°_{Ag} - 0.05916 \log(1/a_{Ag+})$

$= E°_{Ag} - 0.05916 \log\{a_{Cl^-}/K_{sp}\}$

$= E°_{Ag} + 0.05916 \log(K_{sp}) - 0.05916 \log(a_{Cl^-})$

$= 0.80 + 0.05916 \log(1.77 \times 10^{-10}) - 0.05916 \log(a_{Cl^-})$

$= 0.80 + 0.05916(-9.752) - 0.05916 \log(a_{Cl^-})$

$= 0.80 - 0.5769 - 0.05916 \log(a_{Cl^-})$

$$= 0.2231 - 0.05916 \log(a_{Cl^-})$$

b) A side reaction that could change the potential of the silver electrode would be the reaction of silver ion with other halides, such as bromide or iodide, or its reaction with ligands, such as ammonia or thiocyanate.

c) $E_{Cell} = E_{AgCl} - E°_{SHE} = 0.222 - 0.05916 \log(0.0025) - 0.0000$

$\quad = 0.222 - 0.05916 (-2.602) - 0.0000$

$\quad = 0.222 + 0.1539 - 0.0000 = 0.3759 = 0.376$ V

82. a) $BiOCl + 2 H^+ + 3 e^- \rightleftharpoons Bi + Cl^- + H_2O$

$Zn^{2+} + 2 e^- \rightleftharpoons Zn(s)$

$E_{Bi} = E°_{BiOCl} - (0.05916/3) \log\{(a_{Cl^-})/(a_{H^+})^2\}$

$E_{Zn} = E°_{Zn} - (0.05916/2) \log\{1/a_{Zn^{2+}}\}$

b) Anything that would add or remove hydrogen ion or that would precipitate the chloride (such as silver ions) would affect the bismuth half-reaction.

c) $E = \{0.16 - 0.01972 \log \{0.00500)/(0.00500)^2 \} - \{-0.76 - 0.02958 \log(1/0.0200)\}$

$\quad = \{0.16 - 0.01972 (2.301)\} - \{ -0.76 - 0.02958 (1.699)\}$

$\quad = \{0.16 - 0.045\} - \{-0.76 - 0.050\}$

$\quad = 0.115 + 0.81 = 0.925 = 0.92$ V

83. A Pourbaix diagram is a plot of potential versus pH that shows which principal form of a particular element is expected under a set of reaction conditions. This plot also shows the locus of points that separate various forms of a substance at different pH values and different potentials. The dominant species at different pH and potential regimes is then made apparent from this diagram.

84. a) Fe^{2+} b) $Fe(OH)_2$ c) Fe^{3+} d) FeO_4^{2-}

85. A formal potential, or conditional potential, is similar to a standard potential in that it represents the expected potential for a given redox couple when the activities of the species undergoing oxidation or reduction are exactly 1.0. However, a formal potential is instead reported for a specific type of solution or electrolyte in which the oxidation–reaction is being examined.

86. a) $E_{Cell} = E°_{Ce} - E°_{Fe} - (0.05916 \text{ V}) \log\{[Fe^{3+}][Ce^{3+}]/[Fe^{2+}][Ce^{4+}]\}$

$= 1.72 \text{ V} - 0.77 \text{ V} - (0.05916 \text{ V}) \log\{(0.176)(0.0987)/(0.0765)(0.0376)\}$

$= 0.95 \text{ V} - (0.05916 \text{ V}) \log(6.039) = 0.95 \text{ V} - 0.046 \text{ V} = 0.90 \text{ V}$

The cathode is the platinum in the cerium solution and the anode is the platinum in the iron solution.

b) $E_{Cell} = E°_{Ce} - E°_{Fe} - (0.05916 \text{ V}) \log\{[Fe^{3+}][Ce^{3+}]/[Fe^{2+}][Ce^{4+}]\}$

$= (1.47 \text{ V} - 0.73 \text{ V}) - 0.046 \text{ V}$

$= 0.74 \text{ V} - 0.046 \text{ V} = 0.69 \text{ V}$

c) $E_{Cell} = E°_{Ce} - E°_{Fe} - 0.05916 \log\{[Fe^{3+}][Ce^{3+}]/[Fe^{2+}][Ce^{4+}]\}$

$= (1.44 - 0.68) - 0.046 = 0.76 - 0.046 = 0.71 \text{ V}$

(Note: This result assumes that the standard potentials for this system are about the same in 0.5 M H_2SO_4 and 1.0 M H_2SO_4.)

87. a) $E = E° - 0.05916/2 \log\{(a_{C_3H_5O_3-})/(a_{C_3H_3O_3-})(a_{H+})^2\}$

b) $E = E° - 0.02958 \log\{1/(a_{H+})^2\} - 0.02958 \log\{(a_{C_3H_5O_3-})/(a_{C_3H_3O_3-})\}$

$= E° - 0.05916(\text{pH}) - 0.02958 \log\{(a_{C_3H_5O_3-})/(a_{C_3H_3O_3-})\}$

$= E° - 0.414 - 0.02958 \log\{(a_{C_3H_5O_3-})/(a_{C_3H_3O_3-})\}$

$E^{\circ\prime} = E^\circ - 0.414 = 0.244 - 0.414 = -0.170$ V

Not only is there a large difference in the standard potential (+0.244 V) and the formal potential (-0.170 V), but the algebraic sign changes from positive to negative.

90. a) Terminal carbons: oxidation number = 0 - 3 = -3

   Internal carbons: oxidation number = 0 - 2 = -2

   Average oxidation number = {(-3)(2) + (-2)(6)}/8 = -18/8 = -2.25

   b) Methyl carbon: oxidation number = 0 - 3 = -3

   Carboxylate carbon: oxidation number = 0 + 3 = +3

   Average oxidation number = 0

99. a) $E^\circ = \{(1)(0.77\text{ V}) + (2)(-0.41\text{ V})\}/(1 + 2) = -0.01\underline{7}$ V $= -0.02$ V

   b) $E^\circ = \{(2)(0.34\text{ V}) - (1)(0.52\text{ V})\}/(2-1) = +0.16$ V

CHAPTER 11: GRAVIMETRIC ANALYSIS

1. A gravimetric analysis is an analytical method that uses only measurements of mass and information on reaction stoichiometry to determine the amount of an analyte in a sample. Gravimetric analysis is often based on a measurement of the mass of a solid derivative of the analyte: the measurement of a change in mass can also be utilized. The mass relationship of the weighed product, or the change in mass, allows calculation of the amount of analyte in a sample.

2. Richards used gravimetry to find the atomic mass of silver by measuring the amount of AgCl that was produced from a sample mass that contained a known initial mass of silver. Modern chemists use this same approach to the quantitative analysis of silver by using the known atomic mass of silver and measured mass of AgCl to determine the amount of silver that is present in a sample. Both types of experiments make use of the known stoichiometry of this reaction, in which one mole of Ag^+ reacts with one mole of Cl^- for every mole of AgCl that is formed.

3. Most traditional gravimetric methods determine the mass of an analyte by measuring the mass of a precipitate or related solid that contains this analyte in a known ratio. Two other types of gravimetric analysis methods are combustion analysis and thermogravimetric analysis, in which mass measurements are combined with the formation or loss of volatile chemicals. In combustion analysis, a sample is burned to release gases such as carbon dioxide and water, which are then collected and weighed to determine the carbon and hydrogen content of the original sample. In a thermogravimetric analysis, a sample is

heated in a controlled fashion and its change in mass is measured with temperature as it releases volatile components or reacts with gases in its surrounding atmosphere.

4. A precipitant is a precipitating agent that is added to a sample to form a weighable solid whose mass can be used to calculate the mass of analyte that is present.

5. In combustion analysis, a sample is burned to release gases such as carbon dioxide and water, which are then collected and weighed to determine the carbon and hydrogen content of the original sample. In a thermogravimetric analysis, a sample is heated in a controlled fashion and its change in mass is measured with temperature as it releases volatile components or reacts with gases in its surrounding atmosphere.

6. Advantages: inexpensive and requires only simple equipment, and can give high accuracy and good precision for major sample components. Disadvantages or limitations: requires enough analyte to be present for mass measurements, not good for trace analysis, and can be tedious and time-consuming to perform.

7. Traditional gravimetric methods require that the analyte be placed into a solution in a form that is both soluble and that can be precipitated. In the case of a metal or mineral-based sample, the sample must be converted into metal ions that will be soluble in an aqueous solution.

8. Ashing refers to the pretreatment by dry or wet methods that convert metals in the sample into metal ions in a solution. This approach is needed to place the metals into solution as metal ions that can then be treated with a precipitating agent for a gravimetric analysis.

9. a) Wet ashing is a method often used with organic samples that involves adding a weighed amount of sample to a concentrated acid and heating it to the boiling point of the acid (usually in a porcelain dish). This procedure is done in such a way as to oxidize any organic material so that it is lost as CO_2, while the mineral components of the sample remain behind and dissolve in the acid as metal ions. This acid solution is then diluted with water in a volumetric flask and analyzed for its metal ion content.

b) Dry ashing is a method of sample preparation in which a weighed portion of a sample is heated to high temperatures in a porcelain dish that is open to the air. This procedure burns off any organic material and leaves nonvolatile metal oxides behind in the dish. These metal oxides are usually soluble in a dilute solution of hydrochloric acid.

c) In fusion, a weighed and powdered sample is mixed with a material such as solid sodium carbonate that acts as the fusing agent. (Note: Other fusing agents can also be utilized.) This mixture is placed in a platinum crucible and heated to high temperatures. Under these conditions, sodium carbonate decomposes to form CO_2 and sodium oxide (Na_2O). Molten sodium oxide is extremely basic and will dissolve rocks and silica-based materials by reacting with SiO_2 to form the water-soluble salt Na_2SiO_3. After this fusion process is complete and everything has been cooled, the remaining material is dissolved in a dilute solution of hydrochloric acid and analyzed for its metal ion content.

10. a) Dry ashing would melt the chocolate and burn it to CO_2. This would leave behind the iron as iron oxide, as well as other metal oxides.

b) Sodium carbonate fusion will dissolve the silicate to form water-soluble sodium silicate.

c) Wet ashing will keep the dissolved mercury in solution rather than being volatilized away in the other high-temperature methods.

11. Filtration is a process by which a filter is used to physically separate a solid material from a liquid. This process is often used in gravimetric analysis to separate a precipitate from a solution.

12. a) A slurry is a heterogeneous mixture of a solid suspended in a liquid. For instance, the combination of aqueous solutions of $AgNO_3$ and $NaCl$ will quickly result in a slurry, in which $AgCl(s)$ particles are present in a liquid that also contains soluble ions.

    b) The supernate is the liquid remaining after a precipitation has occurred, or the liquid that is in contact with a precipitate. This would be the liquid that remains after an $AgCl(s)$ precipitate has formed and settled to the bottom of the container.

    c) A filtrate is the liquid that passes through a filter during the process of filtration. This would be the solution of soluble ions that passes through a filter as the $AgCl(s)$ particles are collected on a filter with a suitably small pore size.

13. Filter paper is a relatively inexpensive and effective way of isolating many types of precipitates. Once a precipitate has been collected, the filter paper is then burned away to leave only the precipitate or a related solid residue for weighing. As the pore size is decreased, the filter paper is able to isolate precipitates with smaller size particles. However, using a smaller pore size also decreases the rate at which liquid can flow through the filter.

14. a) Medium or course porosity

    b) Fine or medium porosity

    c) Fine or medium porosity

15. Very small particles plug the pores in filter medium, making it very difficult to complete the filtration. Using ripening to produce larger particles will make it easier to collect these particles on filter paper and will allow the use of larger porosity paper, making it possible to conduct the filtration more rapidly.

16. Ashless paper is a special paper that has been treated with acid to dissolve out mineral materials, leaving pure cellulose. When this paper is burned away, it then leaves only a small amount of solid residue behind. This type of paper provides a convenient means for collecting a precipitate, but also allows for removal of the paper prior to a mass measurement.

17. The paper is first folded in half and then into quarters. It is opened up and inserted into a glass funnel. The suspension is poured into the funnel where the liquid passes through and the solid is retained. The precipitate is rinsed with a solution containing only volatile material. The damp filter paper is pressed into a porcelain crucible, where it is heated with a burner until it is dry, and then the paper is burned away leaving only the precipitate.

18. Ignition is the burning away of the filter paper and heating of the precipitate until it is a single, pure compound.

19. A Meker burner is similar to a Bunsen burner, but it has a larger top surface to provide a

uniform hot flame over a relatively wide area. This type of burner is often used during the ignition step of a gravimetric analysis.

20. $2\ Mg(NH_4)PO_4 \cdot 6\ H_2O \rightarrow Mg_2(P_2O_7) + 7\ H_2O + 2\ NH_3$

21. $Sn(C_6H_5O_2N_2)_4 + 26\ O_2 \rightarrow SnO_2 + 24\ CO_2 + 4\ N_2 + 10\ H_2O$

22. Porous-bottomed crucibles are made of either glass or porcelain. The suspension is poured through them usually with the help of an aspirator. Glass crucibles can be dried at oven temperatures, but they will melt at the temperature of a Meker burner. Porcelain crucibles can, on the other hand, be heated to the full heat of a Meker burner. Each must be carefully cleaned after each use.

23. a) An aspirator is attached to a water faucet and, when the water is turned on, creates a low pressure in a side-arm flask that holds a porous-bottomed crucible. The resulting difference in pressure can be used to help pull liquid through a filter that has been placed on the flask.

    b) A sintered glass crucible in a gravimetric analysis when the collected precipitate needs to be dried rather than ignited.

24. Usually, the analysis is desired to be done on a dry sample. As a result, the sample must be dried prior to analysis by some method that only removes water without changing any other aspect of the sample. The precipitate also must be pure and dry to allow calculation of the percent of the sample that is the analyte. Drying of precipitates in sintered crucibles is

usually done in an oven, typically set at a temperature of about 110 °C. Samples are left in such an oven for a suitable length of time and allowed to cool in a dry storage container.

25. A desiccator is a container used to store samples in a dry environment. This container typically has a large opening at the top, which can be covered to make an airtight seal. The bottom portion is filled with some chemical (called a desiccant), such as calcium chloride, that adsorbs water from the air within the container. Samples and filtered precipitates are kept in the desiccator to prevent them from absorbing water from the air and gaining weight.

26. A desiccant is a material that absorbs water vapor from the air. This material is used to maintain dry conditions inside a desiccator. Calcium chloride, phosphorous pentoxide, magnesium perchlorate, and calcium sulfate are four examples of desiccants.

27. A hygroscopic substance is one that absorbs or attracts moisture of the air. This type of substance can be difficult to measure during a gravimetric analysis because it will tend to gain weight when exposed to air.

28. Because we desire a high level of accuracy when using a gravimetric method, all of the associated mass measurements should be conducted using an analytical balance. An analytical balance can measure masses up to 100 or 160 g, with readings to the nearest 0.0001 g. Other types of balances that are sometimes used in a gravimetric analysis involving small samples or precipitates are a semi-micro balance (which can weigh to the nearest 0.00001 g) or a microbalance (which can weigh to the nearest 0.000001 g).

29. Gravimetric iron determinations weigh iron in the form of Fe_2O_3, so the mass of the precipitate would be as follows.

   Mass of precipitate = (0.1253 g)(0.0187 g Fe/g precipitate)(159.7 g Fe_2O_3/mol)

   (1 mol Fe_2O_3/2 mol Fe)/(55.85 g Fe/mol)

   = 0.00335 g

   In order to know this result to an uncertainty of no more than 2 parts-per-thousand, the balance must be able to weigh to the nearest 0.0000067 g, which is beyond the capability of a standard analytical balance. This illustrates the fact that high accuracy can only be accomplished if the sample is large enough for use on an available balance or if the analyte is a major component of the sample.

30. The final solid being weighed must have a mass greater than 0.0500 g, and the original sample must have a mass of at least 1.40 g to give an error less than 2 parts per thousand, as indicated below:

   Mass unknown = (0.0500 g Fe_2O_3)(1 g Fe/1.4297 g Fe_2O_3)(100 g unknown/2.50 g Fe)

   = 1.40 g

31. Precipitation from a homogeneous solution is a technique in which a precipitating agent is formed slowly in the solution after it has been stirred and made homogeneous. This approach eliminates the presence of high, local concentrations of the precipitating agent and results in larger, more filterable precipitates than standard precipitation methods.

32. a) Use the hydrolysis of urea ($H_2N-CO-NH_2$) to gradually increase the pH of a solution that contains some hydrogen phosphate (or the more acidic species, phosphoric acid, etc.).

As the pH increases, more phosphate species will be present as PO_4^{3-} for use in precipitation.

$$H_2N-CO-NH_2 + H_2O \rightarrow CO_2 + 2\ NH_3$$

$$NH_3 + H_2O \rightleftarrows NH_4^+ + OH^-$$

$$HPO_4^{2-} \rightleftarrows PO_4^{3-} + H^+$$

b) Hydrolysis of sulfamic acid: $\quad NH_2SO_3H + H_2O \rightarrow NH_4^+ + H^+ + SO_4^{2-}$

c) Hydrolysis of dimethyloxalate: $\quad (COOCH_3)_2 + H_2O \rightarrow 2\ CH_3OH + (COOH)_2$

33. mol OH^- needed = $(1.5 \times 10^{-3}$ mol Al^{3+}/L$)(0.100$ L$)(3$ mol OH^-/mol $Al^{3+})$

$\qquad = 4.5 \times 10^{-4}$ mol

mol OH^- possible = $(0.500$ g Urea$)(1$ mol Urea/60.0 g$)(2$ mol NH_3/mol Urea$)(1$ mol OH^-/2 mol NH_3)

$\qquad = 0.0083$ mol OH^-

The final term is due to the fact that at pH 9.5, half of the ammonia will be in the form of ammonium, as occurs when pH = pK_a for this acid–base system.

34. mol dmg = $(2)\{(0.0587$ g Ni$)/(58.69$ g Ni/mol$)\}(2$ mol dmg/mol Ni$)$

$\qquad = 0.00400$ mol dmg

Therefore, 0.0040 mol biacetyl and 0.0080 mol hydroxylamine are needed.

35. Peptization is the conversion of a solid precipitate into a colloidal suspension. The presence of a colloidal suspension during a gravimetric analysis can create a slurry that is extremely difficult to filter.

36. To minimize peptization, the precipitate should be washed with a dilute solution of an ionic material. This approach will help prevent the formation of a colloidal suspension and make it easier to filter the final resulting precipitate.

37. Ostwald ripening is a technique for obtaining precipitates with both larger and purer particles in which a precipitate is heated in its original solution to a temperature that is near the boiling point of the solution. This method causes the smallest particles of the precipitate to dissolve and to reprecipitate onto the larger particles.

38. Ostwald ripening causes the smallest particles of the precipitate to dissolve and to reprecipitate onto the larger particles. The resulting precipitate will be purer and easier to filter.

39. A gravimetric factor is a conversion factor often employed in calculations for a gravimetric analysis, during which the conversion factor is used to multiply the measured mass of a precipitate to obtain the mass of desired analyte. This factor makes it easy to convert from the mass of the measured precipitate to the mass of the desired analyte during a gravimetric analysis.

40. a) Gravimetric factor = (137.327 g Ba)/(233.391 g $BaSO_4$) = 0.5884

    b) Gravimetric factor = (63.546 g Cu)/(121.629 g Cu(SCN)) = 0.5225

    c) Precipitates of cupferron are of indefinite composition, so they are ignited to metal oxide for weighing.

    Gravimetric factor = (118.710 g Sn)/(150.709 g SnO_2) = 0.7877

    d) $Mg(NH_4)(PO_4)$ is ignited to $Mg_2P_2O_7$ for weighing.

Gravimetric factor = $(94.9713 \text{ g PO}_4^{3-})/(222.553/2 \text{ g Mg}_2\text{P}_2\text{O}_7) = 0.8534$

41. As CaCO$_3$: Gravimetric factor = $(40.078 \text{ g Ca})/(100.087 \text{ g CaCO}_3) = 0.4004$

    As CaO: Gravimetric factor = $(40.078 \text{ g Ca})/(56.077 \text{ g Ca}) = 0.7147$

42. Desired reaction: $\text{Ag}^+ + \text{Cl}^- \rightleftarrows \text{AgCl(s)}$

    Undesired reactions:

    $2 \text{ AgCl} \xrightarrow{\text{light}} 2 \text{ Ag} + \text{Cl}_2$ (Minimize by conducting method in dim light)

    $2 \text{ Ag}^+ + 2 \text{ OH}^- \rightarrow \text{Ag}_2\text{O} + \text{H}_2\text{O}$ (Minimize by keeping pH low)

    $\text{Ag}^+ + 2 \text{ Cl}^- \rightleftarrows \text{AgCl}_2^-$ (Minimize by not adding huge excess of Cl$^-$)

43. Mass of Ag = $(0.2865 \text{ g AgCl})(107.87 \text{ g Ag}/143.32 \text{ g AgCl}) = 0.2156 \text{ g Ag}$

    Purity of original silver metal = $100 (0.2156 \text{ g})/(0.2365 \text{ g}) = 91.18\%$

44. Amount of HCl = $2(0.035 \text{ mol/L})(0.050 \text{ L}) = 0.0035 \text{ mol}$

    Volume HCl = $(0.0035 \text{ mol})/(0.15 \text{ mol/L}) = 0.0233 \text{ L}$ (or 23.3 mL)

45. Mass of NaCl + mass of NaBr = 0.3654 g

    Mass of AgCl + mass of AgBr = 0.8783 g

    (mol of Cl$^-$)(58.44 g/mol) + (mol of Br$^-$)(102.89 g/mol) = 0.3654 g

    (mol of Cl$^-$)(143.32 g/mol) + (mol of Br$^-$)(187.77 g/mol) = 0.8783 g

    Solve the preceding equations simultaneously to give the following results:

    Mass of NaCl = 0.3368 g   Mass of NaBr = 0.0286 g

46. a) $Ag^+ + Cl^- \rightleftarrows AgCl(s)$; the reaction must be carried out in the presence of excess Ag^+.

b) $[Cl^-]$ = {(3.295 g AgCl)/(143.32 g AgCl/mol)}/(0.100 L) = 0.2299 M

or Conc. Cl^- = (3.295 g AgCl)(58.44 g NaCl/143.32 g AgCl)/(0.100 L) = 13.44 g/L

47. Mass of NaCl + mass of NaBr = 0.8764 g

Mass of AgCl + mass of AgBr = 1.8758 g

(mol of Cl^-)(58.44 g/mol) + (mol of Br^-)(102.89 g/mol) = 0.8764 g

(mol of Cl^-)(143.32 g/mol) + (mol of Br^-)(187.77 g/mol) = 1.8758 g

Solving the preceding equations simultaneously and converting from moles to grams gives the mass of NaCl = 0.4407 g, and the mass of NaBr = 0.4357 g.

48. $Fe^{3+} + 3\ OH^- \rightarrow Fe(OH)_3$ $\qquad\qquad$ $2\ Fe(OH)_3 \rightarrow Fe_2O_3 + 3\ H_2O$

Possible error: Some of the $Fe(OH)_3$ is reduced to Fe(II) during the ignition.

To fix: Treat ignited solid with HNO_3 and reignite.

Possible error: Precipitate is difficult to filter.

To fix: Digest the precipitate by keeping it on a hot plate for several minutes before filtering.

Possible error: Other metals such as Al or Cr are also precipitated.

To fix: Reprecipitate.

49. Mass Fe_2O_3 = (0.257 mol Fe)(1 mol Fe_2O_3/2 mol Fe)(159.7 g Fe_2O_3/mol) = 20.5 g

50. Dimethylglyoxime (dmg) is an organic reagent used for the gravimetric analysis of nickel ions. This reagent can be formed by the slow reaction of biacetyl with hydroxylamine in the presence of nickel ions. This is followed by the reaction of two dmg molecules with one nickel ion to form an insoluble, red-colored complex.

51. Two dmg molecules in their anionic forms can react with a nickel ion to form a square, planar complex in which the nickel is chelated through both pairs of nitrogen atoms. The molecule is further stabilized by hydrogen bonding between the oxygens on the two dmg species. This complex is very insoluble and is quite specific for nickel.

52. As these anionic ligands coordinate to a metal ion, they cause the overall charge to become zero. This neutral charge, plus the larger size of the complex compared to the metal ion, results in a decrease in solubility in an aqueous solution.

53. $[Ni^{2+}]$ = {(0.0658 g Ni(dmg)$_2$)/(288.91 g Ni(dmg)$_2$/mol)}/(0.01000 L) = 0.0228 M

54. a) Gravimetric factor = (26.98 g Al)/(459.41 g Al(ox)$_3$) = 0.05873

    b) Mass of Al = (0.1653 g)(0.05873) = 0.009708 g

55. Mass of Ni = (6.0797 g)(0.2033) = 1.235 g

    % Ni = 100 (1.235 g Ni)/(4.945 g) = 25.0%

56. A combustion analysis is a method in which the combustion of a sample is used to measure the relative amount of carbon, hydrogen, and other elements in a sample. This approach typically involves totally burning an organic sample to form CO_2 and H_2O, trapping and weighing these compounds, and using these masses to measure the carbon and hydrogen

content in the sample. This approach is similar to precipitation-based gravimetry in that a product in a stoichiometric relationship to the analyte is converted to a weighable form prior to weighing. These approaches differ in that the weighable material in a combustion analysis is formed from the gas phase products of combustion rather than by precipitation from a solution.

57. Ascarite absorbs CO_2 and magnesium perchlorate absorbs water.

   $2\ NaOH + CO_2 \rightarrow Na_2CO_3 + H_2O$ Ascarite is NaOH absorbed onto asbestos.

   $Mg(ClO_4)_2 + 2\ H_2O \rightarrow Mg(ClO_4)_2 \cdot 2\ H_2O$

58. a) Mass of C = (0.1362)(12.01)/44.01 = 0.03717 g (mol C = 0.00309 mol)

   Mass of H = (0.04953)(2.014)/18.014 = 0.00554 g (mol H = 0.00550 mol)

   Mass of O = 0.0537 − (0.03717 + 0.00554) = 0.0110 g (mol O = 0.00069 mol)

   b) Relative mol of C = 0.00309/0.00069 = 5.0

   Relative mol of H = 0.00550/0.00069 = 8.0

   Relative mol of O = 0.00069/0.00069 = 1.0

   Empirical formula: C_5H_8O

59. Mass CO_2 = {(0.1753 g)(386.7 g/mol cholesterol)}(27 mol CO_2/mol cholesterol)(44.01 g CO_2/mol)

   = 0.5387 g

   Mass H_2O = {(0.1753 g)(386.7 g/mol cholesterol)}(23 mol H_2O/mol chol)(18.01 g H_2O/mol)

   = 0.1878 g

60. A CHN Analyzer is a device that burns organic compounds and uses the resulting combustion gases to measure the amount of carbon, hydrogen, and nitrogen in the original

sample. Such a device usually assumes that any mass not represented by these three elements in the sample is a measure of the mass of oxygen in the sample.

61. Thermogravimetric analysis is a technique in which the mass of a sample is measured as the temperature of the sample is varied. This method involves measuring the mass of a sample as the sample is heated. As gaseous components are lost (or sometimes gained), the change in mass helps to characterize the sample.

62. A thermobalance is an instrument for performing a thermogravimetric analysis, which includes a high-quality analytical balance along with a furnace for heating the sample in a controlled fashion.

63. A thermogravimetric curve is a graph made by thermogravimetric analysis, in which the measured mass of the sample is plotted versus temperature. This type of plot can be used to identify and measure the components of a sample that are lost or gained during heating.

64. Mass of FeO(OH) =

$$\{(0.0097 \text{ g } H_2O)(18.01 \text{ g/mol})\}(2 \text{ mol FeO(OH)/mol } H_2O)(88.85 \text{ g/mol})$$

$$= 0.0953 \text{ g FeO(OH)} \quad (\text{or } 37.3\% \text{ (w/w) of the sample})$$

Mass of Fe_2O_3 = 0.2564 − 0.0953 = 0.1609 g (or 62.7% (w/w) of the sample)

65. The low-temperature loss is $(NH_4)_2CO_3 \rightarrow 2 NH_3 + H_2O + CO_2$, so the sample contained 0.0574 g of ammonium carbonate. The high-temperature loss is $Na_2CO_3 \rightarrow CO_2 + Na_2O$. Therefore, the mass of sodium carbonate = (0.0124 g)(105.99)/44.011 = 0.0298 g. This leaves 0.0965 g − (0.0574 + 0.0298) = 0.0093 g NaCl.

67. a) Perchloric acid, when hot and concentrated, can react explosively with organic compounds. To avoid such explosions, always wet ash with a mixture of perchloric and nitric acids and raise the temperature gradually. That way the nitric acid destroys the easily oxidized portion and leaves only the more refractory components for the perchloric acid.

b) Hydrofluoric acid can produce severe burns on skin and acts as a temporary pain killer so the unlucky victim may not notice the burn until it has gotten very severe. Always wear gloves and safety glasses when working with hydrofluoric acid and wash your hands and arms carefully when you have finished.

c) Sodium carbonate itself is benign at typical laboratory temperatures. However, when this chemical is heated to melt, the resulting liquid is extremely alkaline and hot, so this melt must be treated carefully and cooled gradually.

69. a) Number of silver ions = $\{(0.0500 \times 10^{-6} \text{ g})/(143.3 \text{ g/mol})\}(6.02 \times 10^{23}) = 2.1 \times 10^{14}$

An equal number of chloride ions would be present.

b) Number of chloride ions per side = $(2.1 \times 10^{14})^{1/3} = 5.9 \times 10^{4}$

The volume of a cube is the third power of the side length, so the side length expressed in number of ions is the cube root of the number of ions in the cube. The area of the face of a cube is the square of the side length, and there are six faces to a cube.

Number of chloride ions on surface = $6 (5.9 \times 10^{4})^{2} = 2.1 \times 10^{10}$

This number of chloride ions represents 0.010% of the total chloride ions in the crystal.

70. a) Total moles of chloride = (0.020 mol/L)(0.150 L) = 0.0030 mol

Total moles of silver = (0.1543 g AgCl)/(143.3 g/mol) = 0.0010 mol

Therefore, total moles of chloride remaining = 0.0030 − 0.0010 = 0.0020 mol

Equilibrium chloride concentration = (0.0020 mol)/(0.150 L) = 0.013 M

Equilibrium silver ion concentration = K_{sp}/[Cl$^-$] = $(1.77 \times 10^{-10})/(0.013) = 1.36 \times 10^{-8}$ M

b) Percent error = $100 \cdot (1.36 \times 10^{-8}$ $M)(0.150$ L$)/(0.0010$ mol$) = 2.0 \times 10^{-4}$ %

71. AgBr is less soluble than AgCl, so it would precipitate quantitatively if enough silver is added to precipitate the AgCl. AgBr also has a higher formula weight (187.8 g/mol) than AgCl (143.3 g/mol).

$$\text{Mass of precipitate} = \text{mass of AgCl} + \text{mass of AgBr}$$
$$= \text{mass of AgCl} + \text{mass of AgCl}(0.010)(187.8/143.3)$$
$$= \text{mass of AgCl } (1 + 0.013) \quad \text{(or a 1.3\% error)}$$

73. Moles of Na$_2$SO$_4$ = (0.1273 g)/(142.043 g/mol) = 8.9621×10^{-4} mol

Mass of Pb^{2+} needed to react with this sulfate = $(8.9621 \times 10^{-4}$ mol)(atomic mass of Pb)

The modern value of lead's atomic mass is 207.2 g/mol, which would give 0.18569 g of Pb^{2+} that is needed in the preceding equation. The addition of this Pb^{2+} would produce 0.27178 g of PbSO$_4$(s).

If the atomic mass of lead is off by 0.01 g/mol, the same type of calculation, using an atomic mass for lead of 207.21 g/mol, would give a required mass of 0.18570 for Pb^{2+} and a mass for the resulting lead sulfate of 0.27179 g. The difference in the mass of the precipitate would be 0.00001 g.

76. Moles ^{141}Ba = (0.0000024 g)/(141 g/mol) = 1.7×10^{-8} mol

[^{141}Ba] = $(1.7 \times 10^{-8}$ mol$)/(1.50$ L$) = 1.13 \times 10^{-8}$ M

After adding H_2SO_4:

$$[Ba^{2+}] = K_{sp}/[SO_4^{2-}] = (1.08 \times 10^{-10})/(0.10 \text{ mol}/1.50 \text{ L}) = 1.62 \times 10^{-9} \text{ M}.$$

Therefore, $100 \cdot (1.62 \times 10^{-9})/(1.13 \times 10^{-8}) = 14.3\%$ of the barium remains in solution.

If 0.020 mol normal barium is added, the sulfate doesn't distinguish between the barium isotopes, so the concentration of dissolved barium is $(1.08 \times 10^{-10})/(0.08 \text{ mol}/1.50 \text{ L}) = 2.0\underline{2} \times 10^{-9}$ M. Of this total concentration, $(1.13 \times 10^{-8})/(0.020) = 5.5 \times 10^{-4}$ (or 0.055%) is the fraction of the dissolved barium, which is actually ^{141}Ba. The concentration of ^{141}Ba in solution $= (2.0\underline{2} \times 10^{-9})(5.5 \times 10^{-4}) = 1.1 \times 10^{-12}$. Therefore, the fraction of ^{141}Ba remaining in solution is $(1.1 \times 10^{-12})/(1.13 \times 10^{-8}) = 1.0 \times 10^{-4}$ (or 0.010%).

77. $K_{sp} = [Ca^{2+}][Ox^{2-}] = 2.32 \times 10^{-9}$ $pK_{a1} = 1.23$ $pK_{a2} = 4.19$

At pH = 5.00, the fraction of oxalate is Ox^{2-} = 0.866 (from fraction of species calculations).

$[Ox^{2-}]$ at equilibrium = $(0.866)(0.00800 \text{ mol} - 0.00300 \text{ mol})/0.100 \text{ L} = 0.00433$ M

$[Ca^{2+}] = K_{sp}/[Ox^{2-}] = (2.32 \times 10^{-9})/(0.00433) = 5.36 \times 10^{-7}$ M

Fraction of calcium not precipitated = $(5.36 \times 10^{-7} \text{ M})(0.100 \text{ L})/(0.00300 \text{ mol})$

$$= 1.8 \times 10^{-5} \text{ (or } 0.0018\%\text{)}$$

Fraction precipitated = $1.000000 - 0.000018$

$$= 0.999982 \text{ (or } 99.9982\%\text{)}$$

CHAPTER 12: ACID–BASE TITRATIONS

1. A titration is a procedure in which the quantity of an analyte in a sample is determined by adding a known quantity of a reagent that reacts completely with the analyte in a well-defined manner. An acid–base titration is a special type of titration in which the reaction of an acid with a base is used for measuring an analyte.

2. The titrant is the reagent added to the sample in a titration. NaOH is the titrant in Figure 12.4.

3. A titration curve is a plot of a quantity that is proportional to the concentration of remaining analyte versus the volume of titrant that has been added to. The equivalence point is the exact point in the titration at which the stoichiometrically required amount of titrant has been added.

4. The equivalence point occurs exactly when a stoichiometric amount of titrant has been added to the sample containing the analyte. The end point is when the end of the titration is detected, such as when an acid–base indicator changes color. If the end point does not match the equivalence point, there will be an error in calculating the amount of analyte in the sample.

5. Titration error = 17.32 mL − 18.05 mL = -0.73 mL

6. First student: Titration error = 25.12 mL − 25.15 mL = -0.03 mL
   Second student: Titration error = 25.17 mL − 25.15 mL = +0.02 mL

7. A volumetric analysis is a method in which volume measurements are used for characterizing a sample. When volume measurements are used during a titration, such as in Figure 12.2, the resulting method is sometimes called a volumetric titration.

8. A gravimetric titration (or weight titration) involves measuring the mass of titrant necessary to react stoichiometrically with an analyte. The type of titration shown in Figure 12.2 makes use of volume measurements and is a volumetric titration.

9. Advantages: Use of simple, inexpensive equipment, and capable of providing high accuracy

    Disadvantages: Not all analytes are suited for such analyses, and the analyte must have a sufficiently high concentration for measurement by a titration.

10. A standard solution of a strong base, such as NaOH (if the analyte is an acid), or a strong acid such as HCl (if the analyte is a base), is added to the sample until an end point is observed. When this is accomplished, the moles of acid and the moles of base can be related to each other through their reaction stoichiometry. For instance, in the titration of a monoprotic acid/base with a monoprotic base/acid, the following relationships will be true at the equivalence point,

$$n_{Analyte} = n_{Titrant} \quad \text{or} \quad C_{Analyte} V_{Analyte} = C_{Titrant} V_{Titrant}$$

    where n = moles, C = molar concentration, and V = volume (in L).

11. Mass of aspirin = (0.04276 L NaOH)(0.1354 mol NaOH/L)(180.16 g aspirin/mol)

    = 1.043 g

12. $[NH_3]$ = (0.04656 L)(0.2034 M HCl)/(0.01000 L) = 0.9470 M

    C = (0.9470 mol NH_3)(17.03 g NH_3/mol)/(980 g soln/L) = 1.64% (w/w)

13. If a known mass of a pure acid or a base is titrated, the molar mass can be calculated using the following equation for an analyte and titrant that react in a 1:1 ratio.

    $$MW_{Analyte} = m_{Analyte}/(M_{Titrant} V_{Titrant})$$

    Terms in this equation include the molar concentration of the titrant ($M_{Titrant}$), the volume of titrant needed to reach the equivalence point ($V_{Titrant}$), and the known mass of the analyte ($m_{Analyte}$), which makes it possible to then find the analyte's molar mass ($MW_{Analyte}$).

14. Molar mass = (0.05465 g HA)/(0.01265 mol NaOH/L)(0.02455 L) = 176.0 g/mol)

15. a) Molar mass base = (3.576 g base)(0.02500 L/0.2500 L)/(0.1380 M HCl)(0.01315 L)

    = 197.0 g/mol

    The other two titrations give molar masses of 196.0 and 196.5 g/mol. The combined results give an average molar mass of 196.5 g/mol with a standard deviation of 0.5 g/mol.

    b) If the base were diprotic, then the true molar mass would be 2 · (196.5 g/mol) = 393.0 g/mol.

    c) The diprotic value makes more sense if the base is composed only of C, H, O, and N, which have near-integral atomic weights. In most organic compounds, there is no way to have a molar mass so far from an integer value unless the compound contains an element, such as chlorine, that has an atomic mass with a value that is about half-way between integers.

16. A titration can provide the number of moles of an acid or a base in a sample. If we know

the molar mass of the acid or the base, we can then determine the total mass of this chemical in the sample. If we also know the total mass or amount of the sample, we can then determine the purity of the acid or the base in this sample.

17. Mass of absorbic acid = (0.2378 M NaOH)(0.03455 L NaOH)(176.13 g/mol)

    = 1.447 g absorbic acid

    Percent ascorbic acid = 100 · (1.447 g ascorbic acid)/(2.0654 g)

    = 70.06% (w/w)

18. Molar mass THAM = 121.14 g/mol

    Results for the first sample (0.1367 g):

    Purity of THAM = 100 · (0.01945 L HCl)(0.05794 M HCl)(121.14 g THAM/mol)/(0.1367 g)

    = 99.86% (w/w)

    The other two samples give purities for the THAM of 99.87% and 99.86%, respectively. The average purity of THAM based on all these results is 99.87%, and the standard deviation of this result is 0.006%.

19. If only an end point is determined, one can learn the purity of a known acid or base and can also determine the molar mass of a pure acid or a base. If the pH of the titration is also measured, one can measure the acid dissociation constant or base ionization constant for the analyte.

20. In a direct titration, the titrant is added to the sample until an end point is determined or the titration is otherwise completed. An example is the titration of vinegar with NaOH to learn the percent of acetic acid in the vinegar sample.

21. In a back titration, an excess of titrant is added to the sample and then the end point is approached by adding a titrant that neutralizes the excess original titrant. An example is the titration of sodium carbonate, in which excess HCl is added, and then the end point is achieved by adding standard NaOH.

22. $NaHCO_3 + HCl \rightarrow NaCl + H_2O + CO_2$

    Mol $NaHCO_3$ = mol HCl − mol NaOH

    $= (0.1028\ M)(0.02500\ L) − (0.0565\ M)(0.00253\ L)$

    $= 0.002570\ mol − 0.000143\ mol = 0.002427\ mol$

    Purity of baking soda $= 100 \cdot (0.002427\ mol)(84.01\ g/mol)/(0.2087\ g)$

    $= 97.69\%\ (w/w)$

    This calculation assumes that none of the other components of the baking soda are acidic or basic.

23. In the digestion step, the sample is decomposed so that all of the nitrogen becomes ammonium ions in a sulfuric acid solution. Treatment with NaOH neutralizes the sulfuric acid and converts the ammonium to ammonia. The distillation step removes the volatile ammonia from the NaOH solution and traps it in water, a boric acid solution, or a standard HCl solution. The final step is an acid–base titration, in which the amount of required titrant is related to the amount of ammonia that was produced from the sample and, therefore, to the amount of nitrogen in the original sample.

24. Mol NH_3 = mol HCl − mol NaOH

    $= (0.6000\ M)(0.05000\ L) − (0.1000\ M)(0.00784\ L) = 0.02922\ mol$

% Nitrogen = 100 · {(0.02922 mol N)(14.007 g N)/(1.00 oz)(28.349 g/oz))} = 1.44% (w/w)

% Protein = (1.44 g N/g cereal)(100 g protein/17.5 g N) = 8.23% (w/w)

25. mol NH_3 = mol HCl − mol NaOH

    = (0.1234 M)(0.05000 L) − (0.08736 M)(0.03559 L)

    = 0.006170 mol − 0.003109 mol = 0.003061 mol

    Mass of N = (0.003061 mol)(14.007 g/mol) = 0.04287 g

    % N = 100·(0.04287 g)/(0.248 g) = 17.3% (w/w)

26. The number of moles of titrant needed to reach an end point must be known. This means that the molarity of the titrant must be known accurately. A standard solution is a solution that has an accurately known concentration and can be used through standardization to find the true concentration of a titrant solution.

27. A primary standard is a substance that can be weighed and dissolved to give an accurately known amount of that material. Commonly used primary standard acids are potassium hydrogen phthalate (KHP) and sulfamic acid; primary standard bases are sodium carbonate and tris-hydroxymethyl aminomethane (THAM, or TRIS).

28. A primary standard must be pure, dry, and stable, and it must have the desired chemical property. Solid NaOH always contains some sodium carbonate and some water. Additionally, it absorbs more water from the atmosphere. HCl is a gas that is difficult to work with, so it is usually sold as an aqueous solution of approximately 37% (w/w). This solution fumes dramatically when opened, so it does not maintain a known concentration.

29. a) First titration:

$$[NaOH] = (0.8127 \text{ g KHP})/[(204.221 \text{ g/mol})(0.03986 \text{ L NaOH})] = 0.09984 \ M$$

Second titration:

$$[NaOH] = (0.7549 \text{ g KHP})/[(204.221 \text{ g/mol})(0.03703 \text{ L NaOH})] = 0.09982 \ M$$

Third titration:

$$[NaOH] = (0.8650 \text{ g KHP})/[(204.221 \text{ g/mol})(0.04244 \text{ L NaOH})] = 0.09980 \ M$$

Average: $[NaOH] = 0.09982 \ M$ Standard deviation = 0.00002 M

b) This comparison can be made by using a Student's t-test.

Standard deviation of the mean = $(0.00002 \ M)/(3)^{1/2} = 1.\underline{2} \times 10^{-5} \ M$

$t = (0.10000 \ M - 0.09982 \ M)/(1.\underline{2} \times 10^{-5} \ M) = 15$

At 95% confidence level, the critical value for t_c is 4.304 at $n - 1 = 2$ degrees of freedom. Because t (15) > t_c (4.304), we can say the results are not the same at the 95% confidence level.

30. For the smallest relative error, one should use about 40 mL of a 50 mL buret for the analysis. Concentrated HCl is about 12 M, so the diluted HCl is about 12(10/250) = 0.48 M. The mass of TRIS needed for analysis is 2.3 g.

31. An increase in titrant concentration will cause the equivalence point to be reached with a smaller volume of titrant. The concentration of titrant will also have some effect on the extent to which the pH changes near the equivalence point. One generally wishes to use about 80% of the volume of the buret to have the smallest relative error without the concern of using more than the buret can hold.

32. For a strong acid titrated with NaOH, more titrant will be required to reach the end point if the analyte concentration is higher. The pH at and after the equivalence point will be unchanged.

    For a weak acid titrated with NaOH, the pH throughout the buffer region will be the same for a dilute versus a more concentrated analyte, unless the weak acid concentration is so great that the assumptions regarding $C > 10^3 K_a$ are invalid. In that case, the rapid rise in pH in the first part of the buffer region will be absent. Typical concentrations are 0.01 to 0.10 M, although more concentrated or more dilute samples can also be titrated.

33. An acid with a pK_a less than about 2.0 has a titration curve similar to a strong acid. If the pK_a is less than 10.0 (and preferably less than 8.0), the acid can often be titrated successfully. If the pK_a is greater than 10.0, the weak acid cannot be successfully titrated under typical conditions in an aqueous system. Similar guidelines apply to weak bases and their pK_b values.

34. A buret contains and delivers the titrant into the titration vessel. If it is not clean, a buret will often hold up drops of titrant on the walls above the meniscus and have an interior volume that is less than what is stated on the buret. Under these conditions, the buret will give less-than-expected volumes for the amount of delivered titrant.

35. To properly read the liquid level in a buret, the buret should be exactly vertical and your eye should be at the same level as the liquid in the buret. Parallax error occurs if your eye is above or below the meniscus because the bottom of the meniscus is in the center of the cylinder and the graduation lines are on the perimeter. Hold a white card behind the buret

and read the bottom of the meniscus. You should be able to estimate the volume to the nearest 0.01 mL in a 50 mL buret, which has graduation lines every 0.10 mL.

36. If one monitors the pH during a titration, it is easy to see if a sample is a mixture of acids/bases or a polyprotic acid/base. It is also possible to use pH measurements during an acid–base titration to estimate the pK_a or pK_b of the sample. These useful features are not available if a visible indicator is instead used to detect the end point.

37. An acid–base indicator is a member of a weak acid/base system, in which one form has a different color than another, and this color changes with pH. A small amount of an acid–base indicator can be added to a titration vessel to monitor a titration by having a color change as the indicator goes from one form to the other over a given pH range. This indicator must be chosen so that its pK_a is close to the expected pH at the equivalence point for the titration.

38. An acid–base indicator is a member of a weak acid/base system, in which one form has a different color than the other, and this color or form changes with pH. When the pH reaches a value of about one pH unit away from the pK_a of the indicator, its color begins to change until the pH is about one unit on the other side of its pK_a value.

39. One key advantage of using acid–base indicators is their simplicity. A well-chosen indicator will show an end point that can be a very accurate measure of the equivalence point. A poorly chosen indicator will give a gradual color change that is a poor measure of the equivalence point. A major disadvantage is that if only a visible indicator is used, one cannot notice the presence of mixtures, or polyprotic acids or bases, and one cannot obtain a

value of the pK_a or pK_b for the analyte, as is possible when using pH measurements.

40. Titration error = 38.08 mL – 38.65 mL = -0.57 mL

41. Thymol blue changes from yellow to blue over a pH range of 8.0–9.6. This is the same approximate pH range for the color change of phenolphthalein, so this student's error should be no worse than that of a student using phenolphthalein. One possible problem is that phenolphthalein is a one-color indicator, which means a smaller fraction of the basic form can be seen. The student using thymol blue may tend to slightly overshoot the equivalence point.

42. The derivative plot for an acid–base titration shows a sharp maximum at the equivalence point, which can be read more confidently than judging the position of the maximum slope in a more traditional pH versus volume plot.

43. The titration curve and plot of the first derivative are shown on the next page.

    The plot of the first derivative gives an end point at 18.38 mL.

    (0.1034 M)(0.01838 L) = [KHP](0.01000 L)

    [KHP] = 0.1900 M

44. A Gran plot is made by graphing a function containing [H^+] instead of pH versus the volume of added titrant, resulting in a linear response for the titration curve near the equivalence point. This type of plot drops to near 0 at and after the equivalence point. Such a plot is used to more easily determine the volume of titrant that is needed to reach the equivalence point.

Figures for Question 43 (see previous page)

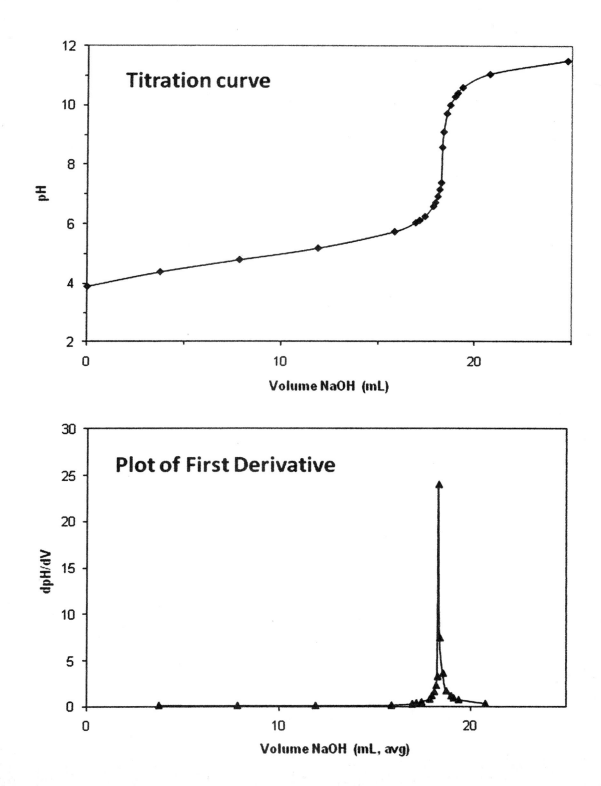

45. The Gran plot is given below. As we saw for the first derivative plot in Problem #43, the Gran plot gives an end point at 18.31 mL.

(0.1034 M)(0.01831 L) = [KHP](0.01000 L)

[KHP] = 0.1893 M

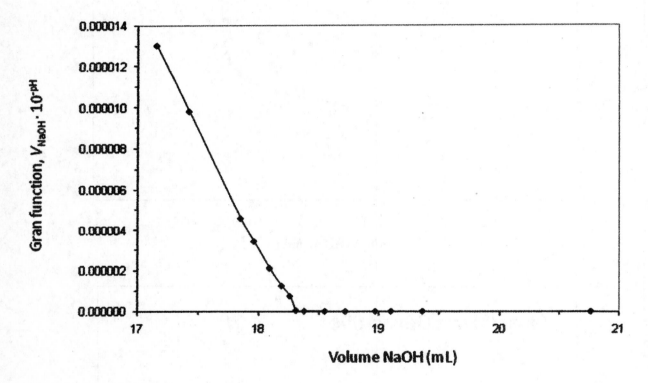

46. a) The equivalence point occurs when $y = 0$

   $x = (1.16 \times 10^{-5})/(3.61 \times 10^{-4}) = 0.03213$ L (or 32.13 mL)

   b) Slope = $-K_a = 3.61 \times 10^{-4}$ $pK_a = 3.44$

47. General reaction for the titration of a strong acid with a strong base in water:

$$H_3O^+ + OH^- \rightleftharpoons 2\,H_2O \qquad (\text{or} \quad H^+ + OH^- \rightleftharpoons H_2O)$$

This reaction is the reverse of the autoprotolysis constant for water and has an equilibrium constant of 1.0×10^{14} at 25 °C. This is the same reaction as for a strong base titrated with a strong acid, only the identity of the analyte and titrant would now be reversed.

48. The equilibrium constant for a weak acid being titrated with a strong base has the following form.

$$K = [A^-]/[HA][OH^-] = K_a/K_w$$

The weaker the acid, the smaller the equilibrium constant will be for this reaction. Similarly, for a weak base, the titration equilibrium constant will be given by the following expression.

$$K = K_b/K_w$$

49. Concentration of weak acid = $(0.1234\ M)(0.01677\ L)/(0.01000\ L) = 0.2069\ M$

$K_a = [H^+][A^-]/[HA] = (4.47 \times 10^{-6}\ M)[(0.01075\ L)/(0.01677\ L - 0.01075\ L)]$

$= 7.9\underline{8} \times 10^{-6} = 8.0 \times 10^{-6}$ (or $pK_a = 5.10$)

50. a) $K = [NH_4^+][F^-]/[NH_3][HF] = (K_{b,NH_3}\ K_{a,HF})/K_w$

b) $K = 1.2 \times 10^6$, which is much smaller than K of HF titrated with NaOH $(K_{a,HF}/K_w = 6.8 \times 10^{10})$

51. Figure 12.12 summarizes the four general regions in the titration of a strong acid with a strong base and the equations that can be used to determine the pH in each of these regions.

52. a) $[HNO_3] = (0.5365\ M)(0.03687\ L/0.02500\ L)(0.10000\ L/0.00500\ L) = 15.82\ M$

b) Nitric acid is a strong acid, so the following is true.

$$[H^+] = 15.82 \, M \, (0.00500 \, L/0.1000 \, L) = 0.791 \, M$$

$$pH = 0.10$$

At the equivalence point the pH is expected to be 7.00.

c) At 10.00 mL: $[H^+] = \{(0.791 \, M)(0.02500 \, L) - (0.5365 \, M)(0.01000 \, L)\}/(0.03500 \, L)$

$$= 0.412 \, M$$

$$pH = 0.38$$

At 25.00 mL: $[H^+] = \{(0.791 \, M)(0.02500 \, L) - (0.5365 \, M)(0.02500 \, L)\}/(0.05000 \, L)$

$$= 0.127 \, M$$

$$pH = 0.89$$

At 40.00 mL: $[OH^-] = \{(0.5365 \, M)(0.04000 \, L - 0.03687 \, L)\}/(0.06500 \, L)$

$$= 0.02583 \, M$$

$$pOH = 1.59 \quad \text{or} \quad pH = 14.00 - 1.59 = 12.41$$

53. a) mol HCl = mol NaOH = $(0.1000 \, M)(0.01608 \, L) = 1.608 \times 10^{-3}$ mol

Mass of HCl = $(1.608 \times 10^{-3} \, \text{mol})(36.45 \, \text{g/mol}) = 0.0586$ g

$[HCl] = (1.608 \times 10^{-3} \, \text{mol})/(0.05000 \, L) = 0.03216 \, M$

b) pH = -log(0.03216) = 1.49

c) At 5.00 mL:

$[H^+] = \{1.608 \times 10^{-3} \, \text{mol} - (0.1000 \, M)(0.00500 \, L)\}/(0.05500 \, L) = 0.0201 \, M$

pH = 1.69

At 10.00 mL:

$$[H^+] = \{1.608 \times 10^{-3} \text{ mol} - (0.1000\ M)(0.01000\ L)\}/(0.06000\ L) = 0.0101\ M$$

pH = 1.99

At 20.00 mL:

$$[OH^-] = \{(0.1000\ M)(0.02000\ L - 0.01608\ L)/(0.07000\ L) = 0.0056\ M$$

pOH = 2.25   pH = 14.00 − 2.25 = 11.75

54. Figure 12.13 summarizes the four general regions in the titration of a strong base with a strong acid and the equations that can be used to determine the pH in each of these regions. In general, these are the same as those used for the titration of a strong acid with a strong base, but with the identities of the analyte and titrant now being reversed. When [OH⁻] is calculated using these equations, this value is then converted into the corresponding value of pH.

55. a) At 0% titration: [OH⁻] = 0.08000 M   or   pOH = 1.10

   pH = 14.00 − 1.10 = 12.90

   At 100% titration: [OH⁻] = [H⁺] = 1.00 × 10⁻⁷ M

   pH = 7.00

   b) At 50% titration: [OH⁻] = (1/2){0.08000 M (0.02000 L)/(0.02000 L + 0.00800 L)

   = 0.0286 M   or   pOH = 1.46

   pH = 14.00 − 1.46 = 12.54

56. a) [OH⁻] = (0.2500 M)(0.03896 L)/(0.10000 L) = 0.0974 M

   pOH = 1.01   pH = 14.00 − 1.01 = 12.99

b) After 10.00 mL:

$[OH^-] = (0.2500\ M)(0.03896\ L - 0.01000\ L)/(0.1100\ L) = 0.06581\ M$

pOH = 1.18   pH = 14.00 − 1.18 = 12.82

After 20.00 mL:

$[OH^-] = (0.2500\ M)(0.03896\ L - 0.02000\ L)/(0.1200\ L) = 0.03950\ M$

pOH = 1.40   pH = 14.00 − 1.40 = 12.60

After 30.00 mL:

$[OH^-] = (0.2500\ M)(0.03896\ L - 0.03000\ L)/(0.1300\ L) = 0.01723\ M$

pOH = 1.75   pH = 14.00 − 1.76 = 12.24

c) At equivalence point: pH = 7.00 since it is a strong acid/strong base titration

After 40.00 mL:

$[H^+] = (0.2500\ M)(0.04000\ L - 0.03896\ L)/(0.1400\ L) = 0.001857\ M$

pH = 2.73

d) Mass of CaO = $(0.2500\ M)(0.03896\ L)(1/2)(56.078\ g\ CaO/mol) = 0.2731\ g$

Percent CaO = $100 \cdot (0.2731\ g)/(0.5654\ g) = 48.3\%$ (w/w)

57. Figure 12.14 summarizes the four general regions in the titration of a weak acid with a strong base and the equations that can be used to determine the pH in each of these regions. However, the calculation after the equivalence point is the same as it is for the titration of a strong acid. The three earlier sections on the titration curve must consider the acid-dissociation constant of the weak acid.

58. If [HF] in a sample is $5.00 \times 10^{-3}\ M$, the volume of NaOH titrant needed to reach the equivalence point is 16.67 mL.

K_a for HF = 6.8×10^{-4} (pK_a = 3.17)

Original sample: $[H^+] = \{(6.8 \times 10^{-4})(5.00 \times 10^{-3})\}^{1/2} = 0.001844\ M$ pH = 2.73

or $[H^+] = 1.53 \times 10^{-3}\ M$ pH = 2.81 (when solved using the more detailed quadratic approach)

At 50% titration: pH = pK_a = 3.17

At 100% titration:

$[OH^-] = \{(1.01 \times 10^{-14}/6.8 \times 10^{-4})(5.00 \times 10^{-3})(0.02500\ L)/(0.02500\ L + 0.01667\ L)\}^{1/2}$

$= \{(1.47 \times 10^{-11})(0.0029998)\}^{1/2} = 2.10 \times 10^{-7}\ M$ or   pOH = 6.68

pH = 14.00 − 6.68 = 7.32

At 30.00 mL of titrant:

$[OH^-] = (7.50 \times 10^{-3})(0.03000\ L - 0.01667\ L)/(0.03000\ L + 0.02500\ L)$

$= 0.00182\ M$ or   pOH = 2.74

pH = 14.00 − 2.74 = 11.26

Using the quadratic formula gives the same answers.

59. Volume of titrant at equivalence point = (0.3654 g)/(46.0 g/mol)(0.1086 M)

$= 0.07314$ L or 73.14 mL

[HA] = {(0.3654 g)/(46.0 g/mol)}/(0.05000 L) = 0.1588 M

At 0% titration: $[H^+] = \{(1.77 \times 10^{-4})(0.1589)\}^{1/2} = 0.00530\ M$ pH = 2.28

or $[H^+] = 0.005215\ M$ pH = 2.28 (when solved using the more detailed quadratic approach)

At 25% titration: $[H^+] = (1.77 \times 10^{-4})(0.75/0.25) = 5.31 \times 10^{-4}$ M

pH = 3.27

At 50% titration: $[H^+] = (1.77 \times 10^{-4})(0.50/0.50) = 1.77 \times 10^{-4}$

pH = 3.75

At 75% titration: $[H^+] = (1.77 \times 10^{-4})(0.25/0.75) = 5.90 \times 10^{-5}$

pH = 4.23

At 100% titration: $[OH^-] = \{(1.01 \times 10^{-14}/1.77 \times 10^{-4})(0.1589)(50.00/123.14)\}^{1/2}$

$= 1.92 \times 10^{-6}$

pOH = 5.72   pH = 14.00 − 5.72 = 8.28

At 110% titration: $[OH^-] = (0.1086)(7.31/131.00) = 0.00606\ M$

pOH = 2.22   pH = 14.00 − 2.05 = 11.78

Using the quadratic formula gives the same values for pOH and pH.

60. $K_a = (10^{-5.87})(10.87\ \text{mL})/(37.65\ \text{mL} - 10.87\ \text{mL}) = 5.48 \times 10^{-7}$

61. Figure 12.15 summarizes the four general regions in the titration of a weak base with a strong acid and the equations that can be used to determine the pH in each of these regions. However, the calculation after the equivalence point is the same as it is for the titration of a strong base. The three earlier sections on the titration curve must consider the base ionization constant of the weak base.

62. The equivalence point occurs at $(0.0445\ M)(25.00\ \text{mL})/(0.07000\ M) = 15.89\ \text{mL}$.

At 0 mL titrant: $[OH^-] = \{(4.47 \times 10^{-4})(0.0445)\}^{1/2} = 0.00446\ M$

pOH = 2.35   pH = 14.00 − 2.35 = 11.65

or $[OH^-] = 0.00424\ M$

pOH = 2.37   pH = 14.00 − 2.37 = 11.63

(when solved using the more detailed quadratic approach)

At 10 mL titrant: $[OH^-] = (4.47 \times 10^{-4})(5.89/10.00) = 2.63 \times 10^{-4}\ M$

$pOH = 3.58 \qquad pH = 14.00 - 3.58 = 10.42$

At 20 mL titrant: $[H^+] = (0.07000)(4.11/45.00) = 0.00693\ M$

$pH = 2.19$

At 15.89 mL titrant: $[H^+] = \{(2.24 \times 10^{-11})(0.0445(25/40.89))\}^{1/2} = 7.80 \times 10^{-7}\ M$

$pH = 6.11$

63. Molar mass = $(0.0356\ g)/(0.01000\ M)(0.02267\ L) = 157.0$ g/mol

    Conc. base [B] = $\{(0.0356\ g)/(157.0\ g/mol)\}/(0.02500\ L) = 0.009070\ M$

    At 0 mL titrant added:

    $pOH = 14.00 - 11.42 = 2.58$ so $[OH^-] = 2.6\underline{3} \times 10^{-3}\ M$

    $K_b[B] = [OH^-]^2 = 6.9\underline{2} \times 10^{-6}$

    $K_b = (6.9\underline{2} \times 10^{-6})/(9.070 \times 10^{-3}) = 7.6\underline{3} \times 10^{-4} = 7.6 \times 10^{-4}$

    At 9.45 mL titrant added:

    $pH = 11.02 \quad pOH = 14.00 - 11.03 = 2.97$

    $[OH^-] = 1.0\underline{7} \times 10^{-3}\ M$

    $K_b = (1.0\underline{7} \times 10^{-3})(9.45/(22.67 - 9.45)) = 7.6\underline{5} \times 10^{-4} = 7.6 \times 10^{-4}$

    At 16.03 mL titrant added:

    $pH = 10.49 \quad pOH = 14.00 - 10.49 = 3.51$

    $[OH^-] = 3.0\underline{9} \times 10^{-4}\ M$

    $K_b = (3.0\underline{9} \times 10^{-4})(16.03)/(22.67 - 16.03) = 7.4\underline{6} \times 10^{-4} = 7.5 \times 10^{-4}$

At equivalence point:

$$\text{pH} = 6.60 \qquad [H^+] = 2.5\underline{1} \times 10^{-7}$$

$$K_a = [H^+]^2/[HB^+] = (6.309 \times 10^{-14})/\{(0.0356/157.0)/(0.02500 + 0.02263)\}$$

$$= 1.3\underline{3} \times 10^{-11}$$

$$K_b = (1.01 \times 10^{-14})/(1.3\underline{3} \times 10^{-11}) = 7.5\underline{9} \times 10^{-4} = 7.6 \times 10^{-4}$$

All four values are in pretty good agreement, but the more trustworthy ones are those from the buffer region, so the best estimate is $K_b = 7.6 \times 10^{-4}$. It is important to recognize that the ionic strength during this titration was never very high, so this value is also a good estimate of the thermodynamic value expected at zero ionic strength.

64. A polyprotic titration curve will have an equivalence point for each acidic or basic group in the sample. If the acid dissociation constants for the different ionizations differ significantly, there will be a step (or break) in the curve at each equivalence point.

65. $[H^+] = (K_{a1} K_{a2})^{1/2}$ or $\text{pH} = (pK_{a1} + pK_{a2})/2$

66. a) 0% titration: $[H^+] = \{(3.71 \times 10^{-5})(0.1000)\}^{1/2} = 0.00193 \ M$

   pH = 2.71

   or $[H^+] = 0.001907 \ M$ and pH = 2.72 (when solved using the more detailed quadratic approach)

   50% titration: $[H^+] = K_{a1} = 3.71 \times 10^{-5} \ M$

   pH = 4.43

   100% titration: $\text{pH} = (pK_{a1} + pK_{a2})/2 = (4.43 + 5.41)/2 = 4.92$

   150% titration: $\text{pH} = pK_{a2} = 5.41$

200% titration: $[OH^-] = \{(10^{-8.58})(0.100)\}^{1/2} = 1.62 \times 10^{-5} M$

pOH = 4.79   pH = 14.00 − 4.79 = 9.21

A titration curve drawn with these values shows a good break at 200% titration, but it shows no break at all at 100% titration because the two pK_a values are too similar.

b) 35% titration: pH = pK_{a1} + log(35/65) = 4.43 + (−0.269) = 4.70

70% titration: pH = pK_{a1} + log(70/30) = 4.43 + 0.368 = 4.80

135% titration: pH = pK_{a2} + log(35/65) = 5.41 − 0.269 = 5.14

170% titration: pH = pK_{a2} + log(70/30) = 5.41 + 0.368 = 5.78

67. a) pK_{a1} = 10.1   pK_{a2} = = 11.0 (from Appendix B)

pK_{b1} = 14.00 − pK_{a2} = 3.0   pK_{b2} = 14.00 − pK_{a1} = 3.9

At 0% titration: $[OH^-] = \{(1.0 \times 10^{-3})(0.1000)\}^{1/2} = 0.010 M$

pOH = 2.0   pH = 14.00 − 2.0 = 12.0

or $[OH^-] = 9.51 \times 10^{-3} M$

pOH = 2.02   pH = 14.00 − 2.02 = 11.98

(The latter answers are obtained when using the more detailed quadratic approach.)

At 50% titration: pH = pK_{a2} = 11.0

At 100% titration: pH = (pK_{a1} + pK_{a2})/2

= (11.0 + 10.1)/2 = 10.55

At 150% titration: pH = pK_{a1} = 10.1

At 200% titration: $[H^+] = (K_{a1}C)^{1/2} = \{(1.0 \times 10^{-11})(0.100)\}^{1/2}$

= $1.0 \times 10^{-6} M$ pH = 6.0

(The same answer is obtained by using the more detailed quadratic approach.)

b) These two base ionization constants are too close together to allow a good break at the first equivalence point to be seen. Accurate calculations would not begin with the approximation that the two ionization steps can be treated separately.

c) This titration curve starts at high pH and shows a good break at the second equivalence point, but no break at the first equivalence point. In this regard, it is similar to the titration curve for adipic acid.

68. a) Concentration of base = (0.1147 M)(37.58 mL/50.00 mL) = 0.0862 M

b) At pH = 10.25, pOH = 14.00 − 10.25 = 3.75 or [OH⁻] = $1.7\underline{8} \times 10^{-4}$ M

K_b = [OH⁻](V_{tit}/(V_{ep} − V_{tit})) = ($1.7\underline{8} \times 10^{-4}$ M)(20.48 mL − (37.58 mL − 20.48 mL)

= 2.13×10^{-4}

pK_b = 3.67

69. Acids and bases in mixtures behave much as they would if not mixed together. Therefore, the same equations can be used. The stronger acids react with the base titrant at the beginning of a titration. The weaker acids react after the stronger acids have been neutralized.

70. 0% titration: In the presence of at least 0.2 M HCl, the phosphoric acid ionization will be negligible, so it can be ignored in the calculation.

[H⁺] = 0.2356 M or   pH = 0.63

The first end point will be the neutralization of the HCl and the first proton from phosphoric acid. This will occur when 34.31 mL of NaOH has been added. The pH will be governed by the two forms of phosphoric acid, as shown in the following equation.

$$\text{pH} = (\text{p}K_{a1} + \text{p}K_{a2})/2 = (2.184 + 7.198)/2 = 4.691 = 4.69$$

The second end point will occur when 10.75 mL more of the titrant has been added and will occur at pH = $(\text{p}K_{a2} + \text{p}K_{a3})/2 = (7.198 + 12.375)/2 = 9.786 = 9.79$.

The third proton of phosphoric acid is so weak that no clean break or end point will be seen for its ionization and titration.

71. If the K_a values of the components of the mixture differ by at least 10^3-fold, separate breaks for the two components will be seen. The first break is for the stronger acid and the second is for the weaker acid. If the two K_a values are too similar, only a single break will be observed, which will allow calculation of the total concentration of the two components.

72. The first end point occurs after both protons are titrated from the sulfuric acid and the second measures the second ionization of oxalic acid.

    At first end point: mol NaOH = 2 mol H_2SO_4 + mol oxalic

    (0.1200 M)(0.01744 L) = 0.002093 mol NaOH = 2 mol H_2SO_4 + mol oxalic

    At second end point: mol NaOH = mol oxalic

    (0.1200 M)(0.02198 L – 0.01744 L) = 5.448×10^{-4} mol

    2 mol H_2SO_4 = 0.002093 mol – 0.0005448 mol = 0.001548 mol

    mol H_2SO_4 = 0.0007741 mol

    [H_2SO_4] = (0.0007741 mol)/(0.02500 L) = 0.03096 M

    [Oxalic acid] = (5.448×10^{-4} mol)/(0.02500 L) = 0.02179 M

73. First break is for HCl:

   mol HCl = mol NaOH = (0.1000 M)(0.02166 L) = 0.002166 mol

   Second break is for HOCl:

   mol HOCl = (0.1000 M)(.03017 L − 0.02166 L) = 0.000851 mol

   Fraction reduced = (0.002166 mol)/(0.002166 mol + 0.000851 mol) = 0.718

74. At first end point:

   (0.1000 M)(0.03789 L) = 2 [H_2SO_4](0.02500 L) + [H_2SO_3](0.02500 L)

   At second end point:

   (0.1000 M)((0.04357 L − 0.03789 L) = [H_2SO_3](0.02500 L)

   [H_2SO_3] = 0.02272 M

   [H_2SO_4] = {(0.03789 L − 0.02500 L (0.02272 M)}/(0.05000 L)

   = 0.06442 M

75. The fraction of titration is defined as the ratio of the moles of titrant that have been added at any given point in the titration versus the moles of analyte that were originally present in the sample. Equations 12.27 and 12.38 give the expressions used for the fraction of a titration for both the titration of an acid with a base and the titration of a base with an acid.

76. See Table 12.5 for the derivation of this equation.

77. The derivation of a fraction of a titration equation for the titration of a strong base with a strong acid is essentially the same as the derivation shown in Table 12.5 for the titration of a strong base with a strong acid. The only difference is in the last step, in which the

equation is rearranged to give $F = (C_A V_A)/(C_B V_B)$ instead of $F = (C_B V_B)/(C_A V_A)$, as it is in Table 12.5. This is also the reason that the fraction of titration equation shown in Table 12.6 for the titration of a strong base with a strong acid is just the reciprocal of the fraction of titration equation shown for the titration of a strong acid with a strong base.

78. a) At midpoint of phenolphthalein range (pH = 8.8):

$[H^+] = 1.58 \times 10^{-9}$ M and $[OH^-] = 6.31 \times 10^{-6}$ M

$F = \{1 - ((1.58 \times 10^{-9} - 6.31 \times 10^{-6})/0.1000\}/\{1 + ((1.58 \times 10^{-9} - 6.31 \times 10^{-6})/0.1000\}$

$F = \{1 + 6.308 \times 10^{-5}\}/\{1 - 6.308 \times 10^{-5}\} = 1.00013$

Titration error = (1.00013)(25.00 mL) – 25.00 mL = +0.0032 mL

This error is too small to detect when using a buret.

At start of range (pH = 8.0):

$[H^+] = 1.00 \times 10^{-8}$ M and $[OH^-] = 1.00 \times 10^{-6}$ M

$F = \{1 - (1.00 \times 10^{-8} - 1.00 \times 10^{-6})/0.1000\}/\{1 + (1.00 \times 10^{-8} - 1.00 \times 10^{-6})/0.1000\}$

$F = \{1 + 9.90 \times 10^{-6}\}/\{1 - 9.90 \times 10^{-6}) = 1.000020$

Titration error = (1.000020)(25.00 mL) – 25.00 mL = 5×10^{-4} mL

This error is too small to detect using a buret.

b) At midpoint of methyl red range (pH 5.3):

$[H^+] = 5.01 \times 10^{-6}$ M and $[OH^-] = 1.99 \times 10^{-9}$ M

$F = \{1 - (5.01 \times 10^{-6} - 1.99 \times 10^{-9})/.1000\}/\{1 + (5.01 \times 10^{-6} - 1.99 \times 10^{-9})\}$

$F = \{1 - 5.008 \times 10^{-5}\}/\{1 + 5.008 \times 10^{-5}\} = 0.99990$

Titration error = (0.99990)(25.00 mL) – 25.00 mL = -0.0025 mL

This error is too small to detect when using a buret.

At midpoint of thymolphthalein range (pH 10.0):

$[H^+] = 1.00 \times 10^{-10}$ M and $[OH^-] = 1.00 \times 10^{-4}$ M

$F = \{1 - (1.00 \times 10^{-10} - 1.00 \times 10^{-4})/.1000\}/\{1 + (1.00 \times 10^{-10} - 1.00 \times 10^{-4})/0.1000\}$

$F = (1 - (-9.99999 \times 10^{-4})\}/\{1 + (-9.99999 \times 10^{-4}\} = 1.0020$

Titration error = $(1.0020)(25.00$ mL$) - 25.00$ mL $= 0.050$ mL

This error is just barely large enough to read from a buret.

Phenolpthalein and methyl red both give end points with only small titration errors, but thymolphthalein gives a slightly noticeable and late end point.

79. a) Methyl red will be 50% in each form at its pK_a of 4.95, where $[H^+] = 1.1\underline{22} \times 10^{-5}$ M and $[OH^-] = 8.9\underline{12} \times 10^{-10}$ M.

$\alpha = (1.1\underline{22} \times 10^{-5})/(1.1\underline{22} \times 10^{-5} + 5.6\underline{2} \times 10^{-10}) = 0.9999\underline{5}$

$F = [0.9999\underline{5} + \{(1.1\underline{22} \times 10^{-5} - 8.9\underline{12} \times 10^{-10})/0.1000\}]/\{1-(1.1\underline{22} \times 10^{-5} - 8.9\underline{12} \times 10^{-10})/0.1000\}$

$= 1.000\underline{06}$

Titration error = $(1.000\underline{06})(25.00$ mL$) - 25.00$ mL $= 0.001\underline{55}$ mL

This error is too small to distinguish when using a buret.

Methyl red will be 10% in the acid form and 90% in the base form at pH = $4.95 + \log (90/10) = 5.90$.

$[H^+] = 1.2\underline{46} \times 10^{-6}$ M and $[OH^-] = 8.1\underline{06} \times 10^{-9}$ M

$\alpha = (1.2\underline{46} \times 10^{-6})/(1.2\underline{46} \times 10^{-6} + 5.6\underline{2} \times 10^{-9}) = 0.999\underline{51}$

$F = [0.999\underline{5}1 + \{(1.24\underline{6} \times 10^{-6} - 8.10\underline{6} \times 10^{-9})/0.1000\}]/\{1 - (1.24\underline{5} \times 10^{-6} - 8.10\underline{6} \times 10^{-9})/0.1000\}$

$= 0.995\underline{4}$

Titration error $= (0.995\underline{4})(25.00 \text{ mL}) - 25.00 \text{ mL} = -0.113 \text{ mL}$

This error is well within the readability of a buret.

b) The equivalence point will be at 25.00 mL, so the methyl red will be 50% in each form at $(25.00 \text{ mL})(1.00017) = 25.004 \text{ mL}$. The resulting titration error of 25.004 mL − 25.00 mL = 0.004 mL is smaller than the readability of a buret.

80. One can use pH to solve for $[H^+]$ and $[OH^-]$, and then use these values to calculate F. The value of F can then be used directly to generate a titration curve by plotting pH on the y-axis and F on the x-axis. Alternatively, F can be converted to the equivalent volume of titrant, which can then be used to make a plot of pH versus volume of added titrant.

86. a) I $OH^- + H^+ \rightarrow H_2O$ Occurs during red–colorless phenolphthalein end point

   II $PO_4^{3-} + H^+ \rightarrow HPO_4^{2-}$ Occurs during red–colorless phenolphthalein end point

   III $CO_3^{2-} + H^+ \rightarrow HCO_3^-$ Occurs during red–colorless phenolphthalein end point

   IV $HPO_4^- + H^+ \rightarrow H_2PO_4^-$ Occurs during red–colorless phenolphthalein end point

   V $HCO_3^- + H^+ \rightarrow H_2CO_3$ Occurs during blue–yellow bromocresol green end point

   VI $H_2CO_3 \rightarrow CO_2 + H_2O$ Occurs mainly during heating and N_2 bubbling

   VII $H_2PO_4^- + H^+ \rightarrow H_3PO_4$ Occurs as the pH stays yellow on adding excess HCl

The back reaction will measure only the phosphate species because CO_2 has been bubbled out into the air and the hydroxide has been converted to water.

b) At phenolphthalein end point:

$$\text{mol HCl} = \text{mol Na}_2\text{CO}_3 + 2 \text{ mol Na}_3\text{PO}_4 + \text{mol NaOH}$$

$$= (0.1234 \, M)(0.02510 \text{ L}) = 0.003097 \text{ mol}$$

At bromocresol green end point:

$$\text{mol HCl} = \text{mol Na}_2\text{CO}_3 + \text{mol Na}_3\text{PO}_4$$

$$= (0.1234 \, M)(0.03845 \text{ L} - 0.02510 \text{ L}) = 0.001647 \text{ mol}$$

At first back titration end point:

$$\text{mol NaOH} = \text{excess mol HCl} + \text{mol Na}_3\text{PO}_4$$

$$= (0.1456 \, M)(0.00424 \text{ L}) = 0.0006173 \text{ mol}$$

At second back titration end point:

$$\text{mol NaOH} = \text{mol Na}_3\text{PO}_4$$

$$= (0.1456 \, M)(0.01214 \text{ L} - 0.00424 \text{ L}) = 0.001150 \text{ mol}$$

Final results:

Amount $Na_3PO_4 \cdot 12 \, H_2O$ = 0.001150 mol

or   (0.001150 mol)(380.16 g/mol) = 0.4373 g

Amount Na_2CO_3 = 0.001647 mol − 0.001150 mol = 0.000497 mol

or   (0.000497 mol)(106 g/mol) = 0.0527 g

Amount NaOH = 0.003097 mol − (0.000497 mol + 2 (0.001150) mol) = 0.000300 mol

or   (0.000300 mol)(40.00 g/mol) = 0.01200 g

CHAPTER 13: COMPLEXOMETRIC AND PRECIPITATION TITRATIONS

1. A complexometric titration involves titrating metal ions with a solution containing a multidentate ligand to form a complex. A water hardness titration is actually a titration of calcium and magnesium ions with a solution of EDTA, in which EDTA is used as a titrant to form complexes with both the calcium and magnesium ions.

2. A precipitation titration involves a titration reaction that forms an insoluble product. The equilibrium in this reaction is controlled by a solubility product and not by a formation constant, as is the case with a complexometric titration.

3. Such a graph is usually made by plotting the negative logarithm of the analyte concentration versus volume of added titrant. If the analyte is a metal ion, the quantity "pM" is used to describe the negative logarithm of the metal ion concentration.

4. A strong acid–strong base titration makes use of a Brønsted–Lowry acid–base reaction. Complexometric titrations are based on Lewis acid–base reactions, and precipitation titrations are based on the formation of insoluble products. In each case, the negative log of the analyte concentration is usually plotted versus volume of added titrant.

5. EDTA reacts with metal ions in a 1:1 reaction with a large equilibrium constant, so the number of moles of a standard solution of EDTA necessary to reach an equivalence point can be used to find the moles of metal ion in a sample when using a complexometric titration.

6. The general titration reaction between EDTA and a metal ion is shown below.

$$M^{n+} + EDTA^{4-} \rightarrow MEDTA^{(4-n)-}$$

This reaction has a 1:1 stoichiometry and a large equilibrium constant for most metal ions. These properties make it relatively easy to detect a good end point when using EDTA to titrate a metal ion.

7. a) $Ca^{2+} + EDTA^{4-} \rightarrow CaEDTA^{2-}$

   b) $[Ca^{2+}] = (0.02752 \text{ mol/L})(0.00746 \text{ L})/(0.05000 \text{ L}) = 4.10 \times 10^{-3} \, M$

   c) $[Ca^{2+}] = (4.10 \times 10^{-3} \text{ mol/L})(1 \text{ mol } CaCO_3/\text{mol } Ca^{2+})(100.09 \times 10^3 \text{ mg/mol})$

   $= 411$ ppm

8. a) $Cu^{2+} + EDTA^{4-} \rightarrow CuEDTA^{2-}$

   b) Mass Cu = $(0.02785 \, M)((0.00756 \text{ L})(63.55 \text{ g Cu/mol}) = 0.01338$ g Cu

   % Cu = $100 \cdot (0.01338 \text{ g Cu})/(2.589 \text{ g ore}) = 0.517$% Cu (w/w)

9. a) $Ag^+ + Cl^- \rightarrow AgCl(s)$

   b) $[Ag^+] = (0.01500 \text{ mol } Cl^-/\text{L})(15.20 \text{ mL } Cl^-)/(25.00 \text{ mL } Ag^+) = 0.00912 \, M$

10. Typically, the chloride would now be in the sample and the flask, and a standard solution of silver ions would be delivered from the buret as a titrant. The indicator could be chromate, as is used in the Mohr method; however, such an indicator would not have worked in Problem 9 because silver chromate would have precipitated immediately.

11. EDTA forms complexes with most metal ions and reacts with these metal ions in a 1:1 ratio. The stability constants for these complexes are typically quite large and these complexes usually form quite rapidly. Na_2H_2EDTA is available as a primary standard material for use in EDTA titrations.

12. EDTA represents a polyprotic acid–base system that can accept as many as six hydrogen ions. Therefore, titrations must be done at a pH, at which the desired metal ions will bind strongly to EDTA, which is a process that generally involves the most basic form of EDTA. Some metal ions may also precipitate at a high pH as metal hydroxides, a side reaction that should be avoided. For these reasons, the pH should be controlled when using EDTA as a titrant.

13. The most important side reaction is the precipitation of the metal ions as metal hydroxides. The solution pH must be low enough to prevent the precipitation of metal hydroxides for the metal ion that is the desired analyte.

14. a) pH 3.0–5.25

    b) pH 2.0–4.1, or pH 10.4–10.6 in presence of an auxiliary agent

    c) pH 8.2–10.2 (pH 8.2–9.1 in presence of an auxiliary agent)

15. According to Table 13.1, a pH of 1.25 to 5.1 can be used for the direct titration of Fe^{3+} with EDTA. This range can be extended to about pH 6.1 in the presence of auxiliary agents. EDTA can be used to directly titrate Cu^{2+} between pH 3.0 and 7.5 and with auxiliary agents up to pH 10.5. EDTA can be used to directly titrate Ni^{2+} between pH 3.1 and 8.6 and with auxiliary agents up to pH 12.1. From this information, it would be possible to selectively titrate Fe^{3+} by using a pH of 1.25 up to a pH just under 3.0. A selective titration of Ni^{2+} can be made directly with EDTA between pH 7.5 and 8.6. Similarly, the amount of Cu^{2+} can then be determined by titrating it along with the Fe^{3+} at pH 3.0 to 5.1 and by using the previously determined concentration of Fe^{3+} to find the concentration of Cu^{2+} in the sample.

Other schemes are also possible.

16. First titration: $[Ca^{2+}] + [Mg^{2+}] = (0.06500\ M)(0.01854\ L)/(0.05000\ L)$

$= 0.02410\ M$

Second titration: $[Ca^{2+}] = (0.06500\ M)(0.01382\ L)/(0.05000\ L)$

$= 0.01797\ M$

$[Mg^{2+}] = 0.02410\ M - 0.01797\ M = 0.00614\ M$

17. Mass $Na_2H_2EDTA \cdot 2\ H_2O = (0.5000\ L)(0.0800\ mol/L)(372.24\ g/mol)$

$= 14.8896\ g$

18. Concentration = $\{(3.576\ g)/(372.24\ g/mol)\}/(0.5000\ L) = 0.01921\ M$

19. Monodentate ligands form a series of complexes with metal ions that have a variety of stoichiometries. These complexes also tend to have a much lower overall formation constants than chelating agents like EDTA. The result is that it is difficult to obtain a well-defined end point when using a monodentate ligand for a complexometric titration.

20. a) DCTA (or CDTA) has two carbons of a cyclohexane ring between the two amino groups instead of the ethylene group found in EDTA. This difference causes metal complexes with DCTA to be more stable than those for EDTA by a factor of about 1000. This property makes DCTA useful as a titrant for metal ions that form weak complexes with EDTA.

b) Trien (or TETA) is a polyethyleneamine that has four amino nitrogen atoms. It forms strong complexes with most transition metal ions, but it does not complex alkaline earth ions. This property makes trien useful in titrating transition elements in the presence of a large excess of calcium ions or magnesium ions.

c) Tetren is a polyethyleneamine that has five amino nitrogen atoms. It forms even stronger complexes than trien with most transition metal ions and also does not bind to calcium ions or magnesium ions to any appreciable extent.

d) EGTA has a $-CH_2CH_2-O-CH_2CH_2-$ group between two nitrogen atoms, but it is otherwise similar in structure to EDTA. The ratio of formation constants of calcium ions to magnesium ions is greater with EGTA ($\sim 10^6$) than with EDTA ($\sim 10^2$), which makes EGTA useful in titrating calcium ions in the presence of magnesium ions.

21. $[Mg^{2+}] + [Ni^{2+}] = (0.04765\ M)(0.01786\ L)/(0.02500\ L) = 0.03404\ M$

    $[Ni^{2+}] = (0.05643\ M)(0.00674\ L)/(0.02500\ L) = 0.01521\ M$

    $[Mg^{2+}] = 0.03404\ M - 0.1521\ M = 0.01883\ M$

22. $[Cu^{2+}] = (0.05000\ M)(0.0190\ L)/(0.02500\ L) = 0.0380\ M$

23. Many metal ions (such as copper ions) require ammonia as an auxiliary ligand to prevent the precipitation of metal hydroxides, such as $Cu(OH_2)$. Fluoride can be used to mask iron ions during the titration of other metal ions.

24. At the equivalence point, mol $CN^- = 2$ mol Ag^+.

    $[CN^-] = [Ag^+](V_{Ag+})/(V_{CN-}) = 2(0.1000\ M)(0.01864\ L)/(0.02500\ L) = 0.1491\ M$

25. An auxiliary ligand is an additional complexing agent that is added to make the titration of an analyte easier to conduct, such as by reducing side reactions involving the analyte.

26. Ammonia forms fairly stable complexes with nickel ions, and therefore, prevents the precipitation of Ni(OH)$_2$. This is a desirable reaction because the titration of an insoluble material is slow and difficult to carry out. This additional reaction raises the value of pNi (or lowers [Ni^{2+}]) before the equivalence point as ammonia complexes with some of the nickel ions and lowers the conditional formation constant between Ni^{2+} and EDTA. It also prevents the loss of Ni^{2+} through its precipitation with hydroxide ions. The overall result is a titration for Ni^{2+} by EDTA that now can be performed up to pH 12 without problems due to undesired precipitation of the analyte.

27. K_f = [NiEDTA^{2-}]/[Ni^{2+}][EDTA^{4-}]

   = [NiEDTA^{2-}]/{(C_{Ni} α_{Ni})(C_{EDTA} α_{EDTA})

   = C_{NiEDTA}/{(C_{Ni} · 0.056)(C_{EDTA} · 0.041)

   = K'_f/(0.056)(0.041) = K'_f(2.3 × 10^{-3})

   K'_f = K_f(2.3 × 10^{-3}) = (2.5 × 10^{18})(2.3 × 10^{-3}) = 5.8 × 10^{15}

   The titration shown in Figure 13.6 is at pH 8.0. At pH 9.0, more of the EDTA will be in the EDTA^{4-} form, so the conditional formation constant will be greater, and the observed break at the end point will be larger.

28. a) pH 8.0–10.5   b) pH 9.5–10.6   c) pH 7.5–10.5

29. A masking agent is used to bind tightly to a metal ion that we wish not to react with the titrating ligand. Masking is a method to lower the conditional stability constant of a metal ion with the titrating ligand. A good example is the use of fluoride to mask iron during the titration of copper.

Demasking is the addition of a reagent that reacts with the masking agent to keep it from reacting with whatever it was masking. An example is the addition of boric acid to react with fluoride to form BF_4^-.

30. $Fe^{3+} + 3\ F^- \rightleftharpoons FeF_3$ $\qquad K = [FeF_3]/[Fe^{3+}][F^-]^3 = 7.\underline{9} \times 10^{11}$

    $\alpha_{Fe3+} = 1/(1 + K[F^-]^3) = 1/(1 + (7.\underline{9} \times 10^{11})(0.050)^3) = 1/(1 + 9.\underline{9} \times 10^7) = 1.0 \times 10^{-8}$

31. $[Fe^{2+}] + [Mn^{2+}] = (0.05753\ M)(0.01876\ L)/(0.02000\ L) = 0.05396\ M$

    $[Mn^{2+}] = (0.05753\ M)(0.00756\ L)/(0.02000\ L) = 0.02183\ M$

    Therefore, $[Fe^{2+}] = 0.05396\ M - 0.02183\ M = 0.03213\ M$

32. $[Zn^{2+}] + [Ni^{2+}] = (0.05000\ M)(0.03475\ L)/(0.02500) = 0.06950\ M$

    $[Zn^{2+}] = (0.05000\ M)(0.01147\ L)/(0.02500\ L) = 0.02294\ M$

    $[Ni^{2+}] = 0.06950\ M - 0.02294\ M = 0.04656\ M$

33. A metallochromic indicator is a compound capable of forming a complex with metal ions. This material is one color when it is not complexed to a metal ion and is a different color when it is present in a complex with a metal ion.

34. Eriochrome black T is commonly used as an indicator in water hardness titrations with EDTA. Its color varies with pH; it is red below a pH of about 6, blue at pH 6 to 11, and orange above a pH of about 11. It forms a more stable complex with magnesium ions than with calcium ions, so the color change at the end point in a water hardness titration is actually the removal of Mg^{2+} from the red complex of the indicator at pH of 10 to the blue form of the indicator. If calcium ions are to be measured in a sample that does not contain magnesium ions, some magnesium ions are added to the EDTA solution before it is

standardized for use in the titration.

35. The indicator should be about 50% in the complexed and uncomplexed form at the pM value of the equivalence point of the titration. That means that pM at the equivalence point should be about $1/K_{MIn}$.

36. The calcium complex of murexide at pH 8 is CaH_3In, which is red–orange. The nickel complex is yellow, so the color change at the end point is yellow to red–orange. The reaction occurring at the endpoint is as shown below.

$$HEDTA^{3-} + NiH_3In \rightarrow NiEDTA^{2-} + CaH_3In + H^+$$

37. At pH 6, the color change will be orange to yellow–orange.

    At pH 8, the color change will be orange to red–orange.

    At pH 10, the color change will be orange to red.

    None of these offer a good change in color because the colors of the calcium and copper complexes are too similar. At pH 8 and 10 (and possibly pH 6 as well), ammonia will be needed as an auxiliary ligand to prevent precipitation of $Cu(OH)_2$. This agent will add the intense blue color of the $Cu(NH_3)_4^{2+}$ complex to the color of the solution. The color will gradually fade as the copper is converted to $CuEDTA^{2-}$, which is also blue but less intense than the ammonia complex.

38. $EDTA^{4-} + Ca(calcein) \rightleftarrows CaEDTA^{2-} + calcein^{2-}$

    EDTA is the titrant, and Ca(calcein) is the indicator form of calcein that disappears as Ca^{2+} is released from this complex and titrated by EDTA at the end point.

39. If the formation of the EDTA complex of a metal ion is slow, finding the endpoint will become tedious as the reaction will get slower and slower as the concentration of metal ion gets smaller as the endpoint is approached. Chromium(III) is slow to exchange its water of hydration for a ligand such as EDTA, so a back titration is called for in this case. A known amount of EDTA in excess of the amount of Cr(III) is added, allowed to react for several minutes, and then the excess EDTA is titrated with a standard solution of zinc ions.

40. mol Al^{3+} = mol EDTA − mol Zn^{2+}

    mol Al^{3+} = (0.1000 M)(0.0050 L) − (0.02000 M)(0.00808 L)

    $= 0.000500$ mol − 0.000162 mol $= 3.38 \times 10^{-4}$ mol

    $[Al^{3+}] = (3.38 \times 10^{-4}$ mol$)/(0.01000$ L$) = 3.38 \times 10^{-2}$ M

41. Solution treated with ascorbic acid:

    mol Cr(III) = (0.0895 M)(0.02500 L) − (0.0458 M)(.00575 L)

    $= 0.0022375$ mol − 0.00026335 mol $= 0.001974$ mol

    Untreated solution:

    mol Cr(III) = (0.0895 M)(0.02500 L) − (0.0458 M)(0.00973 L)

    $= 0.002238$ mol − 0.000446 mol $= 0.001792$ mol

    [Cr(III)] = (0.001792 mol)/0.05000 L = 3.58×10^{-2} M

    [Cr(VI)] = (0.001974 mol − 0.001792 mol)/(0.0500 L) = 3.64×10^{-3} M

42. A displacement titration can be used when a particular analyte cannot be easily titrated directly. One reason this might occur is that no suitable indicator is available for the direct titration. Another reason might be that a side reaction, such as precipitation, might occur

during a direct titration. To carry out a displacement titration, an unmeasured amount of an EDTA complex of a metal having a smaller formation constant than the analyte (often zinc ions or magnesium ions) is added to the unknown. A metal exchange reaction occurs that gives a stoichiometric amount of the other metal ion, which can then be successfully titrated directly.

43. $2\ NH_2OH + 2\ Hg(EDTA)^{2-} \rightarrow 2\ Hg + 2\ EDTA^{4-} + N_2O + H_2O + 4\ H^+$

    mol NH_2OH = mol $EDTA^{4-}$ = mol $ZnCl_2$ = $(0.0300\ M)(0.00755\ L)$

    $= 2.26\underline{5} \times 10^{-4}$ mol

    $[NH_2OH] = (2.26\underline{5} \times 10^{-4}\ mol)/(0.0500\ L) = 4.53 \times 10^{-3}\ M$

44. In an indirect titration, something other than a metal ion is the analyte. For instance, one can measure the amount of fluoride in a solution by precipitating the fluoride with a known amount of calcium ions to give CaF_2, with the excess calcium ions then being titrated with EDTA.

45. mol sulfate = mol $BaCl_2$ - mol EDTA

    $= (0.0500\ M)(0.01000\ L) - (0.0354\ M)(0.00745\ L)$

    $= 0.000500\ mol - 0.0002637\ mol = 0.000236\underline{3}\ mol$

    $[SO_4^{2-}] = (0.000236\underline{3}\ mol)/(0.02500\ L) = 0.00945\ M$

46. Figure 13.7 shows the four general regions during a complexometric titration, and gives the equations that can be used to predict a titration curve for analyte M that reacts with titrant L to produce a 1:1 complex.

47.   a) $Ca^{2+} + EDTA^{4-} \rightarrow CaEDTA^{2-}$

Volume titrant needed = (0.01250 M)(0.02500 L)/(0.01050 M)

= 0.02976 L or 29.76 mL

b) At 0 mL titrant: pCa = $-\log(0.01250)$ = 1.90

At equivalence point (and pH 10.0):

$K'_f = K_f \alpha_{EDTA4-}$ = (4.47 × 10^{10})(0.39$\underline{2}$) = 1.75 × 10^{10}

$[Ca^{2+}]$ = {[(0.01250 M)(0.02500 L)]/[(0.02500 L + 0.02976 L)(1.75 × 10^{10})]}$^{1/2}$

= 5.71 × 10^{-7} M

pCa = 6.24

At 10.00 mL titrant:

$[Ca^{2+}]$ = {(0.01250 M)(0.02500 L) – (0.01050 M)(0.01000 L)}/(0.02500 L + 0.01000 L)

= 0.005928 M or   pCa = 2.23

At 30.00 mL titrant:

$[Ca^{2+}]$ = {(0.01250 M)(0.02500 L)/(0.05500 L)}/[{(0.01050 M)(0.0300 L) – (0.01250 M)(0.0250 L)}

(1.75 × 10^{10})]

= 7.06 × 10^{-9} or   pCa = 8.15

c) Use the results in Part (b) when preparing a rough plot of pCa versus volume of titrant.

48.   a) $Cu^{2+} + H_2EDTA^{2-} \rightarrow CuEDTA^{2-} + 2H^+$

Volume titrant needed = (0.04500 M)(0.05000 L/0.1011 M)

= 0.02226 L or 22.26 mL

At pH 5.00: K'_f = (6.0$\underline{3}$ × 10^{18})(4.1$\underline{6}$ × 10^{-7}) = 2.5$\underline{1}$ × 10^{12}

b) At 0% titration: pCu = $-\log(0.04500)$ = 1.347

At equivalence point:

$$[Cu^{2+}] = \{(0.04500\ M)(0.05000\ L)/(0.05000\ L + 0.02226\ L)(2.51 \times 10^{12})\}^{1/2}$$

$$= 1.1\underline{1} \times 10^{-7}\ M \quad \text{or} \quad pCu = 6.95$$

c) At 5.00 mL:

$$[Cu^{2+}] = \{(0.04500\ M)(0.05000\ L) - (0.1011\ M)(0.005000\ L)\}/(0.05500\ L)$$

$$= 0.03172\ M \quad \text{or} \quad pCu = 1.50$$

At 10.00 mL:

$$[Cu^{2+}] = \{(.04500\ M)(0.05000\ L) - (0.1011\ M)(0.01000\ L)\}/(0.06000\ L)$$

$$= 0.02065\ M \quad \text{or} \quad pCu = 1.68$$

At 20.00 mL:

$$[Cu^{2+}] = \{(0.04500\ M)(0.05000\ L) - (0.1011\ M)(0.02000\ L)]\}/(0.07000\ L)$$

$$= 0.003257\ M \quad \text{or} \quad pCu = 2.49$$

d) Use the results in Parts (b) and (c) when preparing a rough plot of pCu versus volume of titrant.

49. See the derivation in Table 13.4.

50. The approach is essentially the same as given in Table 13.4 for the titration of a metal ion with a ligand. The only difference is in the final step where the fraction of titration is defined as $(C_M\ V_M)/(C_L\ V_L)$ instead of $(C_L\ V_L)/(C_M\ V_M)$. This means the final result will essentially be the reciprocal of the final equation that is shown in Table 13.4.

51. The equivalence point for this titration occurs then the volume of titrant is $(0.02500\ M)(0.02500\ L)/(0.0100\ M) = 0.0625\ L$ (or 62.5 mL).

For optimum accuracy, the value of $K'_{f,MIn}$ at the end point should be equal to $1/[Ca^{2+}]$, which under the conditions used in this titration will be $1/K'_{f,MIn} = 1/(2.7\underline{5} \times 10^5) = 3.6\underline{4} \times 10^{-6}$, when using a mixture of Eriochrome Black T with Mg^{2+} as the indicator at pH 10.00.

The fraction of titration is then found as follows.

$F = [1 - (3.6\underline{4} \times 10^{-6}/0.02500)]/[(3.6\underline{4} \times 10^{-6})/(0.0100) + (1.74 \times 10^{10})(3.64 \times 10^{-6})/\{1 + 1.74 \times 10^{10})(3.64 \times 10^{-6})\}]$

$= 0.9998\underline{5}/\{3.6\underline{4} \times 10^{-4} + 0.9999\underline{8}\} = 0.9994\underline{9}$

The volume of titrant at this endpoint would be $(62.5)(0.9994\underline{9}) = 62.4\underline{7}$ mL. The resulting titration error would be $62.4\underline{7}$ mL $- 62.5$ mL $= -0.03$ mL.

52. As shown in Problem 48, the volume of EDTA needed to reach the equivalence point in this titration is 22.26 mL, at which pCu = 6.95. Based on the equations given in Figure 13.7, at the beginning of the titration pCu = 1.347; half-way to the equivalence point the volume of added EDTA is 11.13 mL and pCu = 1.648. The same values are obtained by using the fraction of titration expression in Equation 13.15 by placing the known values for C_M, C_L, and $K_{f,ML}$ into this equation. In this case, successive approximations or a spreadsheet could be used to place various values of [M] into Equation 13.15 with the goal of determining the values of [M] and corresponding values of pCu that would result in $F = 0.00$ (at the beginning of the titration), $F = 0.500$ (at the half-way point of the titration), or $F = 1.000$ (at the equivalence point).

53. An argentometric titration is a titration in which silver ions are used as the titrants, although this term is also sometimes used to refer to a titration in which silver ions are the analytes.

54. The Volhard method is a titration for silver ions with thiocyanate to form insoluble silver thiocyanate in the presence of Fe(III). The end point is marked by the red color of Fe(SCN) that is formed when essentially all of the silver has been precipitated. Gay-Lussac's method used NaCl as the titrant and an end point that was marked by a "clear point" when the addition of further NaCl gave no further additional precipitation of AgCl.

55. mol Br^- = mol Ag^+ − mol SCN^-

    = (0.1065 M)(.01000 L) − (0.09875 M)(0.00876 L)

    = 1.99$\underline{95}$ × 10^{-4} mol

    [Br^-] = (1.99$\underline{95}$ × 10^{-4} mol)/(0.05000 L) = 4.00 × 10^{-3} M

56. $3 Ag^+ + PO_4^{3-} \rightarrow Ag_3PO_4$

    3 (mol PO_4^{3-}) = mol Ag^+ = mol SCN^-

    mol PO_4^{3-} = (0.06543 M)(0.01247 L)/3 = 2.720 × 10^{-4} mol

    [PO_4^{3-}] = (2.720 × 10^{-4} mol)/(0.10000 L) = 2.720 × 10^{-3} M

57. The titration curve for a method that uses Ag^+ as the titrant often uses the value of pAg on the y-axis because many of the methods for end point detection in an argentometric titration are based on a change in the concentration of Ag^+. The Volhard method is one example of such a technique, in which the presence of excess Ag^+ from the titrant results in a color change at the end point.

58. One example is the use of Ba^{2+} as a titrant to determine the content of sulfate in a sample through the precipitation of $BaSO_4(s)$.

    $$Ba^{2+} + SO_4^{2-} \rightarrow BaSO_4(s)$$

Another example is the analysis of I^- with Tl^+ to form $TlI(s)$.

$$Tl^+ + I^- \rightarrow TlI(s)$$

59. In a Mohr titration, the end point is signaled when the silver ion concentration in the titration vessel becomes high enough to cause the precipitation of silver chromate. This can only happen if silver ion is the titrant. In a Volhard titration, the end point is signaled when the thiocyanate concentration gets high enough for thiocyanate to react with Fe(III) to form red iron(III) thiocyanate. This can only happen if thiocyanate is the titrant.

60. Mass of Ag = (0.02654 M)(0.02398 L)(107.87 g/mol) = 0.06865 g

    % Ag = 100·(0.06865 g)/(0.06978 g) = 98.38% (w/w)

61. $[Cl^-]$ = (0.05647 M)(0.01604 L)/(0.02500 L) = 0.03623 M

62. The Fajans method uses an end point that is marked by a color change when the indicator, fluorescein or a related compound, is absorbed onto the surface of AgCl and shows a characteristic pink color, as occurs when silver ions are present in excess. This means that silver ions should be in the buret. The Mohr method also requires that the silver ions be in the buret, but the Volhard method requires that the silver ions be in the titration flask.

63. An adsorption indicator is a large, colored molecule that has acid–base properties. When it is in the anionic form it will adsorb to the surface of silver halide particles when those particles have a positive charge from adsorbing excess silver ions. This adsorption will not occur when the surface of the silver halide is in the presence of excess chloride ions.

64. $[Cl^-]$ = (0.1000 M)(0.01865 L)/(0.00500 L) = 0.373 M

65. Mass NaBr = (0.1000 M)(0.03185 L)(102.89 g NaBr/mol) = 0.3277 g

    % NaBr = 100 · (0.3277 g)/(5.00 g) = 6.55% (w/w)

66. Nephelometry involves measuring the intensity of light that is scattered by suspended or colloidal particles in a solution. This method of end point detection is best used when the end point is marked by the formation of a precipitate. A sudden increase in the intensity of scattered light is more easily measured than the cessation of formation of precipitate, such as would be used in the case when the "clear point" method is employed for finding the end point.

67. The student is using visual nephelometry to detect the "clear point" end point. The end point is marked by cessation of the formation of an AgCl precipitate on further addition of chloride ions. For this method to work, the early precipitate must be allowed to settle to give a clear solution above it, making it possible to easily see the formation of AgCl or the lack of such precipitate formation.

68. Figure 13.14 shows the four general regions during a precipitation titration and gives the equations that can be used to predict a titration curve for such a method.

69. a) $Ag^+ + Cl^- \rightarrow AgCl(s)$

    Volume titrant needed = (0.0500 M)(0.05000 L)/(0.02500 M) = 0.10000 L or 100.00 mL

    b) At 0% titration: $[Ag^+] = 0.0500\ M$ pAg = 1.30

      At 50% titration:

      $[Ag^+]$ = {(0.0500 M)(0.05000 L) − (0.02500 M)(0.05000 L)}/(0.10000 L)

      = 0.0125 M

pAg = 1.90

At 100% titration:

$$[Ag^+] = (1.77 \times 10^{-10})^{1/2} = 1.33 \times 10^{-5} \, M$$

pAg = 4.88

At 10.00 mL excess titrant:

$$[Ag^+] = (1.77 \times 10^{-10})/\{(0.02500 \, M)(0.01000 \, L/0.16000 \, L)\}$$

$$= 1.13 \times 10^{-7} \, M$$

pAg = 6.95

70. a) $Ag^+ + Cl^- \rightarrow AgCl(s)$

   Volume titrant needed $= (0.0500 \, M)(0.02500 \, L)/(0.0250 \, M) = 0.05000$ L or 50.00 mL

   b) At 0% titration:

   $[Cl^-] = 0.0500 \, M$ or   pCl = 1.30

   At 25.00 mL titrant:

   $$[Cl^-] = \{(0.0500 \, M)(0.02500 \, L) - (0.0250 \, M)(0.02500 \, L)\}/(0.05000 \, L)$$

   $= 0.0125 \, M$ or   pCl = 1.90

   At 50.00 mL titrant:

   $[Cl^-] = (1.77 \times 10^{-10})^{1/2} = 1.33 \times 10^{-5} \, M$ or   pCl = 4.88

   At 75.00 mL titrant:

   $[Cl^-] = (1.77 \times 10^{-10})/(0.0250 \, M)(0.02500 \, L/0.10000 \, L) = 2.83 \times 10^{-8} \, M$

   pCl = 7.55

   c) At 0.00 mL titrant:

   There is no silver present, so pAg is undefined.

At 25.00 mL titrant:

$$[Ag^+] = (1.77 \times 10^{-10})/(0.0125) = 1.42 \times 10^{-8}\ M \text{ or } pAg = 7.85$$

At 50.00 mL titrant:

$$[Ag^+] = [Cl^-] = 1.33 \times 10^{-5}\ M \text{ or } pAg = 4.88$$

At 75.00 mL titrant:

$$[Ag^+] = (1.77 \times 10^{-10})/(2.88 \times 10^{-8}) = 0.00615\ M \text{ or } pAg = 2.21$$

d) The two plots are approximately mirror images of each other. Either could be used to find the end point. It is more common to use a function of the concentration of the analyte, so when chloride is the analyte, one would usually plot pCl. If the method of end point detection relies on a measurement of silver ions, than a plot that uses pAg on the y-axis might be used instead.

71. See Table 13.5 for an illustration of how mass balance equations and a solubility product expression can be used to obtain a fraction of titration equation for a precipitation titration.

72. The approach is essentially the same as given in Table 13.5, except in the final step where the fraction of titration is defined as $(C_M V_M)/(C_X V_X)$ instead of $(C_X V_X)/(C_M V_M)$. This means the final result will essentially be the reciprocal of the final equation that is shown in Table 13.5.

73. As shown in Problem 69, the volume of titrant needed to reach the equivalence point, in this titration is 100.00 mL, at which pAg = 4.88. Based on the equations given in Figure 13.14, at the beginning of the titration pAg = 1.30. Half-way to the equivalence point the volume of added titrant is 50.00 mL and pAg = 1.90. The same values are obtained by

using the fraction of titration expression in Equation 13.21 by placing the known values for C_M, C_X, and K_{sp} into this equation. In this case, successive approximations or a spreadsheet could be used to place various values of [M] into Equation 13.21 with the goal of determining the values of [M] and corresponding values of pAg that would result in $F = 0.00$ (at the beginning of the titration), $F = 0.500$ (at the half-way point of the titration), or $F = 1.000$ (at the equivalence point).

80. At the time this answer was written, the price of silver on the international market was $19.400 per ounce or $623 per kilogram.

a) The trader would make a profit of 4% for each transaction. To clear $1000 the trader would have to trade $25,000 worth of silver. At $623/kg, 40.1 kg of silver would be needed to make a profit of $1000, given a systematic error of 2%.

b) Prices are quoted to the nearest $0.001, but always seem to have a zero in the thousandths place. If we assume that the digit in the thousandths place is significant, we would needs five significant figures in the mass of silver and in the assay method. However, this type of accuracy is not commonly attainable in such an assay.

CHAPTER 14: AN INTRODUCTION TO ELECTROCHEMICAL ANALYSIS

1. Electrochemical analysis involves the measurement of current or voltage and relating this measurement to the identity or concentration of an analyte. Examples of methods for electrochemical analysis include coulometry, potentiometry, voltammetry, and electrogravimetry.

2. a) Current is the amount of electrical charge that flows through a conducting medium in a given amount of time. Current is expressed in units of amperes (A, amps).

    b) Charge is the integral of an electrical current over time and is expressed in units of coulombs (C).

    c) Electrical potential is a measure of the work that is required to bring a charge from one point to another and is expressed in units of volts (V).

    d) Resistance in an electrical circuit refers to the resistance to the flow of current in the presence of an applied electrical potential and is expressed in units of ohms (Ω).

3. The Faraday constant is the number of coulombs that make a mole of electrons and is equal to 96,485 C/mol. The value of the Faraday constant can be used to determine moles of electrons that are needed to provide a certain charge.

4. The charge (Q) in coulombs for a constant current (I) applied for time t is the product of constant current (in amps, A) multiplied by the time (in seconds), or $Q = I \cdot t$.

5. Keeping in mind that 1 A = 1 C/s, the following relationships can be written to obtain the charge and moles of electrons.

$$\text{Charge } (Q) = (125 \times 10^{-6} \text{ C/s})(500.0 \text{ s}) = 6.25 \times 10^{-2} \text{ C}$$

$$\text{mol electrons} = (6.25 \times 10^{-2} \text{ C})/(96{,}485 \text{ C/mol})$$

$$= 6.48 \times 10^{-7} \text{ mol}$$

6. a) Charge $(Q) = (560 \times 10^{-6} \text{ amp})(2.50 \text{ min})(60 \text{ sec/min}) = 0.0840 \text{ C}$

$$= (0.0840 \text{ C})(1 \text{ F}/96{,}485 \text{ C})$$

$$= 8.71 \times 10^{-7} \text{ mol electrons or } 5.24 \times 10^{17} \text{ electrons}$$

b) Mass Cu $= (8.71 \times 10^{-7} \text{ mol electrons})(1 \text{ mol Cu}/2 \text{ mol electrons})(63.546 \text{ g/mol})$

$$= 2.77 \times 10^{-5} \text{ g}$$

7. Electric potential is a measure of the work that is required to bring a charge from one point to another. The difference in electrical potential between two points (E) is expressed in the unit of volts (V) in the SI system. Electromotive force is another term that can be used in place of electric potential in an electrochemical cell that has no appreciable current flowing.

8. Resistance is the tendency of a circuit to resist the flow of electrons. It is measured in ohms.

9. Conductance is the inverse of resistance and is measured in units of ohm^{-1} (or "mho").

10. Ohm's law relates current, voltage, and resistance in an electrical system, as given by the following equation: $E = I R$. If any two of these quantities are known, the third can be calculated.

11. $I = E/R = (0.140 \text{ V})/(4 \times 10^{12} \text{ ohms}) = 3.5 \times 10^{-14} \text{ A}$

12. $R = E/I = (0.0591 \text{ V})/(8.5 \times 10^{-6} \text{ A}) = 6.95 \times 10^3$ ohms

13. $I = E/R = (0.400 \text{ V})/(2.00 \times 10^8 \text{ ohms}) = 2.00 \times 10^{-9}$ A

14. In a direct current, all the electrons move in the same direction. In an alternating current, the electrons change direction repeatedly. Household electricity is based on alternating current (AC) and batteries supply direct current (DC).

15. A graph of the given direct current on the y-axis and time on the x-axis will be a straight line at 5.0 A with a slope of 0.

    A graph of a 60 Hz, alternating 5.0 A current will give a sine wave for a plot of current versus time, where the current signal is centered at 0 amps and reaches a maximum of 5.0 A every second.

16. Potentiometry is the measurement of potential between two electrodes when the resistance is so great that essentially no current flows. The fluoride ion-selective electrode has a potential related to the activity of fluoride in a solution into which it is immersed. Thus, the use of a fluoride ion-selective electrode to measure fluoride levels in a sample is an example of potentiometry.

17. Amperometry is a method of electrochemical analysis, in which the current passing through an electrochemical cell is measured at a fixed potential. Voltammetry is a method of electrochemical analysis in which the current is also measured, but the potential is now varied over time. Both amperometry and voltammetry measure current as the voltage is controlled, but in amperometry the voltage is set to a specific value, and

in voltammetry the voltage is changed according to a predetermined pattern.

18. Coulometry is a technique that uses a measure of charge for chemical analysis. The use of the Faraday constant makes it possible to calculate the moles of electrons that are involved in an oxidation or reduction process and, in turn, the analyte that is consumed or produced as a result of these processes.

19. An electrochemical cell that is used in potentiometry includes at least two electrodes (i.e., an indicator electrode and a reference electrode), with these electrodes being in contact with either the sample (in the case of the indicator electrode) or a reference solution (in the case of the reference electrode). Usually, there is also some type of salt bridge present to provide contact between these two parts of the electrochemical cell. The circuit is completed by making an electrical contact between the two electrodes, which is also used as a means to measure the difference in potential across the cell. These are the same components as discussed in Chapter 10 for other electrochemical cells, but they are now used under conditions in which the difference in potential is measured in the presence of only a negligible current.

20. If the current were large, the concentration of the analyte would change during the measurement, which would also affect the measured potential. Thus, an essentially zero flow of current is required during potentiometry.

21. The Nernst equation is used to relate a measured difference in potential to the activity or concentration of an analyte that can be related in some way to the oxidation–reduction processes that are occurring in the electrochemical cell.

22. An indicator electrode is the electrode in potentiometry whose potential is related to the activity (or concentration) of desired analyte in the sample. The measured potential of this electrode is used to obtain a signal that is related to this activity or concentration for detection and measurement of the analyte.

23. A measured "negative" potential simply means that the anode and cathode are incorrectly identified when the cell potential is being measured or calculated.

24. The standard hydrogen electrode is inconvenient to use because it requires a supply of gaseous hydrogen bubbling into the electrode solution and over its surface. Also, many things can "poison" this type of electrode and create a systematic error in its measured potential.

25. The design of a silver/silver chloride electrode is shown in Figure 14.3. It consists of a silver wire coated with insoluble AgCl immersed in a solution that contains chloride ions. This type of electrode will have a potential that is related to the activity of chloride ions in the surrounding solution. If the activity of Cl^- is held constant in this solution, this electrode can be used as a reference electrode.

26. A calomel electrode consists of liquid mercury in contact with Hg_2Cl_2 immersed in a chloride-containing solution. This type of electrode will have a potential that is related to the activity of chloride ions in the surrounding solution. If the activity of Cl^- is held constant in this solution, this electrode can be used as a reference electrode.

27. A saturated calomel electrode (SCE) has the Hg and Hg_2Cl_2 immersed in a saturated solution of KCl. An SCE is commonly used as a reference electrode. The design of an SCE is shown in Figure 14.4.

28. Metallic indicator electrode: Pt electrode in a solution of Fe^{3+} and Fe^{2+}

    Class one electrode: Cu electrode in a solution of Cu^{2+}

    Class two electrode: Ag wire covered with AgCl in a solution of Cl^-

    Class three electrode: Hg in the presence of $Hg_2C_2O_4$ and CaC_2O_4 and a solution of Ca^{2+}

29. a) Class one electrode

    b) Metallic indicator electrode

    c) Class two electrode

30. a) $E = 0.465 = 0.73 - 0.05916 \log\{[Fe^{2+}]/[Fe^{3+}]\} - 0.242$

    $0.465 = 0.48\underline{8} - 0.05916 \log\{[Fe^{2+}/Fe^{3+}]\}$

    $\log\{[Fe^{2+}]/[Fe^{3+}]\} = 0.023/0.05916 = 0.039$

    $[Fe^{2+}]/[Fe^{3+}] = 2.45$

    b) $[Fe^{2+}]/(0.0763\ M - [Fe^{2+}]) = 2.45\ M$

    $3.45\ [Fe^{2+}] = 0.187$

    $[Fe^{2+}] = 0.0542\ M$ and $[Fe^{3+}] = 0.0221\ M$

31. A salt bridge can be used to provide contact between the two solutions in an electrochemical cell without allowing mixing of these solutions. The presence of a salt bridge is required to maintain electrical contact between these solutions and to allow the flow of current. A salt bridge can take many forms, but it is often in the form of a U-shaped glass tube. This tube is filled with agar that contains an aqueous solution of

potassium chloride.

32. A junction potential arises when ions move more rapidly across a junction (e.g., such as between two liquids with different contents) in one direction than in the other. The magnitude of this potential is usually not known exactly in an electrochemical cell, but it does contribute to the total measured cell potential.

33. A junction potential can form at the interface between a salt bridge and a sample or reference solution if the contents with the salt bridge differ from those of these solutions. The junction potential created at these interfaces will be the largest when diffusion rates of the cations and anions in the salt bridge are quite different. KCl is recommended for use in a salt bridge whenever possible because the diffusion rates of K^+ and Cl^- are quite similar.

34. A liquid junction potential is the potential that arises from unequal distribution of positive and negative charges between solutions, as results from different rates of diffusion by anions and cations across the interface between these solutions. This type of junction potential arises whenever two solutions are in contact and have different compositions, such as a 0.10 M HCl solution in contact with a 0.10 M NaCl solution.

35. A possible benefit of using a junctionless cell is that there will be no liquid junction potential.

36. Potentiometry can be easy to perform and requires relatively inexpensive equipment. This approach can also be quite selective and provide low limits of detection, as occurs in

the use of a pH electrode to measure the activity hydrogen ions in samples.

37. $-0.0650 = K + (0.05916/2) \log[Ca^{2+}]$

$-0.0477 = K + (0.05916/2) \log\{[Ca^{2+}](50/51) + 0.0850(1/50)\}$

$-0.0173 = 0.02958 \{\log[Ca^{2+}] - \log([Ca^{2+}](50/51) + 0.00170)\}$

$-0.5848 = \log[Ca^{2+}] - \log([Ca^{2+}](50/51) + 0.00170)$

$-0.5848 = \log\{[Ca^{2+}]/([Ca^{2+}](50/51) + 0.00170)$

$0.2601 = \{[Ca^{2+}]/([Ca^{2+}](50/51) + 0.00170)$

$0.2601 ([Ca^{2+}](50/51) + 0.00170) = [Ca^{2+}]$

$[Ca^{2+}](0.255) + 4.42 \times 10^{-4} = [Ca^{2+}]$

$[Ca^{2+}](1 - 0.255) = 4.42 \times 10^{-4}$

$[Ca^{2+}] = 5.93 \times 10^{-4} M$

38. The activity (or concentration) of an analyte or a titrant can be monitored potentiometrically as the titration proceeds. An example is the measurement of pH to follow the course of an acid–base titration.

39. Both flow-injection analysis and liquid chromatography involve flowing solutions of changing composition. Potentiometry can be used to measure the activity (or concentration) of suitable analytes continuously in real time as they pass through these systems.

40. An ion-selective electrode is an indicator electrode that can respond to individual types of anions or cations. A glass-membrane electrode is a type of indicator electrode that uses a thin, glass membrane for selectively detecting the desired ion. Most pH electrodes are

made of glass-membrane electrodes that show a specific response for hydrogen ions.

41. Most modern pH electrodes are combination electrodes in which a single chassis contains both an indicating electrode and a reference electrode, with a suitable salt bridge built in. The indicating electrode is a glass or ceramic membrane of a composition that favors the formation of a junction potential that is related to the activity of hydrogen ions in the surrounding solution.

42. The glass membrane separates the solution being analyzed from a standard solution inside the electrode body. The glass is specially formulated to interact with hydrogen ions. The different activities of H^+ on the two sides of the membrane gives rise to a difference in junction potential at these two interfaces, which is then measured and related to the activity of hydrogen ions in the sample being examined.

43. Each pH electrode is slightly different. There is no standard potential value for this type of electrode, so each pH electrode must be calibrated first, using one or more buffer solutions that have a known pH.

44. Sodium error is caused by Na^+ adsorbing onto a glass membrane in a pH electrode instead of H^+. This error is the greatest at a high pH and when the value of $[Na^+]$ is much greater than $[H^+]$. If pH is to be measured under these conditions, it is useful to raise the pH with KOH instead of NaOH to minimize this type of error.

45. The presence of solid NaCl would raise the ionic strength and change the activity coefficients, but it would not change the hydrogen ion concentration. Because a pH

electrode measures the activity of H^+, the response of this electrode would decrease at the higher ionic strength and the observed pH would increase.

46. Both the pH electrode and sodium ion-selective electrode are glass-membrane electrodes, but the glass is of different composition. The glass in a pH electrode is optimized for the detection of H^+ and the glass in a sodium ion-selective electrode is optimized for the detection of Na^+.

47. At an ionic strength 0.0500, the activity coefficient of Na^+ will be about $0.81\underline{7} = 0.82$. Therefore, $[Na^+] = (0.0674)/(0.81\underline{7}) = 0.082\underline{5}\ M = 0.082\ M$.

48. At pH 7.00, $[H^+] = 1.00 \times 10^{-7}\ M$

    For less than a 10% error, $[Na^+]$ must be greater than $10 \cdot (1.00 \times 10^{-7}\ M)/2 = 5 \times 10^{-7}\ M$.

49. A solid-state, ion-selective electrode uses some solid membrane other than glass to achieve a selectivity response for a particular type of ion. Other than the type of membrane that is used, the general principle of operation is the same for the two types of electrodes.

50. A solid-state, ion-selective electrode contains a sensing element that is a crystalline material or a homogeneous pressed pellet. It is necessary in this type of electrode for the sensing element to have selective adsorption or interactions with the ion of interest, which takes the place of the glass membrane in a pH electrode.

51. The sensing membrane in a fluoride ion-selective electrode is LaF_3 doped with Eu^{2+}. Fluoride ions can be selectively exchanged onto this surface. The different surface

concentrations of membrane on the inside versus outside gives rise to a difference in junction potential that is related to the activity (or concentration) of F⁻ in the sample.

52. If the pH is too low, fluoride exists as HF instead of F⁻. If the pH is too high, OH⁻ adsorbs onto the surface and gives a falsely high reading. Ionic strength is controlled to assure that activity coefficients are the same for the standards and samples. EDTA is added to prevent Fe^{3+} from forming complexes with F⁻.

53. The equation of the best-fit line for the standards is $y = -59.2\,x - 99.42$.

    Substitution of E for the values of the unknowns gives the following results: [Unknown 1] = 1.16×10^{-3} M, [Unknown 2] = 1.26×10^{-5} M, [Unknown 3] = 1.73×10^{-1} M. However, it should be noted that Unknowns 2 and 3 are outside of the range of calibration and their results may have some systematic errors because of this.

54. $[Cu^{2+}] = [CuEDTA]/(K'_f\,[EDTA]) = (0.050)/\{(1.0 \times 10^{14})(0.10)\}$

    $= 5.0 \times 10^{-15}$ M or pCu = 14.3

55. A compound electrode is an ion-selective electrode that has been modified for the measurement of other analytes. Two examples are enzyme electrodes and many types of gas sensing electrodes.

56. A gas sensing electrode is a type of compound electrode that has been modified for the measurement of a particular gas. An example is the CO_2 electrode, which is a pH electrode separated from solution by a thin, plastic membrane that allows gases, such as CO_2, to reach the electrode surface. The change in pH that is created in the solution by

the electrode surface due to the acidity of CO_2 produces a potential that is related to the activity (and concentration) of this dissolved gas.

57. A semipermeable membrane can be placed over the sensing element of a pH electrode that allows only small gaseous molecules, such as ammonia, to reach the electrode. The presence of NH_3 in this region will raise the pH of the solution next to the electrode and produces a potential that is related to the activity (or concentration) of this dissolved gas.

58. CO_2 will only be a gas at a low pH, and NH_3 will only be a gas at high pH. Adjustment of the pH allows the same electrode to be used for either analyte.

59. Mass protein = (0.032 mol NH_4^+/L)(0.100 L)(14.0 g N/mol NH_4^+)(100 g protein/16 g N)

    = 0.280 g

    Percent protein = 100 (0.280 g)/(0.3476 g) = 80.5% (w/w)

    Melamine is 66.7% N (w/w). It would be converted to ammonium by the same treatment and would contribute to the apparent percent protein that is measured during this type of analysis.

60. The carbonate is first converted to CO_2. In an acidic solution, CO_2 molecules can transport through a thin, polymeric membrane covering the indicating electrode. Within the membrane is a solution that becomes more acidic due to the acidic character of CO_2. This lowering of the pH is related to the concentration of CO_2, which in turn, is related to the original carbonate concentration.

61. An enzyme electrode uses an immobilized enzyme to convert a suitable substrate into a product that can give a signal at a gas sensing electrode or other type of electrode.

62. The enzyme urease catalyzes the hydrolysis of urea at the surface of a gas sensing, pH electrode. At a high pH, the ammonia goes through the membrane and gives a pH increase that is proportional to the concentration of urea.

64. The solubility of KCl depends on temperature, so the potential of an SCE will be different at different temperatures. This is not a problem if the concentration of KCl is 0.10 M.

65. Calcium metal reacts rapidly with water to produce H_2 and Ca^{2+}.

66. $E = E^\circ - 0.05916 \log(1/a_{Ag^+}) = E^\circ - 0.05916 \log(a_{Br^-}/K_{sp})$

    $= 0.799 - 0.05916 \log(a_{Br^-}) + 0.05916 \log(K_{sp})$

    $= 0.799 + 0.05916 \log(5.35 \times 10^{-13}) - 0.05916 \log(a_{Br^-})$

    $= 0.799 - 0.726 - 0.05916 \log(a_{Br^-})$

    At $a_{Br^-} = 1.00$: $E = 0.073$ V

67. It would require the use of elemental barium, which reacts rapidly with water to give H_2 and Ba^{2+}.

68. $E = K + 0.05916(pH)$

    $= K + 0.05916(4.56)$

    $= K + 0.2698$

    If there is an error of 0.0010 V, $E = K + 0.2708$

    pH = (0.2708 V)/(0.05916 V) = 4.58

CHAPTER 15: REDOX TITRATIONS

1. A redox titration is one in which the titration reaction is an oxidation–reduction reaction. Either the sample or the titrant can be oxidized, while the other is reduced. A chemical oxygen demand (COD) analysis is a redox method because first dichromate is added to a sample to oxidize dissolved organic compounds; the remaining dichromate is then measured in a back titration that also involves an oxidation-reduction reaction.

2. A redox titration is similar to acid–base and complexometric titrations in that a stoichiometric amount of titrant is added to the sample to initiate a rapid reaction having a favorable equilibrium constant. It is also similar in that either a visual or instrumental method is used to mark the endpoint. A redox titration differs only in the type of chemical reaction that is involved in the titration.

3. The titration reaction must be fast, have no side reactions, and have a large equilibrium constant.

4. A large $E°$ means that the reaction will have a favorable $\Delta G°$ and equilibrium constant. These properties help to provide an end point that is easy to detect.

5. A fast reaction is desirable, so that the reaction reaches equilibrium quickly after even small amounts of titrant have been added. If the titration reaction is slow, it is possible that more titrant than is necessary can accidently be added, thus leading to an overestimation of the amount of titrant that is needed to reach the equivalence point.

6. The actual oxidation of dissolved organic compounds by dichromate is very slow. To

overcome this problem in a COD analysis, an excess of dichromate is added to the sample, the solution is heated, and the amount of dichromate that remains is determined through a back titration with iron (II), which takes place through a very rapid reaction.

7. EDTA titrations invariably have a 1:1 stoichiometry. Acid–base titrations are usually based on a 1:1 or 2:1 stoichiometry, and the two neutralizations of a diprotic analyte are usually well separated. Redox titrations frequently involve reactions that have more complicated stoichiometries.

8. In order for a redox reaction to have 1:1 stoichiometry, the two half-reactions must involve the same number of electrons. This is frequently not the case in a redox titration.

9. Overall titration reaction: $n_T A_{red} + n_A T_{ox} \rightleftarrows n_T A_{ox} + n_A T_{red}$

In this reaction, n_A represents the number of electrons that are released during the oxidation of A_{red} to A_{ox}, and n_T represents the number of electrons that are consumed during the reduction of T_{ox} to T_{red}. The stoichiometry for the balanced redox titration reaction can be used to relate the concentration of the titrant (C_T) and the volume of titrant that has been added at the equivalence point ($V_T = V_E$) to the sample volume (V_A) and concentration of analyte in the sample (C_A).

$$n_T C_T V_T = n_A C_A V_A \quad \text{or} \quad (n_T/n_A)\cdot(\text{mol T}) = \text{mol A}$$

10. Mol dichromate added = $(0.2306\ M)(0.02500\ L) = 0.005765$ mol

Mol iron added = $(0.3165\ M)(0.03424\ L) = 0.01084$ mol

Because 6 moles of iron are oxidized by 1 mole of dichromate, the iron reduced

0.0001806 moles of dichromate, leaving 0.003958 moles of dichromate, which has been reduced in the oxidation of the dissolved organic stuff in the sample.

COD = 3800 mg O_2/L

11. mol Fe(II) in sample = (0.1234 M)(0.01786 L) = 0.002204 mol

    Total mol Fe(II) and Fe(III) = (0.1234 M)(0.02254 L) = 0.002781 mol

    Therefore the mol Fe(III) = 0.002781 mol – 0.002204 mol = 0.000577 mol

    Original concentrations:

    [Fe(III)] = (0.000577 mol)/(0.02500 L) = 0.02310 M

    [Fe(II)] = 0.08816 M

12. $5 AsO_3^{3-} + 2 MnO_4^- + 6 H^+ \rightleftarrows 2 Mn^{2+} + 5 AsO_4^{3-} + 3 H_2O$

    mol AsO_3^- = (2/5)(0.05895 M)(0.03755 L) = 8.854 × 10^{-4} mol

    Mass of As = (8.854 × 10^{-4} mol)(74.92 g As/mol) = 0.06634 g

13. Potassium dichromate standard solutions can be prepared directly from accurate weighing of the primary standard $K_2Cr_2O_7$. This chemical is readily available in a sufficiently pure and stable form that can serve as a primary standard oxidizing agent after drying in an oven to remove surface water.

14. Potassium permanganate and sodium thiosulfate must both be standardized after their solutions have been prepared. $KMnO_4$ can be standardized by titrating primary standard ferrous ethylenediammonium sulfate and $Na_2S_2O_3$ can be standardized by titrating primary standard potassium dichromate.

15. a) Mass of $K_2Cr_2O_7$ = (0.02000 M)(2.0 L)(294.19 g/mol) = 11.7676 g

    b) Volume of H_2SO_4 = (1.0 M)(2.00 L)(98.0 g/mol)(1 mL/1.84 g) = 106 mL

16. [$KMnO_4$] = {(1/5)(1.3657 g)/(392.16 g/mol)}/0.03786 L = 0.018397 M

    [$KMnO_4$] = {(1/5)(1.4498 g)/(392.16 g/mol)}/0.04023 L = 0.018379 M

    [$KMnO_4$] = {(1/5)(1.5108 g)/(392.16 g/mol)}/0.04195 L = 0.018367 M

    Average = 0.018381 M Standard deviation = 0.000015 M

17. If multiple oxidation states of the analyte are present in the sample, it is important to do a pretreatment to get all of the analyte into the same oxidation state if the total amount of the analyte is the goal of the analysis.

18. Preoxidation involves raising the oxidation number of all the analyte to a higher oxidation state prior to analysis, and prereduction involves lowering the oxidation state of all the analyte prior to analysis. For instance, prereduction of Fe(III) to Fe(II) is done prior to a titration with an oxidizing agent to measure total iron content. Preoxidation of chromium Cr(II) and Cr(III) to chromate by hot perchloric acid prior to titration with Fe(II) allows measurement of total chromium.

19. A Jones reductor is a column packed with granular zinc that has been surface amalgamated with mercury (to decrease its reactivity with H^+) through which is passed a solution that will undergo prereduction prior to a redox titration. A Walden reductor is similar but uses granular silver as reducing agent. In a Jones reductor, Fe(III) is reduced to Fe(II); the reaction is $Zn + 2\ Fe^{3+} \rightleftarrows Zn^{2+} + 2\ Fe^{2+}$. In a Walden reductor, the same reduction would

occur as follows: $Ag + Fe^{3+} \rightleftarrows Ag^+ + Fe^{2+}$. The Walden reductor is a milder reducing agent, so it can be selective; for instance, it will reduce Fe(III) but will not reduce Ti(IV), while the Jones reductor will reduce both these species.

20. a) Preoxidize the Cr(III) to Cr(VI)

   b) Prereduce the MnO_2 to Mn^{2+}

   c) Prereduce the NO_3^- to NO_2^-

21. Without prereduction: $[Fe^{2+}] = (0.0254\ M)(17.08\ mL)/25.00\ mL = 0.01735\ M$

    With prereduction: $\{[Fe^{2+}] + [Fe^{3+}]\} = (0.0254\ M)(24.57\ mL)/25.00\ mL$
    $$= 0.02496\ M$$

    Therefore, initial $[Fe^{3+}] = 0.00761\ M$

22. The use of an indicator electrode can be used to follow the change in potential as a result of changes in the chemical contents of the sample/titrant mixture during a redox titration. Before the equivalence point, the potential of the indicator electrode is governed by the potential of the analyte, but after the equivalence point, it is governed by the potential of the titrant. Typically, these are quite different potentials, so there is usually a dramatic change in potential in the immediate vicinity of the equivalence point.

23. The typical redox titration curve is a plot of cell potential versus volume of titrant and is sigmoidal in shape. This curve is similar in appearance to the titration of a weak acid or weak base with a strong base or strong acid. The maximum slope in this curve will occur at the equivalence point.

24. A Gran plot for a redox titration can be prepared by plotting the volume of titrant (V_T) on the x-axis and the function $V_T \, 10^{-n_A \, E_{Cell}/(0.05916 \, V)}$ on the y-axis. The intercept of this plot on the x-axis provides the value of the volume of titrant that is needed to reach the end point.

25. A plot is first made by placing the volume of titrant (V_T) on the x-axis and the function $V_T \, 10^{-n_A \, E_{Cell}/(0.05916 \, V)}$ on the y-axis. This produces a linear region for the data at titrant volume below 40.00 mL, with a best-fit line that crosses the x-axis at 41.79 mL. From this volume of titrant, the concentration of Fe^{2+} in the original sample was (0.04179 L)(0.09533 M)/(0.05000 L) = 0.07968 M.

26. Permanganate is a titrant that serves as its own indicator as long as the rest of the components of the solution are not strongly colored. The first small fraction of a drop in excess of the equivalence point is marked by a pink color due to the intensely dark purple color of the titrant.

27. A redox indicator is a substance that is one color when reduced and a different color when oxidized. Ferron, or iron(1,10-phenanthroline)$_3$, is an example of a redox indicator. This indicator is orange when the iron is present as Fe(II), and this indicator is blue when the iron is present as Fe(III).

28. The indicator must change color at a potential close to that of the equivalence point. This occurs in the vicinity of the standard potential for the redox indicator.

29. The $E^{o'}$ for this indicator is 0.85 V and the color change should occur in a range that is (0.05912 V)/2 below and above this value, so it should change over the range of 0.82 to

0.88 V.

30. Its $E^{o'}$ is 1.25 V, so it should change over the range 1.19–1.31 V. The equivalence point of this titration will occur at (0.77 V + 1.72 V)/2 = 1.24 V, so it should be all right with this indicator. However, if the titration is carried out in 1 M H_2SO_4, the equivalence point will occur at (0.68 + 1.44)/2 = 1.06 V, and the color change will be after the equivalence point.

31. Starch forms a blue association complex with iodine in the presence of iodide. Any titration that uses iodine as a titrant or in which iodine is formed by oxidation of iodide can use starch as an indicator. The blue color marks the point at which free iodine is present as triiodide.

32. The potential of a cell is the difference in potential of two electrodes. One electrode is the indicating electrode whose potential depends on the composition of the solution, and the other is a reference electrode whose potential remains constant.

33. A redox reaction involves both an oxidation and a reduction, each of which has a half-reaction.

34. Because the chemical reaction of a titration comes to equilibrium almost immediately after adding titrant, both half-reactions obey the Nernst equation and either can be used, in theory, to estimate the potential during a redox titration.

35. As is the case for the other classes of titrations, the four general regions in a redox titration are at the beginning of the titration, between the initial sample and equivalence point, at the equivalence point, and after the equivalence point.

36. See Figure 15.8 for a summary of how the cell potential can be estimated for each of the four regions in a redox titration curve.

37. At the beginning of the titration, one has not yet added any titrant so the reference electrode potential can only be estimated from the initial ratio of oxidized and reduced forms of the analyte. However, this ratio is not usually known.

38. a) $Fe^{2+} \rightleftarrows Fe^{3+} + e^-$ $\quad Ce^{4+} + e^- \rightleftarrows Ce^{3+}$ $\quad$ Titration reaction: $Fe^{2+} + Ce^{4+} \rightleftarrows Fe^{3+} Ce^{4+}$

   Volume of titrant = (0.0200 M)(100.00 mL)/(0.05000 M) = 40.00 mL

   b) At 0% titration:

   $E_{Ind} = 0.68 - (0.05916/1) \log\{(0.0200\ M)/(1.5 \times 10^{-5}\ M)\} = +0.49\underline{5}$ V

   $E_{Cell} = E_{Ind} - E_{Ref} = 0.49\underline{5}$ V $- 0.222$ V $= +0.27$ V

   After 1% titration (0.40 mL titrant added):

   $E_{Ind} = 0.68 - (0.05916/1) \log(0.99/0.01) = +0.56\underline{2}$ V

   $E_{Cell} = E_{Ind} - E_{Ref} = 0.56\underline{2} - 0.222 = +0.34$ V

   After 10.00 mL titrant added:

   $E_{Ind} = 0.68 - (0.05916/1) \log(30/10) = +0.65$ V

   $E_{Cell} = E_{Ind} - E_{Ref} = 0.65 - 0.222 = +0.43$ V

   c) At equivalence point:

   $E_{Ind} = (1.44\ V + 0.68\ V)/2 = +1.06$ V

   $E_{Cell} = E_{Ind} - E_{Ref} = 1.06 - 0.222 = +0.84$ V

   At 20 mL excess titrant:

   $E_{Ind} = 1.44 - (0.05916/1) \log(40/10) = +1.40$ V

$E_{Cell} = E_{Ind} - E_{Ref} = 1.40 - 0.222 = +1.18$ V

39. a) $Fe^{2+} \rightleftarrows Fe^{3+} + e^-$ $MnO_4^- + 5 e^- + 8 H^+ \rightleftarrows Mn^{2+} + 4 H_2O$

    Titration reaction: $MnO_4^- + 5 Fe^{2+} + 8 H^+ \rightleftarrows Mn^{2+} + 5 Fe^{3+} + 4 H_2O$

    Volume of titrant = (0.2547 M)(10.00 mL)/{5 (0.01543 M)} = 33.01 mL

    b) At 1% titration:

    $E_{Ind} = 0.68 - (0.05916/1) \log(99/1) = +0.56$ V

    $E_{Cell} = E_{Ind} - E_{Ref} = 0.56 - 0.222 = +0.34$ V

    At 30% titration:

    $E_{Ind} = 0.68 - (0.05916/1) \log(70/30) = +0.66$ V

    $E_{Cell} = E_{Ind} - E_{Ref} = 0.66 - 0.222 = +0.44$ V

    At 70% titration:

    $E_{Ind} = 0.68 - (0.05916/1) \log(30/70) = +0.70$ V

    $E_{Cell} = E_{Ind} - E_{Ref} = 0.70 - 0.222 = +0.48$ V

    c) At equivalence point:

    $E_{Ind} = (1/6)(0.68 + 5(1.51)) = +1.37$ V

    $E_{Cell} = 1.37 - 0.222 = 1.15$ V

    At 30% excess titrant:

    $E_{Ind} = 1.51 - (0.05916/5) \log(100/30) = 1.51 - 0.0062 = 1.50$ V

    $E_{Cell} = E_{Ind} - E_{Ref} = 1.50 - 0.222 = +1.28$ V

    d) The plot made with these points will show a sigmoidal-shaped curve, with a sharp change in E_{Cell} in the region of the equivalence point.

40. a) $V^{2+} \rightleftarrows V^{3+} + e^-$ $\hspace{4cm}$ $E° = -0.26$

$V^{3+} + H_2O \rightleftarrows VO^{2+} + 2 H^+ + e^-$ $\hspace{1cm}$ $E° = +0.34$

$Cr_2O_7^{2-} + 6 e^- + 14 H^+ \rightleftarrows 2 Cr^{3+} + 7 H_2O$ $\hspace{0.5cm}$ $E° = +1.36$

This titration will occur in two steps, the first being the oxidation of V^{2+} to V^{3+} and the second taking V^{3+} to VO^{2+}. The two sequential reactions in this titration are given below.

$$Cr_2O_7^{2-} + 6 V^{2+} + 14 H^+ \rightleftarrows 2 Cr^{3+} + 6 V^{3+} + 7 H_2O$$

and

$$Cr_2O_7^{2-} + 6 V^{3+} + 2 H^+ \rightleftarrows 2 Cr^{3+} + 6 VO^{2+} + H_2O$$

Thus, the overall titration to the final product is as follows.

$$2 Cr_2O_7^{2-} + 6 V^{2+} + 16 H^+ \rightleftarrows 4 Cr^{3+} + 6 VO^{2+} + 8 H_2O$$

The half-reaction for the prereduction is the reverse of the two half-reactions shown for the oxidation reactions of vanadium. Frequently, granular zinc (a Jones reductor) is used as the reducing agent. Dichromate is a strong enough oxidizing agent to further oxidize VO^{2+} to VO_2^+. The potential for this half-reaction is 1.00 V. We, however, have decided to use the second end point as the finale of our titration.

b) Volume of titrant = (2/6)(0.0750 M)(25.00 mL)/(0.05746 M) = 10.88 mL

c) At 25% titration:

$E_{Ind} = -0.26$ V

$E_{Cell} = E_{Ind} - E_{Ref} = -0.26 - 0.242 = -0.50$ V

At 50% titration:

$E_{Ind} = \{(-0.26) + (0.34)\}/2 = +0.04$ V

$E_{Cell} = E_{Ind} - E_{Ref} = +0.04 - 0.242 = -0.20$ V

At 75% titration:

$$E_{Ind} = +0.34 \text{ V}$$

$$E_{Cell} = E_{Ind} - E_{Ref} = +0.34 - 0.242 = +0.10 \text{ V}$$

At 100% titration:

$$E_{Ind} = \{0.34 + 1.00)]/2 = +0.67 \text{ V}$$

$$E_{Cell} = E_{Ind} - E_{Ref} = +0.67 - 0.242 = +0.43 \text{ V}$$

At 10.00 mL excess titrant, we are beyond even the third equivalence point. The first $10.88 + 5.44 = 16.32$ mL of dichromate has been reduced to Cr^{3+} and the last $20.88 - 16.32 = 3.56$ mL remains as $Cr_2O_7^{2-}$.

$$[Cr_2O_7^{2-}] = ((3.56 \text{ mL})(0.05746 \text{ } M)/(25.00 \text{ mL} + 20.88 \text{ mL}) = 0.00446 \text{ } M$$

$$[Cr^{3+}] = 2(0.05746 \text{ } M)(16.32 \text{ mL})/(25.00 \text{ mL} + 20.88 \text{ mL}) = 0.04088 \text{ } M$$

$$E_{Ind} = 1.36 - (0.05916/6) \log \{(0.0488)^2/(0.00446)\} = +1.36 \text{ V}$$

$$E_{Cell} = E_{Ind} - E_{Ref} = +1.36 - 0.242 = +1.12 \text{ V}$$

d) The titration curve will have breaks for each change in oxidation state for vanadium.

41. This process is illustrated in Figure 15.4.

42. The approach is the same as given in Table 15.4, but the final fraction of titration equation is now represented by the term $F = (n_T C_A V_A)/(n_A C_T V_T)$. The resulting expression is the reciprocal of the final equation shown in Table 15.4.

43. It was determined in Exercise 15.5 that the equivalence point for the titration in Figure 15.9 occurs when $E_{Cell} = +1.10\underline{5}$ V vs. SHE, or $(+1.105 - 0.222 \text{ V}) = +0.883$ V vs. Ag/AgCl, and $V_T = V_E = 20.00$ mL. We are also told that the given end point occurs at +0.900 V (vs.

Ag/AgCl). This information can be substituted into Equation 15.22 to find F at the given cell potential.

$$F = \cfrac{\cfrac{1}{1 + 10^{-1\,[0.900\,V + 0.222\,V - 0.77\,V]/(0.05916\,V)}}}{\cfrac{10^{-1\,[0.900\,V + 0.222\,V - 1.44\,V]/(0.05916\,V)}}{1 + 10^{-1\,[0.900\,V + 0.222\,V - 1.44\,V]/(0.05916\,V)}}} = 1.000\underline{003}$$

Next, we can use the values for F and V_E to find V_T at the end point and calculate the resulting titration error.

$$V_L = F \cdot V_E = (1.000\underline{003})(20.00\text{ mL}) = 20.000\underline{062}\text{ mL}$$

$$\text{Titration error} = V_{L,\text{End Pt}} - V_L = 20.00\underline{006}\text{ mL} - 20.00\text{ mL}$$

$$= +0.000\underline{062}\text{ mL} \ (< 0.01\text{ mL})$$

This is a very small titration error that is essentially negligible and only corresponds to $+0.062$ μL, or 3.1 ppm for the titration of Fe^{2+} with 0.0050 N Ce^{4+}.

44. We know from Exercise 15.6 that the equivalence point for the titration in Figure 15.9 occurs when $E_{\text{Cell}} = +1.10\underline{5}$ V vs. SHE, or $(+1.105 - 0.222\text{ V}) = +0.883$ V vs. Ag/AgCl, and $V_T = V_E = 20.00$ mL. This would correspond to a value for F of exactly 1.00000 in Equation 15.22. The value of V_T at the half-way point ($F = 0.5000$) would be 10.00 mL, and Equation 15.22 gives the expected value of $E_{\text{Cell}} = +0.548$ V and $E_{\text{Ind}} = +0.77$ V at the half-way point. The same values should be obtained by using the fraction of titration equation by substituting in various values for E_{Cell} and finding which volumes of titrant give $F = 1.00$ for the equivalence point and $F = 0.500$ for the half-way point.

45. Ce(NH$_4$)$_2$(NO$_3$)$_6$ is available as a primary standard. Its solutions under acidic conditions are stable for a long time. Ce(IV) is a strong oxidizing agent that undergoes only a one-electron reduction to Ce(III).

46. The hydroxides of both Ce(IV) and Ce(III) are insoluble in basic solution.

47. The progress of the titration can be monitored potentiometrically or the end point can be seen with a visual indicator such as ferroin (or iron(II) tris-phenanthroline).

48. a) $Sn^{2+} \rightleftarrows Sn^{4+} + 2\,e^-$

   Titration reaction: $2\,Ce^{4+} + Sn^{2+} \rightleftarrows 2\,Ce^{3+} + Sn^{4+}$

   b) $NO_2^- + H_2O \rightleftarrows NO_3^- + 2\,H^+ + 2\,e^-$

   Titration reaction: $2\,Ce^{4+} + NO_2^- + H_2O \rightleftarrows 2\,Ce^{3+} + NO_3^- + 2\,H^+$

   c) Hydroxylamine can be oxidized to either dinitrogen or dinitrogen monoxide. Frequently, a mixture of these two products is produced. The two half-reactions are shown below.

   $$2\,NH_2OH \rightleftarrows N_2 + 2\,H_2O + 2\,H^+ + 2\,e^-$$

   $$2\,NH_2OH \rightleftarrows N_2O + H_2O + 4\,H^+ + 4\,e^-$$

   The titration reactions in this case are as shown below.

   $$2\,Ce^{4+} + 2\,NH_2OH \rightleftarrows N_2 + 2\,H_2O + 2\,H^+ + 2\,Ce^{3+}$$

   $$4\,Ce^{4+} + 2\,NH_2OH \rightleftarrows N_2O + H_2O + 4\,H^+ + 4\,Ce^{3+}$$

49. Volume of titrant = {(0.00654 g)/(151.85 g/mol)}/(0.0543 M)

$$= 0.793 \times 10^{-4} \text{ L or } 0.793 \text{ mL}$$

50. E_{Ind} at the end point will be +1.06 V when ferroin is in the middle of its color change. Assuming that the solution is at standard conditions, the actual value of E_{Ind} at the equivalence point is (1.72 + 0.77)/2 = +1.24 V, so the end point detected by ferroin will be slightly early. The ratio of $[Fe^{2+}]/[Fe^{3+}]$ at this point can be found by using the Nernst equation.

$$1.06 = 0.77 - (0.05916/1) \log ([Fe^{2+}]/[Fe^{3+}])$$

$$\log([Fe^{2+}]/[Fe^{3+}]) = -\{1.06 - 0.77\}/0.05916 = -4.90$$

$$\text{or} \quad [Fe^{2+}]/[Fe^{3+}] = 1.3 \times 10^{-5}$$

The small size of this ratio indicates that the resulting titration error is insignificant. The color change at the end point will be red (or orange) to pale blue, but the latter color may actually appear as green–blue due to the yellow color of the cerium compounds that will be in solution.

51. $[U^{4+}] = \frac{1}{2} \cdot (0.08755 \, M)(0.03125 \text{ L})/(0.02500 \text{ L}) = 0.05472 \, M$

% U = 100·(0.05472 M)(0.02500 L)(238.02 g/mol)/(5.00 g) = 6.51%

52. The manganese in permanganate has an oxidation state of +7, the highest value for manganese. Permanganate is a strong oxidizing agent that can be reduced to +6, +4, or +2, depending on pH and solution conditions. It has an intense purple color and can serve as its own indicator.

53. In an acidic solution, permanganate is reduced to the soluble species Mn^{2+} in a five-electron change. In a neutral or basic solution, permanganate is reduced to the insoluble, black precipitate MnO_2 through a three-electron change.

54. The purple color of permanganate is so intense that a small fraction of a drop added in excess to the equivalence point gives an otherwise colorless solution a pink color. Thus, permanganate is its own indicator.

55. a) $H_2O_2 \rightleftarrows O_2 + 2\,H^+ + 2\,e^-$

   Titration reaction: $2\,MnO_4^- + 5\,H_2O_2 + 6\,H^+ \rightleftarrows 2\,Mn^{2+} + 5\,O_2 + 8\,H_2O$

   b) $2\,Br^- \rightleftarrows Br_2 + 2\,e^-$

   Titration reaction: $2\,MnO_4^- + 10\,Br^- + 16\,H^+ \rightleftarrows 2\,Mn^{2+} + 5\,Br_2 + 8\,H_2O$

   c) $H_3AsO_3 + H_2O \rightleftarrows H_3AsO_4 + 2\,H^+ + 2\,e^-$

   Titration reaction: $2\,MnO_4^- + 5\,H_3AsO_3 + 6\,H^+ \rightleftarrows 2\,Mn^{2+} + 5\,H_3AsO_4 + 3\,H_2O$

56. With permanganate, the half-reaction involves $[H^+]$ to the 8th power. The cerium half-reaction does not involve hydrogen ions. Ce^{4+} is only a one-electron oxidant, but permanganate (in acidic solution) is a five-electron oxidant. All of these issues affect the calculation of cell potential for these titrations.

57. a) $Fe^{2+} \rightleftarrows Fe^{3+} + e^-$ $\qquad\qquad MnO_4^- + 8\,H^+ + 5\,e^- \rightleftarrows Mn^{2+} + 4\,H_2O$

   Titration reaction: $MnO_4^- + 5\,Fe^{2+} + 8\,H^+ \rightleftarrows Mn^{2+} + 5\,Fe^{3+} + 4\,H_2O$

   Volume of titrant = (1/5)(0.01500 M)(50.00 mL)/(0.01000 M) = 15.00 mL

b) At 50% titration:

$$E_{Ind} = 0.68 \text{ V}$$

$$E_{Cell} = E_{Ind} - E_{Ref} = 0.68 - 0.222 = +0.46 \text{ V}$$

c) At equivalence point:

$$E_{Ind} = \{6(1.51) + 0.68\}/7 = +1.39 \text{ V}$$

$$E_{Cell} = E_{Ind} - E_{Ref} = 1.39 - 0.222 = +1.17 \text{ V}$$

d) After a 10.00 mL excess of titrant:

$$E_{Ind} = 1.51 - (0.0591/5) \log(15.00/10.00) = 1.50\underline{8} \text{ V}$$

$$E_{Cell} = E_{Ind} - E_{Ref} = 1.50\underline{8} - 0.222 = 1.28\underline{6} = +1.29 \text{ V}$$

58. a) $H_3AsO_3 + H_2O \rightleftarrows H_3AsO_4 + 2 H^+ + 2 e^-$ $MnO_4^- + 8 H^+ + 5 e^- \rightleftarrows Mn^{2+} + 4 H_2O$

Titration reaction: $2 MnO_4^- + 5 H_3AsO_3 + 6 H^+ \rightleftarrows 2 Mn^{2+} + 5 H_3AsO_4 + 3 H_2O$

Volume titrant = (2/5)(0.113 M)(10.00 mL)/0.0200 M) = 22.60 mL

b) At 50% titration: $E_{Ind} = +0.58 \text{ V}$

$$E_{Cell} = E_{Ind} - E_{Ref} = +0.58 - 0.222 = +0.36 \text{ V}$$

c) $7 E_{Ind} = 5 E°_{MnO4} + 2 E^0_{H3AsO4} = 5 (+1.51 \text{ V}) + 2 (0.58 \text{ V}) = +8.71 \text{ V}$

$E_{Ind} = (+8.71 \text{ V})/7 = +1.24 \text{ V}$

$E_{Cell} = E_{Ind} - E_{Ref} = +1.24 - 0.222 = +1.02 \text{ V}$

d) $E_{Ind} = E°_{MnO4} - 0.01183 \log(22.60/10.00) = +1.50\underline{6} \text{ V}$

$E_{Cell} = E_{Ind} - E_{Ref} = +1.50\underline{6} - 0.222 = +1.28 \text{ V}$

59. Potassium dichromate is available as a primary standard. It is a fairly strong oxidizing agent whose solutions are stable for a long time. Chromium(VI) is carcinogenic, so its use

is now discouraged.

60. Dichromate is invariably used in acidic solutions because the pH dependence of its electrode potential makes it a dramatically weaker oxidizing agent at a higher pH. If it is used in neutral or basic solution, the Cr(III) product precipitates as $Cr(OH)_3$.

61. The orange/yellow color of $Cr_2O_7^{2-}$ and the green color of Cr^{3+} are not intense enough for dichromate to serve as a self-indicator. Therefore, a separate indicator such as diphenylamine sulfonate is used. A potentiometric method can also be used to detect the equivalence point when using dichromate as a titrant.

62. a) $Sn^{2+} \rightleftarrows Sn^{4+} + 2\ e^-$

   Titration reaction: $Cr_2O_7^{2-} + 3\ Sn^{2+} + 14\ H^+ \rightleftarrows 2\ Cr^{3+} + 3\ Sn^{4+} + 7\ H_2O$

   b) $2\ S_2O_3^{2-} + \rightleftarrows S_4O_6^{2-} + 2\ e^-$

   Titration reaction: $Cr_2O_7^{2-} + 2\ S_2O_3^{2-} + 14\ H^+ \rightleftarrows 2\ Cr^{3+} + S_4O_6^{2-} + 7\ H_2O$

   c) $UO^{2+} + H_2O \rightleftarrows UO_2^{2+} + 2\ H^+ + 2\ e^-$

   Titration reaction: $Cr_2O_7^{2-} + 3\ UO^{2+} + 8\ H^+ \rightleftarrows 2\ Cr^{3+} + 3\ UO_2^{2+} + 4\ H_2O$

63. Calculations of dichromate cell potentials are more complicated than those used with permanganate because the $[Cr^{3+}]$ term is squared. Such calculations are more complicated than those for cerium for that reason, as well as the fact that a hydrogen ion term (to the 14th power) is absent in the cerium calculation but present in the permanganate calculation.

64. Ce(IV) would require the greatest volume because it is only a one-electron oxidizer. Dichromate would require the least because it is a six-electron oxidizer.

65. a) $Fe^{2+} \rightleftarrows Fe^{3+} + e^-$ $Cr_2O_7^{2-} + 14\,H^+ + 6\,e^- \rightleftarrows 2\,Cr^{3+} + 7\,H_2O$

   Titration reaction: $Cr_2O_7^{2-} + 14\,H^+ + 6\,Fe^{2+} \rightleftarrows 2\,Cr^{3+} + 6\,Fe^{3+} + 7\,H_2O$

   Volume of titrant = $(1/6)(0.01500\,M)(50.00\,mL)/(0.01000\,M) = 12.50\,mL$

   b) $E_{Ind} = E°_{Fe} = 0.68\,V$ $E_{Cell} = 0.68 - 0.222 = 0.46\,V$

   c) $7\,E_{Ind} = E°_{Fe} + 6\,E°_{Cr} - 0.05916\,\log\{[Fe^{2+}][Cr^{3+}]^2/[Fe^{3+}][Cr_2O_7^{2-}][H^+]^{14}\}$

   At the equivalence point:

   $[Fe^{2+}] = 6\,[Cr_2O_7^{2-}]$, $[Fe^{3+}] = 3[Cr^{3+}]$, and $[H^+]$ can be regarded as 1.0 even though sulfuric acid is diprotic.

   Therefore: $7\,E_{Ind} = 0.68 + 8.16 - 0.05916\,\log\{2[Cr^{3+}]\}$

   $[Cr^{3+}] = (6/2)[Fe^{2+}](50.00/(50.00 + 12.50)) = 0.036$,

   so $E_{Ind} = (1/7)\{8.84 - 0.05916\,\log(0.072)\} = (1/7)\{8.84 + 0.0676\} = 1.27\,V$

   d) $E_{Ind} = E°_{Cr_2O_7} - 0.05916\,\log\{[Cr^{3+}]^2/[Cr_2O_7^{2-}]\}$

   $[Cr^{3+}] = 2[Cr_2O_7^{2-}]_0\{12.5/(12.50 + 50.00 + 10.00)\} = 0.03448\,M$

   $[Cr_2O_7^{2-}] = [Cr_2O_7^{2-}]_0(10.00/(50.00 + 12.50 + 10.00)) = 0.001379\,M$

   $E_{Ind} = 1.36 - 0.00986\,\log\{(.03448)^2/(0.001379)\} = 1.36 - 0.00986\,\log(0.8621)$

   $= 1.36 - 0.00986(-0.0644) = 1.36 + 0.00063 = 1.36\,V$

   $E_{Cell} = 1.36 - 0.222 = 1.14\,V$

66. mol dichromate reduced by iron = $(1/6)(0.100\,M)(0.01723\,L) = 2.872 \times 10^{-4}\,mol$

   mol dichromate present originally = $(0.100\,M)(0.0050\,L) = 5.00 \times 10^{-4}\,mol$

mol dichromate reduced by organics = $5.00 \times 10^{-4} - 2.87 \times 10^{-4} = 2.13 \times 10^{-4}$ mol

67. An iodimetric titration is a titration in which iodine, usually in the form of triiodide, is either the titrant or the analyte. Direct iodimetric methods are those in which a reducing agent is titrated with a standard solution of iodine. Indirect iodimetric methods are those in which an oxidizing agent reacts stoichiometrically to form iodine that is titrated with thiosulfate.

68. Thiosulfate is a mild reducing agent that reacts cleanly with triiodide to form iodide and tetrathionate ion. Thiosulfate is frequently the titrant after triiodide has been formed by the reaction of iodide with an analyte that is an oxidizing agent.

69. Starch, and especially amylase, is a helical compound into which the linear triiodide ion can insert to form a "starch–iodide complex" that is blue. This complex will only form when free triiodide is present in solution with the starch. Either the formation or the disappearance of the blue color can be used to indicate an end point in an iodimetric titration.

70. Iodine is formed stoichiometrically by the reaction of excess iodide with hydrogen peroxide. This iodine forms triiodide with some of the excess iodide that is then titrated with thiosulfate.

$$H_2O_2 + 2\,I^- + 2\,H^+ \rightarrow I_2 + 2\,H_2O \qquad I_2 + I^- \rightarrow I_3^-$$

$$I_3^- + 2\,S_2O_3^{2-} \rightarrow 3\,I^- + S_4O_6^{2-} \qquad I_3^- + Starch \rightarrow Starch \cdot I_3^-$$

71. a) $H_3AsO_3 + H_2O \rightleftarrows H_3AsO_4 + 2e^- + 2H^+$ $\quad I_2 + 2e^- \rightleftarrows 2I^-$

   Titration reaction: $H_3AsO_3 + I_2 + H_2O \rightleftarrows H_3AsO_4 + 2I^- + 2H^+$

   b) $N_2H_4 \rightleftarrows N_2 + 4e^- + 4H^+$ $\quad I_2 + 2e^- \rightleftarrows 2I^-$

   Titration reaction: $N_2H_4 + 2I_2 \rightleftarrows N_2 + 4I^- + 4H^+$

   c) $CH_2O + H_2O \rightleftarrows HCO_2H + 2H^+ + 2e^-$ $\quad I_2 + 2e^- \rightleftarrows 2I^-$

   Titration reaction: $CH_2O + I_2 + H_2O \rightleftarrows 2I^- + HCO_2H + 2H^+$

72. $O_3 + 6I^- + 6H^+ \rightleftarrows 3I_2 + 3H_2O$

   mol $S_2O_3^{2-}$ = (0.00456 mol O_3/L)(0.100 L)(3 mol I_2/mol O_3)(2 mol $S_2O_3^{2-}$/mol I_2)

   $\qquad$ = 0.002736 mol

   Volume S_2O_3 = (0.002736 mol)/(0.123 mol/L) = 0.02224 L or 22.24 mL

73. $2Cu^{2+} + 4I^- \rightleftarrows 2CuI + I_2$

   $[Cu^{2+}]$ =

   (0.01784 L)(0.0548 mol $S_2O_3^{2-}$/L)(1 mol I_2/2 mol $S_2O_3^{2-}$)(2 mol Cu^{2+}/mol I_2)/0.00500 L

   = 0.1955 M

74. The Karl Fischer method is used to determine, by titration, the amount of water in a sample. The amount of water present in the sample limits the amount of iodine that is reduced by SO_2 to form iodide. Therefore, the remaining iodine can be measured and allows the amount of water to be calculated.

75. Concentration of titrant = (0.0525 g H_2O/10.62 mL titrant) = 0.00494 g/mL

   % H_2O = 100·(7.84 mL titrant)(0.004943 g H_2O/mL)/(5.00 g flour) = 0.775% (w/w)

76. Iodine can add across a carbon–carbon double bond. An excess of iodine is reacted with a fat sample and the excess iodine is titrated with thiosulfate to calculate the number of double bonds in the sample. Bromine is often used because it reacts more rapidly and completely with double bonds. The excess bromine is then reacted with iodide to form bromide and iodine. The resulting iodine is then titrated with thiosulfate. Bromine is not titrated with thiosulfate because it is too strong an oxidizing agent and forms a mixture of bromine-containing products.

77. mol bromine added = (0.350 M)(0.02500 L) = 0.00875 mol

   mol $S_2O_3^{2-}$ = (0.576 M)(0.00487 L) = 0.002805 mol

   mol iodine = (0.002805 mol)/2 = 0.001402 mol

   mol iodine reacted = 0.00875 mol − 0.001402 mol = 0.007348 mol

   Iodine number = 100·(0.007348 mol I_2)(253.0 g/mol I_2)/(11.54 g fat)

   = 16.1

78. $2\ BrO_3^- + 6\ I^- + 6\ H^+ \rightarrow 3\ I_2 + 2\ Br^- + 3\ H_2O$

   $I_2 + 2\ S_2O_3^{2-} \rightarrow 2\ I^- + S_4O_6^{2-}$

   [BrO_3^-] = (0.0453 mol $S_2O_3^{2-}$/L)(0.00724 L)(1 mol I_2/2 mol $S_2O_3^{2-}$)(2 mol BrO_3^-/3 mol I_2)/(0.0250 L)

   = 0.00437 M

83. a) Half reaction: $MnO_4^- + e^- \rightleftarrows MnO_4^{2-}$

   b) $C_3H_8O_3 + 4\ MnO_4^- + 20\ OH^- \rightleftarrows 3\ CO_3^{2-} + 14\ H_2O + 14\ MnO_4^{2-}$

   $5\ Fe^{2+} + MnO_4^- + 8\ H^+ \rightleftarrows 5\ Fe^{3+} + Mn^{2+} + 4\ H_2O$

$$4\,Fe^{2+} + MnO_4^{2-} + 8\,H^+ \rightleftarrows 4\,Fe^{3+} + Mn^{2+} + 4\,H_2O$$

mol Fe^{2+} needed if all the MnO_4^- is titrated with Fe^{2+}:

$$5\,(0.01000\,M)(0.05000\,L) = 0.002500\,mol\,Fe^{2+}$$

mol Fe^{2+} needed if all MnO_4^- is reduced to MnO_4^{2-}:

$$4\,(0.01000\,M)(0.05000\,L) = 0.002000\,mol\,Fe^{2+}$$

mol Fe^{2+} actually needed = $(0.1235\,M)(0.01913\,L) = 0.002362\,mol\,Fe^{2+}$

mol MnO_4^- + MnO_4^{2-} = $(0.1000\,M)(0.05000\,L) = 0.005000\,mol$

$0.002362\,mol = 5\,(mol\,MnO_4^-) + 4\,(0.005000\,mol - mol\,MnO_4^-)$

mol MnO_4^- remaining = $0.002362\,mol - 0.002000\,mol = 0.000362\,mol$

mol glycerine = $(1/14)(0.005000\,mol - 0.000362\,mol) = 3.31 \times 10^{-4}\,mol$ (or 0.0304 g)

84. a) Volume of titrant at equivalence point:

$$(0.01000\,M)(0.70)(0.1000\,M)/(0.02500\,L) = 0.0280\,L \text{ (or 28.0 mL)}$$

b) $E_{Cell} = (E^o - 0.05916\,\log\{[Fe^{2+}]/[Fe^{3+}]\}) - (+0.222\,V)$

$\quad = (0.77 - 0.05916\,\log\{0.70/0.30\}) - (+0.222\,V)$

$\quad = +0.53\,V$

c) $E_{Cell} = (0.77 - 0.05916\,\log\{0.35/0.65\}) - (+0.222\,V) = +0.56\,V$

(Note: If there had been no Fe^{3+} in the original sample, the potential would have been $0.77 - 0.222 = +0.55\,V$.)

d) $2\,E_{Cell} = (E^o_{Fe} + E^o_{Ce} - 0.05916\,\log\{[Fe^{2+}][Ce^{3+}]/[Fe^{3+}][Ce^{4+}]\}) - (+0.222\,V)$

$2\,E_{Cell} = (0.77 + 1.70 - 0.05916\,\log\{[Fe^{2+}]([Fe^{3+}] + 0.00700)/[Fe^{3+}][Fe^{2+}]\}) - (0.222)$

$2\,E_{Cell} = 2.47 - 0.05916\,\log\{([Fe^{3+}] + 0.00700)/[Fe^{3+}]\} - 0.222$

$2\,E_{Cell} = 2.47 - 0.05916\,\log\{(0.0100 + 0.00700)/0.0100\} - 0.222$

$2\ E_{Cell} = 2.47 - 0.0136 - 0.222 = +2.23$ V

$E_{Cell} = +1.18$ V

e) $E_{Cell} = (1.70 - 0.05916 \log \{28.00/48.00\}) - (+0.222$ V$) = +1.49$ V

CHAPTER 16: COULOMETRY, VOLTAMMETRY, AND RELATED METHODS

1. Electrogravimetry is a method that involves weighing a solid that has deposited on an electrode as a result of the reduction or oxidation of an analyte.

2. This method could be used to reduce these metal ions from an aqueous solution and onto a platinum gauze electrode. The increase in mass of the electrode would then be measured and used to determine the amount of these metal ions that were present in the original sample or solution.

3. Mol Cu^{2+} reduced = $(5.0 \times 10^{-3}$ A$)(456$ s$)(1$ mol $e^-/96,485$ C$)(1$ mol $Cu^{2+}/2$ mol $e^-)$

    $= 1.1\underline{8} \times 10^{-5} = 1.2 \times 10^{-5}$ mol Cu^{2+}

4. Mass Cu^{2+} reduced = $(0.150$ L$)(0.0764$ mol/L$)(63.546$ g Cu^{2+}/mol$)$

    $= 0.728$ g

5. Mol Cu = $(24.5673$ g $- 23.9854$ g$)(1$ mol Cu/63.546 g$) = 9.157 \times 10^{-3}$ mol

    $[Cu^{2+}] = (9.157 \times 10^{-3}$ mol$)/(0.250$ L$) = 3.66 \times 10^{-2}$ M

    Mol Pb = $(10.9858$ g $- 10.6489$ g$)(1$ mol $PbO_2/239.2$ g$)(1$ mol Pb/1 mol $PbO_2)$

    $= 1.408 \times 10^{-3}$ mol Pb

    $[Pb^{2+}] = (1.408 \times 10^{-3}$ mol$)/(0.250$ L$) = 5.63 \times 10^{-3}$ M

6. % Cu = $100 \cdot (3.7618$ g$)/(4.5631$ g$) = 82.44\%$ (w/w)

    % Zn = $100.00\% - 82.44\% = 17.56\%$ (w/w)

7. % Cu = $100 \cdot (16.0467$ g $- 15.7649$ g$)/(1.2764$ g$) = 22.08\%$ (w/w)

8. Increase in mass of cathode = (28.7654 g − 27.8645 g) = 0.9009 g

   Both this mass and the volume of the 25.00 mL pipet are known to four significant figures, so it is also possible to calculate a concentration from these values that has four significant figures.

9. Controlled potential electrolysis means that in doing electrogravimetry the potential of the electrolysis is kept at a set value. This condition assures that only analytes more easily reduced than the set potential will actually be reduced.

10. A potentiostat is a device that sets and maintains a known potential between two electrodes. This is an important piece of equipment in many methods for electrochemical analysis.

11. In coulometry, the current and time necessary for complete reduction or oxidation of an analyte are measured. The product of current and time makes it possible to calculate the number of moles of electrons that were used in the reaction. This information can then be used to determine how much analyte was reduced or oxidized.

12. The term "100% current efficiency" means that all of the applied current is being used in the desired oxidation or reduction reaction. This condition is necessary in coulometry because it is assumed, when calculating the amount of analyte that was oxidized or reduced, that all the electrons passed through the system (or all of the current) were used to oxidize or reduce the analyte. This same condition is not needed in electrogravimetry because in this method the mass of the final product is instead measured and used to determine the amount of analyte that was in the sample.

13. Direct coulometry is the reduction or oxidation of the analyte itself. This approach is in contrast with a coulometric titration, in which a substance is formed coulometrically that reacts quantitatively with the analyte. Constant current coulometry is an approach that requires the current be maintained at a constant value even though this will cause the potential to change. The coulometric reduction of silver ions is an example of both of these methods.

14. No, Ag^+ requires one electron per ion to be reduced to silver metal, while Ni^{2+} requires two electrons per ion to be reduced to nickel metal. Therefore, the original concentration of Ag^+ must have been double the original concentration of Ni^{2+} for this situation to occur.

15. Coulombs = (0.100 mol Ag/L)(0.100 L)(96,485 C/mol Ag) = 9.65×10^2 C

16. In a typical coulometric titration, the titrant is formed *in situ* by a coulometric reduction or oxidation process. This titrant then reacts with the analyte. In a volumetric titration, a titrant solution is added to the analyte and sample from a buret.

17. Mass vitamin C = {(0.03000 amp)(384 s)/96485}(1 mol vitamin C/2 mol e^-)(176.1 g/mol)

    = 0.105 g

    % vitamin C = 100·(0.0105 g)/(0.250 g) = 4.20% (w/w)

18. Mol EDTA = mol Cu^{2+} = (0.01000 C/s)(198.5 s)(1 mol Cu^{2+}/2 F)(1 F/96,485 C)

    = 1.029×10^{-5}

    $[Cu^{2+}]$ = (1.029×10^{-5} mol)/0.100 L) = 1.03×10^{-4} M

19. $E = E^\circ - (0.05916/n) \log\{[\text{Red}]/[\text{Ox}]\}$

    As red and ox change in concentration (or activity) due to the coulometry, the value of the potential E must also change if the given oxidation–reduction reaction follows the Nernst equation.

20. In constant potential coulometry, the potential is maintained at a predetermined value, while coulometry is carried out. This approach causes the current to change during the analysis. The advantage of this method is its selectivity; substances that are reduced at slightly higher potential values will not be reduced if the potential is kept constant at a value below these higher potentials. The disadvantage of this method is that the computation of the number of coulombs required is now more difficult than simply taking the product of current and time. The integral of a current-versus-time plot is instead required.

21. The current is sufficient to reduce Ni^{2+}, but not Cd^{2+}. Thus, nothing more can be said about the value of $[Cd^{2+}]$, but the value of $[Ni^{2+}]$ is as follows.

    $[Ni^{2+}] = (458\ \text{C})(1\ \text{mol Ni}^{2+}/2\ \text{F})(1\ \text{F}/96{,}485\ \text{C})/(0.250\ \text{L}) = 9.49 \times 10^{-3}\ M$

22. Voltammetry is a method in which a current is measured as the potential is changed as a function of time.

23. A solute can arrive at the surface of an electrode during voltammetry by means of convection, migration, or diffusion.

24. In DC voltammetry, the potential is gradually made more negative. When this potential becomes sufficiently low for an analyte to be reduced, the current associated with this

reduction after analyte diffusion to the electrode is measured. This current is then used along with the results for standards to determine how much analyte was present in the original sample.

25. a) A voltammogram is a plot of current versus applied potential that is obtained in voltammetry.

b) The limiting diffusion current is the current when 100% of the analyte diffusing to the cathode becomes reduced (as is measured at the plateau of a voltammogram).

c) The half-wave potential is the potential in a voltammogram when the current is half of the diffusion current and is related to the E^o of the material that is undergoing the oxidation–reduction process.

26. Faradaic current results from the reduction of an analyte and is related to the concentration of the analyte. Charging current results from the movement of nonelectroactive ions to the cathode and is not related to the concentration of the analyte.

27. Only the relative levels of the three soluble oxidation states can be determined in this case (because no data for standards is provided), but these results do indicate that CrO_4^{2-}, Cr^{3+}, and Cr^{2+} were all initially present at identical concentrations. This comparison is given below.

The reduction of Cr(VI) to Cr(III) yields 34.5 mA and is a 3 e^- reduction.

The reduction of Cr(III) to Cr(II) yields 46.0 − 34.5 = 11.5 mA and is a 1 e^- reduction.

The reduction of Cr(II) to Cr(0) yields 69.0 − 46.0 = 23.0 mA and is a 2 e^- reduction.

These results indicate that the three oxidation states are present in a relative concentration ratio of (34.5/3):(11.5/1):(23.0/2) = (11.5):(11.5):(11.5). Therefore, the initial concentrations were all equal.

28. The first wave is related to the reduction of O_2 to H_2O_2 and the second is the reduction of H_2O_2 to H_2O. The concentration of dissolved O_2 at 20 °C in air-saturated water is (0.04339 g/L)(0.21)(1 mol/32.0 g) = 2.85×10^{-4} M. The current associated with reduction of the original H_2O_2 is 104.6 mA – 2 (43.5 mA) = 17.6 mA.

$[H_2O_2]$ = $[O_2]$(17.6 mA/43.5 mA) = $(2.85 \times 10^{-4}\,M)(0.4046) = 1.15 \times 10^{-4}\,M$

29. Silver has a higher electrode potential than copper and so is reduced first. The 12.4 mA wave is silver (1 e⁻) and the 34.2 mA wave is copper (2 e⁻).

$[Ag^+]/[Cu^{2+}]$ = (12.4)/((34.2 – 12.4)/2) = 1.14

30. LOD for current = 3 · 0.065 mA = 0.195 mA

LOD for $[Cd^{2+}]$ = $(2.5 \times 10^{-2}\,M)$(0.195 mA/56.8 mA) = $8.6 \times 10^{-5}\,M$

31. The working electrode is the electrode at which the analyte is oxidized or reduced. The reference electrode provides a constant potential against which the working electrode potential can be compared. The auxiliary electrode serves to carry the current associated with the complementary oxidation or reduction process; this electrode is used so that the reference electrode is not forced to carry any significant current, which would change the potential of the reference electrode over time.

32. Half-reactions that involve H^+ or OH^- will show a strong dependence of their half-wave potential on pH, so control of the pH is necessary when such half-reactions are examined by voltammetry. The solution must also have a high enough ionic strength, as supplied by an inert electrolyte, to keep IR potential from being too great and to keep activity coefficients constant.

33. Amperometry is a method for electrochemical analysis in which the potential between two electrodes is fixed and the current associated with analyte reduction or oxidation is measured.

34. The reaction in the Karl Fischer method is as shown below.

$$H_2O + SO_2 + I_2 \rightarrow 2\,HI + SO_3$$

The titrant is iodine, which reacts quickly with water to form iodide. After a stoichiometric reaction with water, the excess iodine (that is now present with the iodide that has been generated by the titration) will cause a sudden increase in potential of a polarized pair of electrodes (i.e., biamperometry). This change in potential is used to signal the end point of the titration.

35. Oxygen is reduced in two steps, both of which are two-electron half-reactions. The potential is set at a value where the four-electron process is complete. The electrode pair is separated from the solution by a membrane through which only gas molecules, such as O_2, can pass. The current under these situations is a measure of the O_2 concentration.

36. The top 15 feet of the lake are in equilibrium with atmospheric oxygen. This no longer occurs below the "thermocline" mark at 15 feet. Water below that level is more highly

reduced and has less oxygen. This effect becomes more pronounced as the depth increases from 15 to 75 feet below the surface.

37. In anodic stripping voltammetry, the analytes are reduced onto an inert electrode for a predetermined time (i.e., typically many seconds). Following this step, the potential is reversed and scanned to a more positive value, so that the reduced analytes are now oxidized sequentially in the order of their redox potentials. The total current for each analyte is measured as this species is oxidized and used with the results for standards to determine the amount of analyte that was present in the original sample.

38. a) Mass of cadmium = $(4.0 \times 10^{-6}$ C$)(1$ F$/96{,}485$ C$)(1$ mol Cd$/2$ F$)(112.41$ g Cd/mol$)$

$$= 2.3 \times 10^{-9} \text{ g}$$

b) Charge = $(4.0 \times 10^{-6}$ C$)(1000$ s$/500$ s$) = 8.0 \times 10^{-6}$ C

c) Fraction of original cadmium that was reduced and later reoxidized =

$$(2.07 \times 10^{-11})/(4.5 \times 10^{-8} \ M)(0.100 \text{ L}) = 0.0046$$

39. $[Pb^{2+}] = (5.0 \times 10^{-8} \ M)(17.5 \times 10^{-7}$ A·s$)/(27.8 \times 10^{-7}$ A·s$) = 3.15 \times 10^{-8} \ M$

Total mol Pb^{2+} in sample = $(3.15 \times 10^{-8} \ M)(0.100 \text{ L}) = 3.15 \times 10^{-9}$ mol

Mol Pb^{2+} reduced = $(17.5 \times 10^{-7}$ A·s$)(1$ F$/96{,}485$ C$)(1$ mol $Pb^{2+}/2$ F$) = 9.07 \times 10^{-12}$ mol

Fraction Pb^{2+} reduced = $(9.07 \times 10^{-12}$ mol$)/(3.15 \times 10^{-9}$ mol$) = 0.00288$ (or 0.288%)

40. In electrogravimetry, the entire amount of analyte must be plated onto the electrode. This amount must be enough to weigh, so generally, it needs to be greater than 0.100 g. In coulometry or voltammetry, the amounts of analyte are much smaller, so a small electrode will suffice.

41. No error occurs in coulometry because the number of moles of analyte is not changed by minor dilution. In voltammetry, the signal measures the concentration of analyte, which is changed by dilution.

42. The power plant discharges heat energy into the water and raises its temperature. The solubility of O_2 in water is smaller at high temperatures, so oxygen leaves the water immediately downstream. By 2000 meters, the temperature has returned to the value of the upstream water and enough time has lapsed for the oxygen concentration to nearly reequilibrate with the atmosphere.

CHAPTER 17: AN INTRODUCTION TO SPECTROSCOPY

1. Spectroscopy is the field of science that deals with the measurement and interpretation of light that is absorbed or emitted by a sample. A spectrometer is an instrument that is designed to electronically measure the amount of light that occurs in a spectrum at a particular spectral band or group of bands.

2. Remote sensing refers to the use of an analytical instrument to examine a distant sample. An example of remote sensing is when a satellite uses spectroscopy and light that is being reflected or emitted from the surface of Earth to learn about the chemical or physical properties of the region that is being monitored.

3. A spectrum is a pattern that is observed when light is separated into its various colors, or spectral bands, and that is usually shown as a graph of absorbance or emission intensity versus wavelength. This information can be useful in identifying some of the components of a sample and in measuring the amounts of these components that are present.

4. The wavelength at which this complex has its strongest absorption of light is 510 nm. This is the same wavelength that would be used in most situations to measure this complex by absorption spectroscopy.

5. The strongest absorption occurs at 525 nm, and the weakest absorption occurs at 700 nm.

6. The use of spectroscopy to identify a sample or measure chemicals in a sample is called "spectrochemical analysis". The use of spectroscopy to measure a spectrum is known as "spectrometry".

7. a) Atomic spectroscopy and molecular spectroscopy

   b) Ultraviolet-visible spectroscopy and infrared spectroscopy

8. Spectroscopy can be used alone for either the measurement or identification of chemicals. For instance, the change in color of an acid–base indicator can be used to help determine when the endpoint has been reached in a titration. It is also sometimes possible during a titration to determine the endpoint by observing the change in color and spectrum for the analyte or titrant as they combine to form a product. Another way spectrometers are often used is as detectors for other analytical methods (e.g., liquid chromatography and electrophoresis).

9. Modern scientists define light as "electromagnetic radiation", which is a wave of energy that propagates through space with both electrical and magnetic field components. Visible light refers to the part of the electromagnetic spectrum that extends from approximately 380 to 780 nm.

10. Some aspects of light are best explained in terms of it being a wave, and other aspects are best explained in terms of it being composed of particles called photons. It behaves both ways and cannot be said to be just one or the other. This is what is meant by the wave-particle duality of light.

11. a) A crest is the maximum value in the electric field vector in a wave of light.

    b) A trough is the minimum value in the electric field vector in a wave of light.

    c) The amplitude is the intensity or maximum value of the electric field vector in a wave of light.

d) The frequency is the number of crests passing a given point per second in a wave.

e) Wavelength is the distance between adjacent crests in a light wave.

f) Wavenumber is the reciprocal of the wavelength.

12. The speed of light in a vacuum is 2.9979×10^8 meters per second and is given the symbol c.

13. Refractive index is the ratio of the speed of light in a vacuum divided by the speed of light in a given medium.

14. a) $v = (2.9979 \times 10^8 \text{ m/s})/(1.333) = 2.249 \times 10^8$ m/s

    b) $v = (2.9979 \times 10^8 \text{ m/s})/(1.507) = 1.989 \times 10^8$ m/s

    c) $n = (2.9979 \times 10^8 \text{ m/s})/(2.9965 \times 10^8 \text{ m/s}) = 1.0005$

    d) $n = (2.9979 \times 10^8 \text{ m/s})/(2.2063 \times 10^8 \text{ m/s}) = 1.3588$

15. $t = (0.25 \text{ m})/\{(2.9979 \times 10^8 \text{ m/s})/1.0003)\} = 8.3 \times 10^{-10}$ s

16. $n = (2.9979 \times 10^8 \text{ m/s})/(17 \text{ m/s}) = 1.8 \times 10^7$

17. a) $\nu = c/\lambda = (2.9979 \times 10^8 \text{ m/s})/(3.50 \times 10^{-6} \text{ m}) = 8.56 \times 10^{13}$ s^{-1} Infrared light

    b) $1/(635 \times 10^{-9} \text{ m}) = 1.57 \times 10^6$ m^{-1} or 1.57×10^4 cm^{-1} Visible light

    c) $(2.9979 \times 10^8 \text{ m/s})/(2.1 \times 10^{18} \text{ s}^{-1}) = 1.4 \times 10^{-10}$ m = 0.14 nm   X-ray

    d) $(5.5 \times 10^{11} \text{ s}^{-1})/(2.9979 \times 10^{10} \text{ cm/s}) = 18.3$ cm^{-1} Infrared light (far IR)

18. Wavelength = $(2.9979 \times 10^8 \text{ m/s})/(4.5 \times 10^{14} \text{ s}^{-1}) = 6.6\underline{6} \times 10^{-7}$ m = 670 nm

    Wavenumber = $1/(6.6\underline{6} \times 10^{-7} \text{ m}) = 1.50 \times 10^6$ m^{-1} = 1.50×10^4 cm^{-1}

    This is in the red range of visible light.

19. A wavelength of 2.5 μm is in the infrared range and 400 nm is in the visible range. The number of photons at 2.5 μm that would have the equivalent energy of a photon at 400 nm would be 2500/400 = 6.25, but you can't have just a fraction of a photon, so at least seven such photons would be need to supply at least as much energy as one 400 nm photon.

20. Difference in frequency = $(2.9979 \times 10^8 \text{ m/s})(1/500 \times 10^{-9} - 1/501 \times 10^{-9})$

    $= (2.9979 \times 10^8 \text{ m/s})(2{,}000{,}000 - 1{,}996{,}007) = 2.9979 \times 10^8)(3993 \text{ m}^{-1})$

    $= 1.197 \times 10^{12} \text{ s}^{-1}$

    Difference in energy = $(6.6256 \times 10^{-34} \text{ J s})(1.197 \times 10^{12} \text{ s}^{-1}) = 7.93 \times 10^{-22}$ J

21. Wavelength = hc/E

    $= (6.6256 \times 10^{-34} \text{ J·s})(2.9979 \times 10^8 \text{ m/s})(6.02 \times 10^{23} \text{ mol}^{-1})/(145 \times 10^3 \text{ J/mol})$

    $= 8.25\underline{0} \times 10^{-7}$ m

    $= 825$ nm

22. Emission of light occurs when an atom or molecule achieves an excited state and then loses that extra energy as a photon.

23. a) The thermal energy available from the flame causes a 3s electron from a sodium atom to go to the 3p orbital, thus forming an excited sodium atom. A very short time later (~1 ns) the atom returns to its ground state as the excited 3p electron falls back into the 3s orbital with the release of a photon of light having a wavelength of 589 nm.

    b) The 3s orbital is the highest occupied atomic orbital of a sodium atom, so every ground-state sodium atom has a 3s electron as its highest energy valence electron. The 3p orbital is the lowest unoccupied atomic orbital of a sodium atom, and the transition from a

3s to 3p orbital is allowed. Thus, a high temperature can cause this excitation to occur. The return from the 3p to 3s orbital occurs rapidly and has an energy difference that corresponds to a wavelength of 589 nm for the emitted light.

24. Most compounds when put into the high temperature environment of a flame will break apart into their constituent atoms. These atoms, especially for elements that have small excitation energies, can have a valence electron excited to a higher energy state. A very short time later, the excited atom can give up its surplus energy as it returns to the ground state while emitting a photon. This emitted photon will have an energy equal to that of the difference in the energy of an atom's excited and ground states, which were involved in this transition.

25. An emission spectrum is a plot of the intensity of emitted light versus the wavelength of that light. The wavelength of emitted light is characteristic of the identity of the emitting element, and the intensity of that light is a measure of the concentration of that element in the flame.

26. a) 475 nm is in the green–blue region of visible light

   b) 250 nm is in the ultraviolet region

   c) 675 nm is in the red–orange region of visible light

   d) 1000 nm is in the infrared region (i.e., near infrared)

27. Absorption is the transfer of energy from an electromagnetic field (as possessed by light) to a chemical entity (for instance, an atom or molecule). The remaining light that passes through the sample is said to have undergone transmission. Light that falls onto a sample is

either absorbed by the sample or transmitted through the sample (or scattered by the sample).

28. Light absorption requires an external source of light that shines onto the sample to be either transmitted or absorbed. Light emission originates from the sample itself and requires a source of energy to excite the sample atoms or molecules.

29. An absorption spectrum is a plot of the fraction of light that is transmitted by a sample or the amount of light that is absorbed by the sample versus the wavelength of the incident light. The size and position of the peaks in an absorption spectrum can be used to measure and identify the chemical that is causing the absorption of light.

30. Much of the ultraviolet and infrared light coming from the Sun is absorbed by molecules in Earth's atmosphere. Visible light, however, is not absorbed appreciably. As a result, the atmosphere is said to be transparent to visible light.

31. The color that we see transmitted through or reflected off of a sample is the result of the light that is not absorbed by the sample. Thus, a blue object appears blue because orange and red light have been considerably absorbed by the object. This effect means the color of an object will appear as the complementary color to the type of light that is being strongly absorbed.

32. For the first solution, 450–500 nm light is blue to green–blue, so if this light is absorbed, the object will appear as the complementary color, which is red–orange. If the solution absorbs 250–300 nm light, this light is in the ultraviolet region, and the object will appear colorless.

33. Because it appears dark blue, the strongest absorption for light by this colored product will be in the orange region of the visible spectrum, or from about 580–620 nm.

34. Light reflection occurs when a portion of light that is not absorbed by a sample bounces off of the sample's surface. Reflection can be diffuse or specular. Reflection occurs when there is a difference in the index of refraction at the interface between two media.

35. a) Specular reflection is "mirror-like" reflection and occurs off of a smooth surface, allowing the original image of the light to be retained. The angle of reflection is equal to the angle of incidence in this situation.

    b) Diffuse reflection occurs off of an uneven surface in all directions, even if the incident light comes from a distinct angle. The original image is not retained in this situation.

    c) The angle of incidence is the angle between the incoming light beam and the surface of the sample.

    d) The angle of reflection is the angle between the reflected light and the reflecting surface.

36. The Fresnel equation allows calculation of the degree to which light will be reflected at a boundary. This depends on the relative difference in the refractive indices for the two sides of the boundary.

$$\frac{P_R}{P_0} = \frac{(n_2 - n_1)^2}{(n_2 + n_1)^2}$$

37. The refractive indices of the materials at the reflecting surface determine the extent of reflection. This will be dependent on the wavelength because the index of refraction is dependent on wavelength.

38. a) $F = P_R/P_0 = (n_2 - n_1)^2/(n_2 + n_1)^2 = (2.417 - 1.000)^2/(2.417 + 1.000)^2 = 0.1720$

   b) $F = (1.333 - 1.309)^2/(1.333 + 1.309)^2 = 8.25 \times 10^{-5}$

   c) $F = (1.53 - 1.00)^2/(1.53 + 1.00)^2 = 0.044$

   d) $F = (1.00036 - 1.575)^2 / (1.00036 + 1.575)^2 = 0.0498$

39. a) $F = (1.0003 - 1.0000)^2/(1.0003 + 1.0000)^2 = 2.25 \times 10^{-8}$

   This fraction is far too small to be an important issue when using sunlight for remote sensing.

   b) Reflection of light by particulate matter in the atmosphere would be an important factor.

40. Refraction is the change in direction of light as it passes obliquely from one medium to another, which has different values for their refractive index.

41. Snell's law relates the index of refraction of each medium at a reflecting surface to the sine of the angles off the normal of the incident beam and the refracted beam, as given by the following relationship: $n_2 \sin(\theta_2) = n_1 \sin(\theta_1)$.

42. At 30° $\sin(\theta_2) = n_1 \sin(\theta_1)/n_2 = (1.0003)(0.5000)/(1.5171) = 0.3297$ or $\theta_2 = 19.25°$

   At 45° $\sin(\theta_2) = n_1 \sin(\theta_1)/n_2 = (1.0003)(0.7071)/(1.5171) = 0.4662$ or $\theta_2 = 27.79°$

   At 60° $\sin(\theta_2) = n_1 \sin(\theta_1)/n_2 = (1.0003)(0.8660)/(1.5171) = 0.5710$ or $\theta_2 = 34.82°$

43. a) $\sin(\theta_2) = n_1 \sin(\theta_1)/n_2 = (1.0003)(0.7660)/(1.51714) = 0.5050$ or $\theta_2 = 30.33°$

   $\sin(\theta_2) = n_1 \sin(\theta_1)/n_2 = (1.0003)(0.7660)/(1.52430) = 0.5207$ or $\theta_2 = 30.18°$

   $\sin(\theta_2) = n_1 \sin(\theta_1)/n_2 = (1.0003)(0.7660)/(1.65548) = 0.4628$ or $\theta_2 = 27.57°$

   b) $\sin(\theta_2) = n_1 \sin(\theta_1)/n_2 = (1.3325)(0.7660)/(1.51714) = 0.6728$ or $\theta_2 = 42.28°$

   $\sin(\theta_2) = n_1 \sin(\theta_1)/n_2 = (1.3325)(0.7660)/(1.52430) = 0.6696$ or $\theta_2 = 42.04°$

$\sin(\theta_2) = n_1 \sin(\theta_1)/n_2 = (1.3325)(0.7660)/(1.65548) = 0.6165$ or $\theta_2 = 38.06°$

c) The glass that gives the greatest change in direction is the one with the smallest value of θ_2, which is dense flint glass. Ordinary crown glass gives the smallest change in direction. In interpreting these results, it is important to remember that the value of θ_2 is the angle away from the normal, not the angle away from the nondefracted beam of light.

44. Scattering refers to the change in the direction of travel of one particle (such as a photon) due to its collision with another particle (for instance, an atom or molecule). This process differs from reflection in that reflection is caused by a surface, and scattering is caused by particles such as atoms or molecules. Rayleigh scattering is a type of scattering in which the wavelength and energy of the light is not changed as a result of the scattering event.

45. Scattering is important in methods such as Raman spectroscopy and can also interfere with absorbance measurements.

46. a) Diffraction is the spreading or bending of waves as they pass through an aperture or past the edge of a barrier.

    b) Interference is the combination of two waves, so that the resultant wave is the sum of the disturbances of the interfering waves.

    c) A diffraction pattern is the pattern of constructive and destructive interference that is formed by the diffraction of light.

    d) Constructive interference occurs when the electric fields of two waves are in phase with each other, so that the resultant wave has an amplitude equal to the sum of the two interacting waves.

    e) Destructive interference occurs when the electric fields of two waves are exactly out of phase with each other, such that their amplitude becomes 0.

47. Diffraction gratings are commonly used in instruments for spectroscopy to separate different wavelengths of light. X-ray diffraction is a powerful method that used diffraction patterns to learn the structure of chemicals.

48. $n\lambda = 2d\sin(\theta)$, where n is the order of diffraction (usually 1), λ is the wavelength of the X-ray used in the measurement, d is the interplanar spacing between layers of atoms (expressed in the same units of length as the wavelength), and θ is the angle at which constructive interference occurs.

49. $n\lambda = 2d\sin(\theta)$

    $\sin(\theta) = n\lambda/(2d) = (1)(0.63\ \text{Å})/(2)(1.356\ \text{Å}) = 0.2323$

    $\theta = 13.43$ degrees

50. a) $\sin\theta = n\lambda/d = (1)(600\ \text{nm})/(1.000 \times 10^{-3}\ \text{mm})(1.0 \times 10^6\ \text{nm/mm}) = 0.600$

    $\theta = 36.87$ degrees

    b) $\sin\theta = n\lambda/d = 0.610$

    $\theta = 37.59$ degrees

    c) $\sin\theta = $ opposite/hypotenuse $= x/10.0$

    Therefore, 600 nm light will be at 3.687 cm away from normal, and 610 nm light will be at 3.759 cm away. The difference is 0.072 cm.

51. The main components of a simple spectrometer for measuring light emission include a means of atomizing and exciting the sample atoms (usually a flame or plasma that also acts as the "sample cell"), a monochromator to separate the light that is emitted, and a detector to

measure the intensity of the emitted light.

52. $P_E = kC$, where P_E represents the radiant power of the emitted light, C is the concentration of the species that is emitting this light, and k is a proportionality constant.

53. A plot of concentration versus emission intensity for the standards gives an equation a straight line with a best-fit response of $y = 51450\,x + 0.2127$ ($r^2 = 0.9995$). Based on this best-fit response, the concentration of Unknown #1 is 3.94×10^{-4} M, and the concentration of Unknown #2 is 1.10×10^{-3} M.

54. The emission intensity for the diluted sample will be $128(10.00/50.00) = 25.5$, and its concentration will be 4.92×10^{-4} M, as determined by the same equation as used in Problem 53.

55. The main components of a simple spectrometer for measuring light absorption include a source of the light, a monochromator to select the appropriate band of wavelengths for the measurement, a sample holder that must be transparent to the light used in the measurement, and a detector to measure the intensity of the transmitted light.

56. Transmittance is the ratio of the power of light transmitted through a sample to the power of light that is incident upon the sample.

    Percent transmittance is the percentage of the incident light that is transmitted through the sample. It equals 100 times the transmittance.

    Absorbance is the negative base-10 logarithm of the transmittance.

    $$T = P/P_0 \qquad \%T = 100(T) \qquad A = -\log(T)$$

57.

Transmittance	%Transmittance	Absorbance
0.156	15.6	0.807
0.358	35.8	0.446
0.561	56.1	0.251
0.689	68.9	0.162
0.780	78.0	0.108
0.056	5.6	1.250

58. Beer's law relates the absorbance to the concentration of the absorbing material, as well as to the path length that light must follow to be transmitted through the sample. The proportionality constant that relates these terms is called the "molar absorptivity," when the concentration is expressed in units of molarity and is given the symbol epsilon (ε). The value of the molar absorptivity will depend on such things as the identity of the sample, the wavelength, and the sample matrix.

59. $\varepsilon = A/(b\,C) = -\log(0.674)/(5.0 \text{ cm})(3.40 \times 10^{-4}\,M) = (0.1713)/(5.0)(3.40 \times 10^{-4}\,M)$

   $= 101 \text{ L/mol·cm}$

60. $C = A/(b\,\varepsilon) = (0.367)/(1.00 \text{ cm})(6.87 \times 10^{3} \text{ L/mol·cm}) = 5.34 \times 10^{-5}\,M$

61. Absorbance of sample = 0.350 − 0.050 = 0.300

   $\varepsilon = A/(b\,C) = (0.300)/(1.00 \text{ cm})(1.0 \times 10^{-4}\,M) = 3.00 \times 10^{3} \text{ L/mol·cm}$

62. A Beer's law plot is a graph of solution absorbance versus the concentration of absorbing analyte. Such plots are used to learn the concentration of unknown solutions once the measured absorbances are plotted for standards with known concentrations.

63. An absorbance of $A = 0.76$ in this plot corresponds to a concentration for protein A of 38 µg/mL. At 51% T, the value of A will be 0.309, which corresponds to a concentration of 16 µg/mL.

64. This plot gives a linear relationship that provides an unknown concentration of 2.94×10^{-4} M at an absorbance of 0.615.

65. $A_{Total} = A_1 + A_2 = b(\varepsilon_1 C_1 + \varepsilon_2 C_2)$

66. At a low absorbance (high transmittance), there is a loss in precision because it becomes more difficult for the spectrometer to differentiate the amount of light that is transmitted in the presence and absence of the sample (or P is close to P_0). At a high absorbance (low transmittance), there is a large difference between P and P_0, but P is now quite small, making this measurement subject to imprecision due to factors such as noise from the detector.

67. This plot gives the relative error of a spectrometer if the uncertainty in the transmittance measurement is the same over the entire transmittance range. It reaches a minimum at $T = 0.368$ ($A = 0.434$), but it is fairly small within the absorbance range 0.1 to 0.8. More sophisticated spectrophotometers have less uncertainty over the entire range, so trustworthy measurements can be made at both higher and lower absorbances.

68. a) Approximately 0.3 to 0.4 for 2.5% and 0.1 to 1.3 for 3%

    b) Approximately 0.3 to more than 2.0 for 2.5% and 0.20 to more than 2.0 for 3%

69. If the range of wavelengths used in the experiment is too wide, the value of molar

absorptivity will not be constant over the range. This will cause a downward curvature of a Beer's law plot at high concentrations.

If the absorbing species do not behave independently over the concentration range that is being studied, a curvature in the Beer's law plot will occur. The most common kinds of samples in which this can happen are unbuffered weak acids and substances that can dimerize.

70. Monochromatic light consists of only a single wavelength and polychromatic light consists of many different wavelengths. Beer's law assumes that monochromatic light is present, but no light source is perfectly monochromatic. What is actually important is that the band of wavelengths is so narrow that there is no noticeable difference in the value of the molar absorptivity over the range of wavelengths that are being used for the analysis when using Beer's law.

71. Such a plot will curve in a downward fashion. If a wavelength at the top of a peak is used, the range of absorptivities will be small compared to that on the side of a peak and will minimize this effect.

72. a) The absorbance at 465 nm is low and on the side of a peak, while 645 nm is a peak maximum; therefore, a plot made using data obtained at 645 nm will have a larger slope and better linearity than a plot made using data obtained at 465 nm.

b) The absorbance at 645 nm is low, but 660 nm is a peak maximum; therefore, a plot made using data obtained at 660 nm will have a larger slope and better linearity than a plot made

c) The absorbance at 475 nm is low, while 645 nm is a peak maximum; therefore, a plot made using data obtained at 645 nm will have a larger slope and a better linearity and response than a plot made using data obtained at 475 nm.

d) The absorbance at 440 nm is on the side of a peak, but 660 nm is a peak maximum; therefore, a plot made using data obtained at 660 nm will have a larger slope and better linearity than a plot made at 440 nm.

73. Stray light is light that gets to the detector without going through the sample. The presence of stray light will cause the apparent absorbance to be too small and will lead to low results for the analysis.

74. Light can reflect off the surface of the cuvette in four places: at the air–glass interface, at the glass–liquid interface, at the liquid–glass interface, and at the glass–air interface. This reflection will decrease the intensity of light that finally reaches the detector, and it is apt to differ when a blank solution is in the cuvette compared to when an absorbing solution is present. The result will be an error in the measured absorbance.

75. $T = 0.050 = (P + P_S)/(P_0 + P_S) = (0.050 + 0.010)/(1.000 + 0.010) = 0.059$,

or $A = 1.23$ instead of 1.30. The measured absorbance will be 1.23 instead of 1.30, which is a relative error of 5.4%.

76. $T = (P - P_R)/(P_0 - P_R) = (0.050 - 0.010)/(1.000 - 0.010) = 0.0404$,

or $A = 1.393$ instead of 1.30. This change causes an error of $100(0.09)/(1.30) = 6.9\%$.

79. a) $\Delta E \geq h/(4\pi \Delta t) = (6.626 \times 10^{-34} \text{ J·s})/\{(4)(3.14159)(1.0 \times 10^{-8} \text{ s})\}$

$\Delta E \geq 5.27 \times 10^{-27}$ J

b) $E = hc/\lambda = (6.626 \times 10^{-34} \text{ J·s})(2.9995 \times 10^8 \text{ m/s})/(589.3 \times 10^{-9} \text{ m}) = 3.372 \times 10^{-19}$ J so $E = 3.372 \times 10^{-19}$ ($\pm 5.27 \times 10^{-27}$ J) or $3.372 (\pm 0.0000000527) \times 10^{-19}$ J

The upper and lower observed wavelengths in the range created by the uncertainty principle would be as follows.

Upper wavelength: $(6.626 \times 10^{-34}$ J s$)(2.9995 \times 10^8$ m/s$)/(3.3720000527 \times 10^{-19}$ J$)$

$= 5.894 \times 10^{-7}$ m (or 589.4 nm)

A similar calculation for the lower wavelength gives a range of approximately 589.3 ($\pm$ 0.1) nm for the observed light.

c) In order for $\Delta E = 0$, $\Delta \lambda$ would have to be infinite.

CHAPTER 18: MOLECULAR SPECTROSCOPY

1. Molecular spectroscopy can be defined as the examination of the interactions of light with molecules.

2. Molecules can absorb light, emit light, and scatter light. All of these interactions can lead to chemical information.

3. Electronic energy levels – ultraviolet-visible absorption spectroscopy

   Vibrational energy levels – infrared absorption spectroscopy

   Rotational energy levels – microwave spectroscopy

   Nuclear spin energy levels – nuclear magnetic resonance spectroscopy

4. In colorimetry, the analyte is combined with a reagent that will form a colored product. The color of this product is then compared to the color of standards, making it possible to determine the amount of analyte that is present in the sample or to simply see if the analyte is present above a certain level as part of a screening assay. Iron(III) can be measured at low concentrations through the use of colorimetry by the formation of the iron(III) complex with thiocyanate.

5. Ultraviolet-visible spectroscopy is good for measuring low concentrations of species that absorb in the range of 200–800 nm. Infrared spectroscopy is useful for learning the identity of molecular samples, preferably when the sample is quite pure.

6. Ultraviolet-visible spectroscopy is a type of spectroscopy that is used to examine the ability of an analyte to interact with ultraviolet or visible light through absorption. Light in the ultraviolet or visible range has the same amount of energy as occurs between the energy levels for some of the electrons in molecules. Transitions for organic molecules that can

absorb ultraviolet-visible light often involve nonbonded or pi electrons.

7. A chromophore is a functional group within a molecule that is responsible for the absorption of a photon. A chromophore in ultraviolet-visible spectroscopy commonly involves multiple bonds, especially conjugated multiple bonds, or unshared pairs of electrons.

8. In order from easiest to most difficult: d > c > b > a (not possible).

   a) Impossible, because there are no double bonds

   b) Possible, but not very strong absorption because the molecule contains only one double bond

   c) Possible, but still not very strong absorption because the two double bonds are not conjugated

   d) Easiest to examine because the two double bonds are conjugated

9. Beer's law is a relationship between the absorbance and the concentration of an absorbing analyte. This relationship can be used for the measurement of analytes by their absorption of ultraviolet-visible light.

10. $C = A/(\varepsilon\, b) = (0.002)/(5.6 \times 10^5 \text{ L/mol·cm})(1.00 \text{ cm}) = 3.6 \times 10^{-9}\ M$

11. $A = -\log(0.436) = 0.360$

    $\varepsilon = A/(b\, C) = 0.360/(5.00 \text{ cm})(5.7 \times 10^{-3} \text{ mol/L}) = 12.6 \text{ L/mol·cm}$

    Lower limit of detection $= (0.001)/(5.00 \text{ cm})(12.6 \text{ L/mol·cm}) = 1.58 \times 10^{-5}\ M$

12. Upper limit of linear range $= 1.00/(5.00 \text{ cm})(12.6 \text{ L/mol·cm}) = 1.58 \times 10^{-2}\ M$

13. The basic components of an ultraviolet-visible absorbance spectrometer include a light source, a monochromator to separate and select the desired wavelength of light, a cuvette (sample holder) to contain the sample, and a detector and recorder to measure the intensity of light transmitted through the sample.

14. A tungsten lamp emits a continuum of light, most importantly in the visible region. It is heated electrically to a high temperature, causing it to glow white-hot and emit light as a result of blackbody radiation.

15. A hydrogen discharge lamp is a sealed silica tube containing low pressure of hydrogen gas and two electrodes across which is passed a high voltage. The high voltage breaks the H_2 molecule into two H atoms, many of which are excited or ionized to H^+ and a free electron. The excited atoms emit light at specific wavelengths, but the ionized atoms emit a continuum of light in the ultraviolet region when the electrons recombine with the H^+ ions and return to a lower energy state. A deuterium lamp operates in the same manner, except that the gas in the lamp is D_2 instead of H_2; this type of lamp emits more intense ultraviolet radiation than an ordinary hydrogen lamp.

16. Prisms (mostly used in older instruments) and transmission or reflection gratings

17. The sample holder (cuvette) must be transparent in the ultraviolet or visible spectral region, must be inert with respect to the sample, and must maintain a stable geometry. For the visible region, this is most easily accomplished by using glass to construct the sample holder. Glass, however, is opaque in much of the ultraviolet region, so fused silica or fused quartz is used instead when working with ultraviolet light. Some instruments use a cylindrical cross-section tube and other, more expensive instruments use a cuvette with a square cross-section, usually with a 1.00 cm interior path length.

18. A photomultiplier is a vacuum tube with a photodiode that emits an electron when struck by a photon. This electron is amplified and produces an electrical current that is proportional to the number of photons that originally hit the photodiode. A simpler device is a phototube, which is similar to a photomultiplier but does not include the same amplification capability. (Note: A newer device is a photodiode, which is a transistor which causes a current to flow when struck by light but does not require a transparent casing or internal vacuum.)

19. A single-beam instrument is less complicated and less expensive than a double-beam instrument. In a single-beam instrument, there is only one path that light takes through the instrument from the light source to the detector (i.e., as the light travels through a sample, standard, or blank). In a double-beam instrument, both the sample and a reference blank are examined simultaneously. However, in a single-beam instrument, the blank must be read to set up the instrument and then the sample is looked at later. If any instrumental parameters have changed between examining the blank and the sample, a serious error may occur.

20. A diode-array detector is actually a large number of very small photodiode devices placed very close beside each other. That way the entire spectrum of light can be examined simultaneously. In such an instrument, the monochromator usually is placed after the sample rather than before it, which is different than the layout that is used in a standard single-beam or double-beam instrument.

21. Flow injection analysis is a technique by which a series of samples are injected into a flowing stream of liquid containing a reagent that reacts with the analyte to give a colored product. As each sample flows through a suitable detector, its absorbance (and, thus, its

concentration) is measured. Ultraviolet-visible spectroscopy is often used for detection in this method.

22. If the value of the molar absorptivity of a substance is known, its concentration in an unknown solution can be calculated through Beer's law by using $C = A/(\varepsilon\, b)$.

23. a) $[Fe^{2+}] = 0.762/11{,}000$ L/mol·cm)$(1.0$ cm$)\}(50.00$ mL$/20.00$ mL$) = 1.73 \times 10^{-4}$ M

    b) If some other species absorbs at the given wavelength, the measured absorbance will be too high and will be interpreted inaccurately as a higher concentration of Fe^{2+}.

24. The only two values that remain the same are the molar absorptivity and the mass of dye, as given here by the symbol "g".

    $C_1 = g/(1.00$ L$)$ and $C_2 = g/V$, where V is the volume of the pool

    $\varepsilon = A_1/(b_1 C_1) = A_2/(b_2 C_2)$

    $C_2 = A_2 b_1 C_1/(b_2 A_1) = A_2 b_1 (g/1.00$ L$)/(b_2 A_1) = g/V$

    $g/V = ((0.142)(0.100$ cm$)(g/1.00$ L$)/[(10.00$ cm$)(0.768)] = 0.00185$ g

    $V = 1/(0.00185$ g$) = 541$ L

    The most important assumption is that after stirring in the liter of dye, the solution in the pond will be homogeneous. It is also important that no other solute in the pool absorbs at 450 nm.

25. One can calculate the value of ε from all three solutions and get an average value. To do this, subtract the absorbance of the blank from each absorbance and calculate A/C. The three solutions give values, respectively, of 32, 34, and 32, or an average of 32.7. Use of this average value with the results for the unknown gives $C = (0.465 - 0.005)/(32.7)(1.00$ cm$) = 14.1 \times 10^{-3}$ M. (Note: This result assumes that there is no background absorbance.)

    Alternatively one can use a spreadsheet to calculate the slope of a plot of absorbance vs. concentration and use the equation of a straight line to solve for the concentration of the unknown.

26. The molar absorptivity is calculated to be 0.103/0.10 = 1.03, 0.257/.25 = 1.028, 0.515/0.50 = 1.03. (Note: This result assumes that there is no background absorbance.)

$$C = (0.318)/(1.03) = 0.309$$

Actually, this is not the concentration, but is the mass of nitrate-nitrogen in the solution. The corresponding concentration is (0.309 mg)/(0.050 L) = 6.17 mg/L or (6.17 ppm). Alternatively, a spreadsheet graph of A vs. C will yield the same result.

27. In a standard addition measurement, the absorbance of a sample is measured as is a portion of the unknown added to a known amount of analyte. Comparison of the two absorbances will give the concentration of the unknown.

28. $A_1 = k\,[\text{Caffeine}]_1 = 0.243$

$A_2 = k\,[\text{Caffeine}]_2 = 0.387$

$[\text{Caffeine}]_1 = (0.243/0.387)[\text{Caffeine}]_2$

$\qquad = 0.628\{[\text{Caffeine}]_1 + (1.0 \times 10^{-2}\,M)(10/60)\}$

$\qquad = 0.628[\text{Caffeine}]_1 + 0.00104$

$(1 - 0.628)[\text{Caffeine}]_1 = 0.00104$

$[\text{Caffeine}]_1 = (0.00104)/(0.372) = 2.81 \times 10^{-3}\,M$

The original coffee is more concentrated by a dilution factor of (60/50) = $3.37 \times 10^{-3}\,M$.

29. Let x = mol of Fe^{2+} in 10 mL sample

$A_1 = k\,[Fe^{2+}]_1$ $\qquad\qquad\qquad\qquad\qquad A_2 = k\,[Fe^{2+}]_2$

$0.367 = k\,(x)/(0.010\,L)$ $\qquad\qquad\qquad 0.538 = k\,(x + 1.0 \times 10^{-4}\,\text{mol})/(0.015\,L)$

$$k = 0.367\,(0.010\text{ L})/(x) \qquad\qquad k = 0.538\,(0.015\text{ L})/(x + 1.0\times 10^{-4}\text{ mol})$$

Setting these two expressions for k equal to each other makes it possible to solve for x.

$$0.367\,(0.010\text{ L})/(x) = 0.538\,(0.015\text{ L})/(x + 1.0\times 10^{-4}\text{ mol})$$

$$24.467\,x + 2.44\times 10^{-3} = 53.8\,x$$

$$x = 8.3\underline{4}\times 10^{-5}\text{ mol Fe}^{2+}$$

In original sample: $[Fe^{2+}] = (8.3\underline{4}\times 10^{-5}\text{ mol})/(0.010\text{ L}) = 0.0083\underline{4}\,M = 0.0083\,M$

30. An acid–base indicator is a weak acid for which the absorption spectrum of the acid form is different from that of its conjugate base. Methyl red is red in an acidic solution and yellow in a basic solution because its pK_a is about 5.0. Its absorbance maximum in an acid is about 520 nm, and in a base the absorbance maximum is about 425 nm.

31. In a spectrophotometric titration, the absorbance of a solution is measured as a titrant is added. The change (or lack of change) in absorbance before and after the end point is used to determine where the end point occurs in this titration.

32. Beer's law is followed, which is a linear relation between absorbance and concentration of colored species. Thus, as more titrant is added and creates product or excess titrant, the measured absorbance tends to change in a linear fashion with the amount of added titrant.

33. a) A will rise until the end point and then level out.

b) A will decrease until the end point and then level out near 0.

c) A will start at 0 and remain there until the end point, when it will begin to rise.

d) A will start at some non-zero value, rise until the end point, and then level out.

e) A will stay the same throughout the entire titration; no end point can be determined in this case by using absorbance measurements at the given wavelength.

34. a) The absorbance reaches 0.750 at 7.02 mL, as determined from a plot of absorbance versus the volume of trien.

$[Cu^{2+}]\, V_{Cu} = [trien]\, V_{trien}$

$[Cu^{2+}] = (0.0500\ M)(7.02\ mL)/(10.00\ mL) = 3.51 \times 10^{-2}\ M$

b) The analyte by itself has a very small absorbance but the copper-trien complex absorbs quite strongly. Past the endpoint when only more trien is added, there is no increase in absorbance; this result means trien has no measurable absorbance at the given wavelength.

c) $\varepsilon_{Cu2+} = \{0.005/(1.0\ cm)(3.51 \times 10^{-2}\ M)\}(50/10) = 0.71$ L/mol·cm

$\varepsilon_{trien} = 0$ L/mol·cm

$\varepsilon_{Cu(trien)} = \{0.750/(1.0\ cm)(3.51 \times 10^{-2}\ M)\}(50/10) = 106.2$ L/mol·cm

35. Ultraviolet-visible spectroscopy can be used to examine several analytes in a sample by measuring the absorbance of the mixture at least the same number of wavelengths as components that one wishes to measure. The spectrum of each analyte must also be known, and the molar absorptivities of the various components should be different at the wavelengths that are used for the absorbance measurements. Absorbances are additive, so each absorbance that is measured will be the sum of the absorbance of the multiple components. The resulting Beer's law expressions at each wavelength are then solved to obtain the concentrations of the individual components.

36. There must be three measurements of the sample at wavelengths where the molar absorptivities of the three components are significantly different.

37. An isobestic point is a wavelength in an absorption spectrum at which two absorbing species, related through an equilibrium, have the same molar absorptivity and, hence, the same absorbance. An isobestic point should be avoided if the goal is to use multiple

measurements to find the concentrations of several individual species in a sample. However, an isobestic point can also be utilized to provide a wavelength at which the measured absorbance can be used to find the combined concentration of each species without being affected by changes in the relative amount of these species versus one another.

38. $A_{400} = (1.0)\{(570)[P] + (220)[Q]\} = 0.436$

    $A_{600} = (1.0)\{(35)[P] + (820)[Q]\} = 0.644$

    $[P] = \{(0.436) - (220)[Q]\}/ 570$

    $0.644 = \{(35)\{(0.436) - (220)[Q]\}/570 + 820[Q]$

    $0.644 = \{15.26 - 7700[Q]\}/570 + 820[Q]$

    $0.644 = 0.02677 - 13.51[Q] + 820[Q]$

    $0.6172 = [Q](806.49)$

    $[Q] = 7.65 \times 10^{-4}\,M$

    $[P] = \{(0.436) - (220)(7.65 \times 10^{-4})\}/570$

    $[P] = \{0.436 - 0.168\}/570 = 4.69 \times 10^{-4}\,M$

39. $K_a = [H^+][A^-]/[HA] = (10^{-5.20})(3.69 \times 10^{-5})/(1.31 \times 10^{-5}) = 1.78 \times 10^{-5}$

40. Vibrational transitions, with some superimposed rotational transitions, account for the absorption of light in IR spectroscopy. These transitions are of much smaller energy than the electronic transitions that form the basis of ultraviolet-visible spectroscopy.

41. Molecules are always vibrating, even at a temperature of absolute zero. If a molecule absorbs a photon of infrared radiation, it will vibrate more strongly until it loses that excess vibrational energy by bumping into neighboring molecules. The energy, therefore, ends up as greater kinetic energy (i.e., a higher temperature) of the entire sample. Vibrations

include the stretching of covalent bonds, the bending of molecules and more complex motions.

42. IR spectroscopy is most commonly used for the identification of chemical substances. IR spectra have many peaks and their wavelength and intensities give a pattern essentially unique to each substance. Most IR experiments are done on fairly pure samples, so the technique is not used for trace analysis because the complex patterns of other constituents obscure the pattern of a dilute analyte.

43. A typical IR spectrum plots the percent transmittance (% T) on the y-axis and wavenumber or wavelength of light on the x-axis. There are usually a large number of very sharp peaks in this spectrum. The typical wavelength range for an IR spectrum is 2.5 to 16.0 micrometers.

44. The same four components are present, including a light source, monochromator, sample holder, and detector. However, these components are usually present in a different sequence in instruments for ultraviolet-visible spectroscopy and IR spectroscopy. For instance, in ultraviolet-visible spectroscopy, the monochromator is usually before the sample and in IR spectroscopy it is usually after the sample.

45. The light source in IR spectroscopy is a substance that is heated to emit blackbody radiation in the IR region (but not too much in the visible region). Because glass and silica are opaque at wavelengths greater than 2.5 μm, the heated source must not be in a glass envelope that has an inert atmosphere. Instead, this light source must be robust enough to glow in air without burning or decomposing in any way.

46. Both are sources for IR radiation. A Nernst glower is composed of rare earth oxides (e.g., zirconium, yttrium, or thorium oxides) and is heated to about 1200–2000 K. It emits adequate amounts of light at wavelengths ranging from 1–40 μm. A globar is composed of silicon carbide and is heated to about 1300–1500 K, giving usable light at wavelengths of 0.4–20 μm.

47. Both glass and quartz are opaque at wavelengths greater than about 2.5 micrometers and will absorb IR radiation. Thus, a prism made of these materials cannot be used as a monochromator in an instrument for IR spectroscopy.

48. A diffraction grating is the main part of the monochromator in a scanning instrument for IR spectroscopy and is used to separate the various wavelengths of IR radiation from each other. A diffraction grating for IR spectroscopy has fewer lines per centimeter than a diffraction grating for ultraviolet-visible spectroscopy because the wavelengths of light that must be separated are greater. Typically, the monochromator comes after the sample in IR spectroscopy and before the sample in ultraviolet-visible spectroscopy.

49. Sample holders for IR spectroscopy are usually made of NaCl or KBr. If resistance to water is important, CaF_2 or AgCl can be used. These ionic substances are transparent in the IR region.

50. All ordinary solvents have complicated IR spectra, so it is preferable to measure spectra of pure substances rather than a solution, especially a dilute solution. A drop of a liquid sample is often simply put onto a flat plate of NaCl and another similar NaCl flat plate is put on top of it, clamped into place, and a spectrum taken of the resulting film. Solid samples can be mixed with dry KBr and pressed into a thin disk that is put into the beam.

51. IR photons have insufficient energy for the use of a phototube as a detector. A frequently used detection device in IR spectroscopy is a thermocouple that utilizes the temperature increase caused when IR radiation is absorbed by the detector.

52. The spectrum of CO_2 and H_2O in air gives a complex baseline that can be eliminated by using an instrument with a double-beam design. In more modern instruments, this background signal is stored in a computer and subtracted electronically from the signal for a sample.

53. In Fourier transform infrared (FTIR) spectroscopy, an entire set of IR wavelengths falls on the sample simultaneously after going through an interferometer. The resulting signal (as a function of time or position of the mirror in the interferometer) is then converted into a conventional spectrum of signal intensity versus frequency or wavenumber by using a mathematical operation known as a Fourier transform.

54. An experimental IR spectrum can be compared to large sets of spectra in a computer library. If the substance is not in the library, much can still be inferred about the structure of the sample based on knowledge of the wavelengths that are absorbed by various functional groups. Wavelengths of absorption and relative peak heights are compared with known samples for further help in identification. The large number of peaks in an IR spectrum, and their relationship to the various vibrational modes of functional groups, is what makes IR spectroscopy useful in this type of analysis.

55. A correlation chart for IR spectroscopy is a table of wavelengths (or wavenumbers) at which various functional groups will absorb in the IR region. This type of chart is useful in identifying the types of functional groups that can give rise to the peaks seen in an IR spectrum.

56. Double bonds absorb at about 1650 cm^{-1} (6.0 μm), while alkanes do not absorb here.

57. Ketones, such as acetone, show strong absorption near 1700 cm^{-1}. There is no such absorption present in this case, so the solvent must be the hydrocarbon mixture.

58. Triple bonds show a sharp peak at about 2200 cm^{-1} and double bonds at about 1650 cm^{-1}, so this compound must contain a triple bond.

59. Actual books of spectra were formerly used to compare IR spectra, but now most IR spectrometers have extensive libraries in computer files that can be compared rapidly with an experimental spectrum to see which of the reference spectra provide a good match. This approach can be used with IR spectroscopy to quickly identify possible compounds that might be represented by the IR spectrum.

60. Fluorescence is used to describe light that is emitted by a sample after it has become electronically excited by absorbance of a photon, with the light emission being due to a "spin-allowed" transition (such as a singlet–singlet transition). Fluorescence occurs when a molecule is excited electronically, and then the molecule very rapidly loses its recently acquired energy as a photon through this type of spin-allowed transition.

61. Molecules which fluoresce are often fairly rigid, flat molecules containing conjugated multiple bonds or aromatic groups.

62. Electronic excitation usually takes a molecule from the ground state electronic state (either at its ground vibrational state or an excited vibrational level) to a vibrationally excited state of an excited electronic state. The molecule then loses vibrational energy prior to fluorescence, during which it drops to a vibrationally excited state of the electronic ground state through the emission of a photo. Afterwards, the remaining vibrational energy is lost

upon returning to the vibrational ground state of the electronic ground state. Thus, the photon released by means of fluorescence will have less energy than the initially absorbed photon.

63. At low concentrations of analyte, the fluorescence intensity is linearly related to the concentration. At higher concentrations, the fluorescence intensity becomes independent of concentration.

64. Phosphorescence is the emission of a photon from an excited triplet state falling in energy to a ground singlet state.

65. Both involve emission of a photon from an electronically excited molecule that became excited by absorbing a photon of sufficient energy. The difference is that fluorescence involves a molecule in an excited singlet state dropping to a ground singlet state and phosphorescence involves a molecule in an excited triplet state dropping to a ground singlet state. The triplet→singlet transition in phosphorescence is "forbidden", which makes it a much slower process than the fast singlet→singlet transition in fluorescence.

66. The intensity of emitted light is directly related to the concentration of the emitting species in phosphorescence at low analyte concentrations. It should be noted that phosphorescence is much less frequently used in analysis than is fluorescence. Usually phosphorescence is only observed if the sample is cooled to low temperatures (or is immobilized on a solid matrix, in some cases).

67. Chemiluminescence occurs when a chemical reaction yields a product that is in an electronically excited state, and then loses its energy as a photon.

68. Bioluminescence is a special kind of chemiluminescence in which the chemical reaction takes place within a living organism. The most common examples are the light emitted by fireflies and glow worms.

69. A spectrofluorometer has a light source, a grating monochromator to select the excitation wavelength, a sample container (typically made of fused silica on all four sides), a second monochromator (mounted at 90° to the incident light) used to select the wavelength of the fluoresced light, and a detector.

70. A simple fluorometer has a light source, a set of filters to select an appropriate band of excitation wavelengths, a sample holder (which is usually a silica test tube), a slit mounted at 90° to the path of the incident radiation, and a detector. This design is much less expensive than a spectrofluorometer, but it is useful for doing analyses of well-characterized samples.

71. An excitation spectrum results from varying the wavelength of excitation while measuring fluorescence intensity at a fixed wavelength. This spectrum is closely related to an absorption spectrum. An emission spectrum results from fixing the excitation wavelength while scanning the fluoresced wavelengths.

72. Phosphorescence usually occurs only at low temperatures, so the capability of working at low temperatures must be built into the spectrometer. Also, phosphorescence occurs slowly, whereas fluorescence is quite rapid. Both fluorescence and phosphorescence frequently occur simultaneously, so the difference in speed is important in separating these two types of luminescence. To do this, a burst of excitation radiation is placed onto the sample and the luminescence is measured as a function of time. The rapid luminescence is fluorescence and the slower luminescence is phosphorescence.

73. A luminometer is an instrument for measuring chemiluminescence. The reactants in the chemical reaction are mixed rapidly in the vicinity of the detector, and the emitted light intensity is measured as a function of time.

74. Inspection shows that the unknown sample fluoresces in the linear region of the calibration curve, so the simple relation $I_F = k\,C$ can be used to describe this response.

    $k = (32.0)/(8.0 \times 10^{-5}) = 4.0 \times 10^5$

    $C_{unknown} = (25.8)/(4.0 \times 10^5) = 6.45 \times 10^{-5}\,M$

    If the unknown had fluoresced at a level greater than 32.0 units, a nonlinear fit would have to be used to find the concentration of the unknown sample.

79. At 400 s, the reaction has gone to completion because when all the NAD^+ has been converted to NADH, and the absorbance will be $(6000)(1)(2.0 \times 10^{-5}) = 0.120$.

    Lactic acid is in large excess, so its concentration is effectively unchanged during this reaction. LDH is a catalyst, so its concentration doesn't change either. The rate law for this reaction then becomes pseudo-first order and takes on the following form.

    Rate = $(2.4 \times 10^6)(1.0 \times 10^{-3})[LDH][NAD^+] = (2.4 \times 10^3)[LDH][NAD^+]$, where the term 2.4×10^3 has apparent units of $M^{-1}\,s^{-1}$.

    A plot of $\log(A_e - A)$ versus time for this analysis gives a linear plot that has a best-fit equation of $y = -0.0075\,x - 0.9208$.

    The observed rate constant is given the negative value of the slope, or equals $0.0075\,s^{-1}$. Thus, $(2.4 \times 10^3\,M^{-1}\,s^{-1})[LDH][NAD^+] = (0.0075\,s^{-1})[NAD^+]$ or $[LDH] = (0.0075\,s^{-1})/(2.4 \times 10^3\,M^{-1}\,s^{-1}) = 3.1 \times 10^{-6}\,M$

80. The absorbance reaches a maximum when the concentrations of U and R are equal, which indicates that these reactants combine in a 1:1 stoichiometry as they form product P.

CHAPTER 19: ATOMIC SPECTROSCOPY

1. Atomic spectroscopy refers to the measurement of the wavelength or intensity of light that is emitted or absorbed by free atoms. To do this type of measurement, we first need to convert the analyte into atoms from its original state.

2. In atomic emission spectroscopy, an excited atom gives off a photon as it drops to the ground state. In atomic absorption spectroscopy, a photon is absorbed by an atom in the ground state to take it up to an excited state. Both of these methods use spectroscopy and require the analyte to be in the form of free atoms.

3. A molecule can exhibit energy changes related to vibration and rotation in addition to electronic changes, whereas an atom can show only electronic changes. As a result, the spectrum for an analyte in atomic spectroscopy is much simpler than a typical spectrum seen in molecular spectroscopy.

4. Hydrogen atoms contain only one proton and one electron and can undergo only electronic transitions. A hydrogen molecule has two hydrogen atoms covalently bound together that can vibrate and rotate in addition to undergoing electronic transitions.

5. The relative intensities of the several emission lines from excited states of the hydrogen atom reflect the populations of the excited states that, in turn, are controlled by the surface temperature of the star. The light that is emitted by the stars can also be used in a similar manner to determine the types of others elements that are present and their relative composition, thus providing information on the age of the star.

6. Atomic emission spectroscopy and atomic absorption spectroscopy

7. Flame emission spectroscopy is a type of atomic emission spectroscopy in which a flame is used as the sample atomization and excitation source. Shortly after Robert Bunsen had invented the burner that bears his name, he observed different colors of the flame when he put compounds or solutions of different metallic elements into the flame. When mixtures were put into the flame, it became necessary to use a prism to separate the colors. These separated colors could be observed visually or captured on a photographic plate to give a permanent record. This was the beginning of flame emission spectroscopy.

8. The light emitted from each element in a mixture is passed through a monochromator to separate the wavelengths characteristic of each element. This spectrum is recorded and the intensity of the lines of each element is proportional to its concentration in the sample.

9. Rutherford put a radioactive sample into an evacuated tube that could be subjected to a high electrical voltage to excite any atoms in the tube. At first, no emission was seen. As time went by, helium appeared in the tube and was identified by its emission spectrum through the use of atomic emission spectroscopy.

10. Atomization refers to converting the sample, which is usually in an aqueous solution, into free atoms. This process is necessary to any kind of atomic spectroscopy.

11. a) Desolvation refers to the evaporation of solvent from a droplet of solution in the flame to leave behind a cluster of tiny particles containing the analyte and nonvolatile materials from the sample.

    b) Volatization is the thermal formation of a gas-phase form of the analyte.

c) Dissociation is the process in the breaking of ionic or covalent bonds of the analyte to form free atoms.

d) Excitation in atomic spectroscopy is the elevation of a ground state atom to one of its excited states.

e) Ionization in atomic spectroscopy is the loss of an electron from a free atom to form an ion.

12. Desolvation, volatilization, and dissociation are necessary for both atomic emission spectroscopy and atomic absorption spectroscopy. Excitation is necessary for atomic emission spectroscopy, but it is not needed for atomic absorption spectroscopy. Ionization is undesirable for both atomic emission spectroscopy and atomic absorption spectroscopy because it converts some of the atoms into ions and decreases the resulting signal from the atoms.

13. Typically, a fraction of a millisecond is all the time that elapses between when a droplet is aspirated into the flame until it exits from the top of the flame. Desolvation and the processes involved in sample atomization must all occur within this time limit.

14. Droplet volume = $(4/3)(3.14)\{(1.5 \times 10^{-4} \text{ cm})/2\}^3 = 1.77 \times 10^{-12}$ mL or 1.77×10^{-15} L

    Mass of $CaCl_2$ = $(2.5 \times 10^{-5}$ mol/L$)(111$ g/mol$)(1.77 \times 10^{-15}$ L$) = 4.9 \times 10^{-18}$ g

15. Hydrogen has only one electron, so electron-electron repulsion energies are nonexistent, whereas all other atoms have more than one electron.

16. When the atom is heated, its outer 3s electron can be excited into a 3p orbital. This requires energy. After a brief time ($\sim 10^{-8}$ s), the electron returns to the lower 3s orbital to

form a ground state atom again. The drop from 3p to 3s orbital represents a decrease in the electron's energy. This energy is released as a photon having an energy that corresponds to a wavelength at 589 nm.

17. $\Delta E = (6.62 \times 10^{-34} \text{ J·s})(3.00 \times 10^8 \text{ m/s})/(589.0 \times 10^{-9} \text{ m}) = 3.37 \times 10^{-19}$ J

$\upsilon = (3.00 \times 10^8 \text{ m/s})/(589.0 \times 10^{-9} \text{ m}) = 5.09 \times 10^{14} \text{ s}^{-1}$ (or Hz)

18. a) These are all weaker emissions because the initial state in the transition is at a higher energy than the lowest energy (3p) excited state. These higher energy states are less likely to be populated by the temperature of the flame.

b) 4p → 3s: $\upsilon = (3.00 \times 10^8 \text{ m/s})/(330 \times 10^{-9} \text{ m}) = 9.09 \times 10^{14} \text{ s}^{-1}$

$E = 6.02 \times 10^{-19}$ J

5p → 3s: $\upsilon = 1.05 \times 10^{15} \text{ s}^{-1}$

$E = 6.97 \times 10^{-19}$ J

3d → 3p: $\upsilon = 3.67 \times 10^{14} \text{ s}^{-1}$

$E = 2.43 \times 10^{-19}$ J

4s → 3p: $\upsilon = 2.63 \times 10^{14} \text{ s}^{-1}$

$E = 1.74 \times 10^{-19}$ J

c) $\Delta E_{5p-3d} = E_{5p-3s} - (E_{3d-3p} + E_{3p-3s})$

$= \{6.97 - (2.43 + 3.37)\} \times 10^{-19} = 1.17 \times 10^{-19}$ J

19. Some factors that contribute to the intensity of light emission by a sample during atomic spectroscopy are the temperature, the number of atoms in the flame at any one time, the probability of the transition occurring, and the difference in energy that occurs during this transition.

20. The use of more than one type of flame gives greater flexibility in terms of the elements that can be examined by atomic spectroscopy. Most metallic elements can be measured while using an acetylene/air flame. Some elements, however, form refractory oxides if a flame with excess oxygen is used. In these cases, it is advantageous to use nitrous oxide as an oxidant. In any case, the flame must be hot enough to cause excitation of a significant fraction of the atoms.

21. a) 2267 K b) 3342 K c) 3080 K

22. The temperature in different parts of the flame can differ by a large extent. The composition of the flame also differs dramatically in different places. To achieve the greatest sensitivity and repeatability of a measurement in atomic spectroscopy, one must find the best part of the flame to use in this analysis.

23. Figure 19.2 shows the general regions of a flame and gives the general locations of the indicated regions in the flame.

24. The Boltzmann equation relates the population of atoms in an excited state to the temperature and to the difference in energy that is present between the two states that are being compared. The greater the temperature, the greater the population of excited atoms, and the greater the intensity of the light that will be emitted by these atoms.

25. Based on the Boltzmann equation, the fraction of sodium atoms in the 3p versus 3s orbitals at these temperatures will be 1.5×10^{-5} and 8.8×10^{-4}, respectively. As a result, the enhancement in signal intensity due to just this effect will be $(8.8 \times 10^{-4})/(1.5 \times 10^{-5})$, which is a 59-fold increase in intensity.

26. A propane-air flame has a typical temperature of around 2200 K and an inductively coupled plasma has a typical temperature of 7500 K. The much higher temperature of the plasma compared to the flame causes a much greater fraction of atoms to be excited in the plasma, which results in a greater intensity of emitted light. Substituting these temperatures into the Boltzmann equation gives $N_i/N_0 = 1.51 \times 10^{-5}$ at 2200 K and 3.85×10^{-2} at 7500 K, which represents a relative enhancement in signal of 2550-fold when going from 2200 K to 7500 K. Similar results are obtained when using other temperatures in this range (e.g., 2267 K as a maximum temperature for a propane-air flame and 6000–10,000 K as the temperatures within an ICP).

27. Emission, absorption, and fluorescence can also be used to measure analytes through the use of atomic spectroscopy. In the use of emission, the sample atoms give off light that is produced following the thermal excitation of these atoms. The signal is linearly proportional to the number of sample atoms in this case. In the use of absorption, the ground state atoms absorb light at the proper wavelength. The absorption signal is related to the logarithm of the number of sample atoms. In the use of fluorescence, the atoms give off light after these atoms have been excited by absorbing a photon. The fluorescence signal is proportional to the number of sample atoms.

28. $[Cu^{2+}] = (1.00 \text{ ppm})(12.7 - 0.2)/(18.4 - 0.2) = 0.687$ ppm

29. Response ratio is $35.5/46.7 = 0.760$

    Lithium concentration in unknown = $(30.0 \text{ ppm})(5.0 \text{ mL}/55.0 \text{ mL}) = 2.73$ ppm

    $54.5/[Na^+] = 0.760(21.3/2.73)$

    or $[Na^+] = 54.5(2.73/21.3)/(0.760) = 9.19$ ppm in the mixed solution

Concentration of Na^+ in original solution = (9.19 ppm)(55 mL/50 mL) = 10.1 ppm

30. $A_{Unknown} = A_{Std}(C_{Unknown}/C_{Std}) = (0.305)(0.687/1.00) = 0.209$

31.

Sample Zn Conc. (ppm)	% Transmittance	Absorbance
0.00	89.5	0.048
0.50	67.4	0.171
1.00	50.8	0.294
2.00	28.8	0.541
5.00	5.27	1.278
Unknown	35.6	0.448

A plot of absorbance versus zinc concentration is linear and has a best-fit line of $y = 0.246\ x + 0.0482$ ($r^2 = 0.999$). Based on this best-fit line, the concentration of zinc in the sample is $(0.448 - 0.0482)/(0.246) = 1.62$ ppm.

32. $(0.386)/(0.497) = [Zn^{2+}]/([Zn^{2+}] + 0.500) = 0.7766$

$[Zn^{2+}](1 - 0.7766) = 0.7766\ (0.500) = 0.3883$

$[Zn^{2+}] = 1.74$ ppm in the diluted solution and 3.48 ppm in the original solution

33. Atomic fluorescence spectroscopy measures fluorescence that is given off by excited atoms. Atomic fluorescence and atomic emission both involve excited atoms dropping to their ground state while emitting a photon of the energy difference between the two states. The two methods differ in the means by which excitation occurs. In atomic emission, the atoms are excited thermally. In atomic fluorescence, the atoms are excited by absorbing a photon of the energy needed to cause the excitation.

34. While a constant flow of solution containing the analyte is sprayed into the flame, the exciting light can be pulsed on and off rapidly, such as through the use of a signal chopper. When the light beam is blocked, only light given off by thermally excited atoms will detected. When the light beam is not blocked, atoms are excited both thermally and by the absorption of light; they then give off fluorescence that will be measured. The difference in these two signals is the light due to fluorescence.

35. $[Mn]_{Unknown} = [Mn]_{Std} (17.4 - 3.5)/(28.5 - 5.7) = (12.0)(13.9/22.8) = 7.3$ ppm

36. When measuring a very low concentration analyte by atomic fluorescence spectroscopy, the intensity of light when the chopper is open compared to when it is closed is the difference in two small numbers. In contrast, in atomic absorption spectroscopy, very low concentration analytes give intensities that are the difference in two large numbers. The latter is always more prone to noise and will lead to a worse limit of detection.

37. The design of this type of burner is given in Figure 19.4. A laminar flow burner is also called a "premix" burner. The sample solution is aspirated into a nebulizer, from which an aerosol of droplets of sample is dispersed into a mixture of fuel and oxidant. This mixture passes into the actual burner, where it is ignited in a slot burner that is usually 10 cm in length.

38. Flow =

    (1 mL/min)(2 ng/mL)(1 min/60 sec)(1 mol/65.39 g)(10^{-9} g/ng)(6.02 × 10^{23} atoms/mol)

    = 3.0 × 10^{11} atoms/s (or 3.0 × 10^{12} atoms in 10 s)

39. The mixture of fuel and oxidant fills the burner chamber and can explode if the flow rate is insufficient. This problem is avoided by being sure that the flow rate is greater than the rate of combustion. Premix burners are designed with a steel cable connecting the burner head to the chassis, so that if flashback does occur, the burner head won't fly into the air where it can damage itself, the surroundings, or the operator. A blowout plug at the base of the burner further helps to avoid problems from flashback.

40. The general design of a hollow cathode lamp is shown in Figure 19.5. This light source consists of a quartz tube that is filled with a low pressure of an inert gas such as argon or neon. One electrode (the cathode) is a small, open-ended cylinder made from the element that is to be measured. The other electrode is an inert metal. A high voltage ionizes some of the gas in the tube and the resulting ions are accelerated to the electrically negative cathode. These ions strike the cathode's surface and sputter atoms off of the surface. Some of these atoms are also excited by the collision and emit light at their characteristic wavelength. This light exits through the end of the tube and goes to the monochromator, flame, and then on to the detector.

41. a) A multielement hollow cathode lamp is made from either an alloy containing all the desired elements or a cathode made up of strips of the several elements.

    b) Each element emits light characteristic of itself. The many wavelengths in such a lamp are separated by a monochromator, so that each element can be determined separately from the others.

42. Absorption lines of atoms are very narrow, so the emission line from the lamp must be even narrower. In this case, the measurement will fit the assumption of monochromatic

light that is made in Beer's law. The hollow cathode lamp is used to emit light that is exactly at the right energies for absorption by the desired atoms.

43. The Heisenberg uncertainity principle says that there is an unavoidable uncertainty in the measurement of the energy of a system. This results in the emission wavelengths of a group of emitting atoms that are slightly different and that produce a slightly broadened emission line.

   The Doppler effect results from the atoms in a flame being in constant, rapid motion in all directions. Atoms that emit their photon when going toward the detector appear to be of higher energy than those going away from the detector, so the emission line is broadened even more than expected from the uncertainty principle.

44. The general design of a graphite furnace is given in Figure 19.6. This device consists of a cylinder of graphite through which the beam of light passes from the hollow cathode lamp and onto the detector. The cylinder can be heated electrically by resistive heating to atomize a sample placed on a platform in this cylinder.

45. The major advantage of using a graphite furnace is that the furnace contains the sample atoms for a much longer time than does a flame source, giving each atom more opportunities to absorb photons. A much smaller amount of sample is thus consumed and required for the analysis. The major disadvantage of the graphite furnace is that the precision is not as good as it is for a laminar flow instrument, which uses more sample by providing a steady signal for several seconds.

46. An excess of oxidant assures that the fuel (e.g., acetylene) is totally converted into CO_2

and H_2O and does not result in soot, which would make an absorption measurement impossible due to the light scattering off the soot particles. An excess of oxygen, however, can cause some elements to form refractory oxides that will not atomize in the brief time they spend in the flame.

47. Nitrous oxide is sometimes used instead of oxygen as an oxidant for a flame. In this case, excess oxidant will not react with analyte atoms to form refractory oxides. The high temperature of the nitrous oxide flame also helps to break apart refractory oxides.

48. The beam of light must pass through the portion of the flame that has the greatest concentration of atomized analyte. This position will depend on the type of flame that is being used and on the analyte.

49. If flow rate, fuel or oxidant pressure, or element being examined changes, it is best to reposition the light path to optimize the analysis because any of these changes can result in the highest concentration of free atoms to be in a different part of the flame.

    To carry out such an optimization, aspirate a standard solution of the analyte into the flame. Move the burner in all directions until the highest absorbance signal is measured. Then aspirate a blank solution and set the absorbance to zero. Rerun several standards before doing the unknown samples.

50. A refractory oxide is a solid oxide of a metal that does not decompose on heating to the temperature of the flame and in the amount of time that this compound will be present in the flame. Two examples of refractory oxides are Al_2O_3 and MoO_2. If analyte atoms form such refractory compounds, they will not exist as free atoms and will not undergo

absorption as atoms. Low results will occur when such compounds form for an analyte that is to be examined by atomic absorption spectroscopy.

51. After the solvent has evaporated, insoluble and refractory calcium phosphate will form and cause low results. To minimize this problem, one can add La^{3+} (to form an even less soluble lanthanum phosphate) or add EDTA (to form CaEDTA, which prevents precipitation of the calcium phosphate). In the second case, the EDTA burns away in the flame and results in atomic calcium that can then be measured.

52. A releasing agent is a chemical added to a sample or standard and reacts with a precipitating agent, such as phosphate or sulfate, to form an insoluble substance. This allows the analyte to atomize without forming a refractory compound. A classic example is the addition of lanthanum ions during a determination of calcium ions. Lanthanum ions are used because they form a stable phosphate salt. Also, lanthanum is almost never an element of interest in such an analysis.

53. A protective agent is a chemical added to a sample or standard and reacts to prevent the formation of refractory compounds of the analyte on evaporation of a solvent in the flame. EDTA and other organic complexing agents form complexes that can prevent the formation of insoluble, refractory compounds of many elements with phosphate, sulfate, or hydroxide ions.

54. a) $A_{std} = k\,[Ca^{2+}]_{Std}$ so $k = (0.78)/(10.0) = 0.0768$

$A_{Soln} = k\,[Ca^{2+}]$ so $[Ca^{2+}] = (0.543)/(0.0768) = 7.07$

The original calcium concentration in milk = $(7.07)(50/5)(100/10) = 707$ ppm

b) The EDTA was added as a protective agent to assure calcium phosphate would not form in the flame. Milk is rich in phosphate, which is another concern. If EDTA had not been added, there is a good chance that a low result would have been found.

55. Hydrides of some elements are molecular and quite volatile. These hydrides can also easily be destroyed in a flame, thus releasing free atoms of the desired analyte for measurement.

56. The formation of ions from atoms will reduce the number of atoms that are available for measurement by atomic spectroscopy. The higher the temperature of the atomization source, the greater the chance that ionization will occur.

57. An ionization buffer is an element added to a sample or standard that is easily ionized to supply a constant supply of free electrons in a flame or atomization source. This high concentration of free electrons suppresses ionization of atoms for the analyte.

58. Potassium is fairly easily ionized. If lithium is present, the lithium also ionizes to form a constant concentration of free electrons that suppresses the ionization of potassium. This effect means that there are more neutral potassium atoms in the presence of lithium than in its absence, so the intensity of emission by potassium atoms will increase.

59. Light emitted by excited analyte atoms is at the same wavelength as light from the hollow cathode; this emitted light would add to the intensity of the transmitted light. This type of interference can be dealt with by using a signal chopper or other means of signal modulation. Light emitted by the flame itself or other elements in the aspirated solution is of a different wavelength and can be eliminated by using a monochromator.

60. A "signal chopper" is a rotating fan that is placed into the light path. This chopper interrupts the beam of light from the hollow cathode, but it does not interrupt the light emitted by atoms in the flame. The difference in the signal with the beam that is open versus blocked gives the signal that is associated with the transmitted light.

61. See Figure 19.9 for an example of this type of plot. In each of the listed cases, the top of the square wave represents the intensity observed when the chopper is open and is the sum of transmitted and emitted light. The bottom of each square wave occurs when the chopper is closed and corresponds to only the intensity of the emitted light. The height of the square wave is, therefore, the intensity of transmitted light. When the concentration of an analyte is too high, nearly all the light from the lamp is absorbed, so little is transmitted but a large amount of light is emitted. In this situation, the difference is between two large values and is difficult to interpret. When the concentration is too low, very little light is absorbed, so nearly all the light from the lamp is transmitted and little light is emitted. This sounds like a favorable situation, but it is quite similar to the blank solution to which it must be compared to calculate the transmittance and absorbance.

62. $I = 24.5 - 16.6 = 7.9$ $I_0 = 86.5 - 0.4 = 86.1$

    $T = 7.9/86.1 = 0.0917$ $A = -\log(0.0917) = 1.038$

63. The parts that are the same are the nebulizer, flame, monochromator, and detector. The components needed for atomic absorption, but not for emission, are a hollow cathode lamp source and a chopper. It is possible to determine several elements at once by using atomic emission by monitoring wavelengths of each element. However, this approach is not possible in atomic absorption because different wavelengths must be available from

the light source.

64. Inductively coupled plasma atomic emission spectroscopy (ICP–AES) is conducted at a much higher temperature than flame emission spectroscopy. Also, the excited atoms are in an argon environment, so no oxides will form from an atomized sample reacting with excess oxidant. (Note: Some oxides are present due to the water that is also present.)

   The plasma in ICP-AES uses argon gas, while flame emission often uses acetylene and oxygen as the fuel/oxidant mixture. These two methods are similar in that emission from excited electronic states of free atoms is measured.

65. A radio frequency current in a rugged helical conductor is passed around a silica tube through which argon gas is flowing. A spark is struck in the argon and a few atoms are ionized. This initiates the formation of a plasma, a region in which the majority of atoms exist as free ions in the presence of free electrons. The radio frequency current causes movement of these electrons (and, to a smaller extent, the ions), which leads to collisions and an increase in temperature.

66. The ICP-AES has the lowest detection limit because it is by far the highest temperature in this group of methods, which causes the greatest fraction of analyte atoms to be excited for measurement. The graphite furnace is not as hot as the ICP-AES, but the use of this device provides a low detection limit because it contains the analyte atoms for a far greater time than is possible with flame-based instruments.

67. ICP-AES can measure several elements either simultaneously or sequentially. If these elements are to be measured simultaneously, there must be a detector available for each

element (e.g., an array detector). If these elements are to be measured sequentially, the instrument is programmed to monitor one wavelength for a short time and then to move on to the conditions of the next elements. Sequential measurements take a longer time, but they have a major advantage in that any set or order of elements can be detected. Different emission lines of different sensitivities for the same element can be monitored in either mode.

70.   a) $\Delta\lambda_D = (7.16 \times 10^{-7})(589)(3000/23)^{1/2} = 0.0048$ nm

      b) 0.0027 nm

      c) 0.0068 nm for 589 nm   0.0038 nm for 330 nm

CHAPTER 20: AN INTRODUCTION TO CHEMICAL SEPARATIONS

1. A chemical separation is a method that involves the complete or partial isolation of one chemical from another in a mixture of two or more substances. Other common separation methods include precipitation, filtration, centrifugation, and distillation.

2. a) Distillation is a phase-based separation that uses two or more chemical or physical environments to separate chemicals.

    b) Centrifugation is a rate-based separation that uses an approach where chemicals have different rates of travel through a system.

    c) Extraction is a phase-based separation that uses the distribution of analytes between two phases to remove analytes from interferences, to transfer analytes to alternative solvents, or to concentrate analytes prior to their analysis.

    d) Chromatography has components of both phase- and rate-based separations. It is a separation technique in which the components of a sample are separated based on how they distribute between two chemical or physical phases, one of which is stationary and the other of which is allowed to travel through the separation system.

3. a) Chemical separations can be employed to remove an analyte from other agents that might interfere with its measurement; b) a chemical separation can be utilized to place an analyte in a matrix that is more compatible with the analysis method; c) a chemical separation can be used to adjust the sample volume or concentration to a level that is more compatible with the measurement method; and c) a chemical separation can be used to examine several analytes in a sample.

4. a) Filtering sand away from a river water sample for analysis of pollutants in the water will remove sample components that may interfere in the analysis.

    b) Precipitation of barium sulfate from a sample for a gravimetric analysis of barium will isolate barium from the rest of the sample for later analysis.

    c) Evaporation of the water away from a protein solution will make it possible to change the matrix for the remaining protein and to possibly obtain a more concentrated solution of this protein for analysis.

5. An extraction is a separation technique that makes use of differences in the distributions of solutes between two mutually insoluble phases. This technique is often used in sample preparation to remove analytes from interferences, to transfer analytes to alternative solvents, or to concentrate analytes prior to their analysis.

6. A liquid-liquid extraction is an extraction that is conducted with two immiscible liquids, one of which is the sample. The two phases used for the extraction are both liquids. After the analyte has distributed between the two liquids, these phases are allowed to separate. After these two liquids have formed distinct layers, the phase now containing the analyte is removed for later use or analysis.

7. a) The partition constant is an equilibrium constant that gives the ratio of the activities for a solute in two phases, as this solute distributes between these phases at a given pressure and temperature.

    b) The partition ratio is an equilibrium constant that gives the ratio of the concentrations for a solute in two phases, as this solute distributes between these phases at a given pressure and temperature.

c) The distribution ratio is the ratio of the analytical concentrations for a solute in two phases, as this solute distributes between these phases at a given pressure and temperature.

8. Benzene$_{\text{Water}}$ ⇌ Benzene$_{\text{Chloroform}}$

$$K_D° = \frac{a_{\text{Benzene,Chloroform}}}{a_{\text{Benzene,Water}}} \qquad K_D = \frac{[\text{Benzene}]_{\text{Chloroform}}}{[\text{Benzene}]_{\text{Water}}}$$

The values of $K_D°$ and K_D are related by using the activity coefficients for benzene in each of the two phases.

9. For a weak monoprotic acid (HA), the value of D_c in Equation 20.4 as the pH is decreased will approach $K_{D,HA}$. As pH is increased, the value of D_c will become approximately equal to the term $K_{D,HA}[H^+]/K_a$. This change occurs because of the change in the relative size of the term $K_a/[H^+]$ as the pH is altered.

10. $D_c = K_{D,H2A}[H^+]^2/([H^+]^2 + K_{a1}[H^+] + K_{a1}K_{a2})$

    This equation differs from that for a monoprotic acid through the inclusion of additional terms and acid dissociation constants in the denominator, which allow for the loss of several hydrogen ions by the same polyprotic acid.

11. $D_c = K_{D,B}/(1 + K_b[H^+]/K_w)$ or $D_c = K_{D,B}/(1 + K_b/[OH^-])$

12. The value of D_c (or $K_D°$ and K_D) reflects the ability of a solute to enter one phase versus another. The greater the difference in the distribution ratios for two constants (especially if one of these distribution ratios is small and the other is large), the easier it will be to separate these solutes by an extraction.

13. A liquid-liquid extraction uses two immiscible liquids. A liquid and solid are the two phases present during a liquid-solid extraction. A supercritical fluid can also be used as an extracting phase with various samples, including solids, in a supercritical fluid extraction.

14. An extraction based on partitioning involves a chemical entering both of the two phases being employed for the extraction. An extraction that uses adsorption involves the interactions of solutes with the surface of a solid.

15. Single-step extractions use one step to place a sample in contact with the extracting phase.

16. The phase ratio is a ratio that gives the relative amount of one phase versus another in an extraction or chromatographic method. The phase ratio is one factor that will determine the relative moles of an analyte that will be extracted from one phase to another.

17. Two ways the amount of extracted analyte can be varied in a single-step extraction is by changing the phase ratio or by changing the distribution ratio.

18. $$f_{\text{Phase 1,1}} = \frac{1}{1 + 110\,(15.0\text{ mL}/200.0\text{ mL})} = 0.108\underline{1} \text{ (or 10.8\% still in the water sample)}$$

    $$f_{\text{Phase 2,1}} = 1 - 0.108\underline{1} = 0.891\underline{9} = 0.892 \text{ (or 89.2\% extracted)}$$

19. If the volume of pentane is changed to 25.0 mL:

    $$f_{\text{Phase 1,1}} = \frac{1}{1 + 110\,(25.0\text{ mL}/200.0\text{ mL})} = 0.067\underline{8} \text{ (or 6.8\% still in the water sample)}$$

    $$f_{\text{Phase 2,1}} = 1 - 0.067\underline{8} = 0.932\underline{2} = 0.932 \text{ (or 93.2\% extracted)}$$

Volume of pentane need to extract at least 99% of the chloroform:

$$f_{\text{Phase 2, 1}} = 0.99 \quad \text{and} \quad f_{\text{Phase 1, 1}} = 1 - 0.99 = 0.01$$

$$f_{\text{Phase 1,1}} = 0.01 = \frac{1}{1 + 110\,(V_2/200.0 \text{ mL})}$$

$$1/(0.01) = 100 = 1 + 110\,(V_2/200.0 \text{ mL})$$

$$V_2 = (100 - 1)(200.0 \text{ mL}/110) = 180 \text{ mL}$$

20. $f_{\text{Phase 2,1}} = 0.875 \quad \text{and} \quad f_{\text{Phase 1,1}} = 1.000 - 0.875 = 0.125$

$$f_{\text{Phase 1,1}} = 0.125 = \frac{1}{1 + K_{\text{ow}}\,(15.0 \text{ mL}/100.0 \text{ mL})}$$

$$1/(0.125) = 8.00 = 1 + K_{\text{ow}}\,(15.0 \text{ mL}/100.0 \text{ mL})$$

$$K_{\text{ow}} = (8.00 - 1)(100.0 \text{ mL}/15.0 \text{ mL}) = 46.7$$

21. $f_{\text{Phase 2,1}} = 0.89 \quad \text{and} \quad f_{\text{Phase 1,1}} = 1.00 - 0.89 = 0.11$

$$f_{\text{Phase 1,1}} = 0.11 = \frac{1}{1 + D_c\,(15.0 \text{ mL}/80.0 \text{ mL})}$$

$$1/(0.11) = 9.1 = 1 + D_c\,(15.0 \text{ mL}/80.0 \text{ mL})$$

$$D_c = (9.1 - 1)(80.0 \text{ mL}/15.0 \text{ mL}) = 43.2$$

22. If a sample is extracted several times with equal but separate volumes of a second phase, the fraction of the remaining solute that is removed with each extraction step will be the same. However, the overall amount of extracted solute will increase.

23. We first need to find the value of D_c for this extraction using the data given for a single extraction with 15.0 mL of ethyl ether and 100.0 mL of the sample.

$f_{\text{Phase 2,1}} = 0.743$ and $f_{\text{Phase 1,1}} = 1.000 - 0.743 = 0.257$

$$f_{\text{Phase 1,1}} = 0.257 = \frac{1}{1 + D_c\,(15.0\text{ mL}/100.0\text{ mL})}$$

$1/(0.257) = 3.89 = 1 + D_c\,(15.0\text{ mL}/100.0\text{ mL})$

$D_c = (3.89 - 1)(100.0\text{ mL}/15.0\text{ mL}) = 19.3$

We can now use this information to find the degree of extraction when using the same phases, but we now utilize three extractions with three fresh portions of 5.0 mL ethyl ether and a single 100.0 mL portion of the sample.

$$f_{\text{Phase 1,1}} = \{1/[1 + 19.3\,(5.0\text{ mL}/100.0\text{ mL})]\}^3 = 0.132$$

$f_{\text{Phase 2,1}} = 1.000 - 0.132 = 0.868$ (or **86.8% extraction**)

24. a) $f_{\text{Phase 1,1}} = \{1/[1 + 19.0\,(10.0\text{ mL}/50.0\text{ mL})]\} = 0.208$

   $f_{\text{Phase 2,1}} = 1.000 - 0.208 = 0.792$ (or **79.2% extraction**)

   b) $f_{\text{Phase 1,1}} = \{1/[1 + 19.0\,(10.0\text{ mL}/50.0\text{ mL})]\}^2 = 0.043$

   $f_{\text{Phase 2,1}} = 1.000 - 0.043 = 0.957$ (or **95.7% extraction**)

   c) $f_{\text{Phase 1,1}} = \{1/[1 + 19.0\,(10.0\text{ mL}/50.0\text{ mL})]\}^n = 0.01$

   We already know that two extractions will lead to 95.7% extraction. If n is increased to three, the following result is obtained.

   $$f_{\text{Phase 1,1}} = \{1/[1 + 19.0\,(10.0\text{ mL}/50.0\text{ mL})]\}^3 = 0.009$$

   $f_{\text{Phase 2,1}} = 1.000 - 0.009 = 0.991$ (or **99.1% extraction**)

   Thus, at least three extractions are needed under these conditions to give at least 99% extraction into the ether.

25. Side reactions can be used to convert an analyte to a form that is either easier or more difficult to extract. For example, the use of a low pH will make it easier to extract a monoprotic acid by placing this chemical mainly in its neutral HA form. The use of a high pH will make it easier to extract a monoprotic base, which will be mainly in its neutral B form under these conditions.

26. A back extraction is a type of extraction in which a solute is allowed to distribute from its extracting phase back into a fresh portion of its original solvent. As an example, a low pH could be used to extract a weak monoprotic acid into an organic solvent and away from many other sample components. A high pH can later be used to extract this same monoprotic acid back into an aqueous solution.

27. a) It is first necessary to find the value of $K_{D,HA}$ for this weak acid based on the data given for pH 3.00.

$$D_c = K_{D,HA}/(1 + K_a/[H^+])$$

$$75 = K_{D,HA}/\{1 + (10^{-6.21})/(10^{-3.00})\} \text{ or } K_{D,HA} = 75.05$$

At pH 4.00:

$$D_c = (75.05)/\{1 + (10^{-6.21})/(10^{-4.00})\} = 74.6$$

$$f_{\text{Phase 1,1}} = \{1/[1 + 74.6 \,(20.0 \text{ mL}/30.0 \text{ mL})]\} = 0.020$$

$$f_{\text{Phase 2,1}} = 1.000 - 0.020 = 0.980 \text{ (or 98.0% extraction)}$$

b) At pH 8.00:

$$D_c = (75.05)/\{1 + (10^{-6.21})/(10^{-8.00})\} = 1.20$$

$$f_{\text{Phase 1,1}} = \{1/[1 + 1.20 \,(20.0 \text{ mL}/30.0 \text{ mL})]\} = 0.556 \quad (55.6\% \text{ back extraction})$$

c) The overall fraction of the compound in the original sample that will be present in the fresh portion of water after the biochemist has performed the back extraction is $(0.556)(0.980) = 0.545$ (or 54.4% overall back extraction into water).

28. a) At pH 9.00:

$$D_c = K_{D,B}/(1 + K_b[H^+]/K_w)$$

$$= 210/\{1 + (10^{-9.52})(10^{-9.00})/(10^{-14.00})\} = 209.\underline{9} = 210$$

$$f_{\text{Phase 1,1}} = \{1/[1 + 210\,(10.0\text{ mL}/15.0\text{ mL})]\} = 0.007$$

$$f_{\text{Phase 2,1}} = 1.000 - 0.007 = 0.993 \text{ (or 99.3\% extraction)}$$

b) At pH 3.00:

$$D_c = K_{D,B}/(1 + K_b[H^+]/K_w)$$

$$= 210/\{1 + (10^{-9.52})(10^{-2.00})/(10^{-14.00})\} = 0.69$$

$$f_{\text{Phase 1,1}} = \{1/[1 + 0.69\,(10.0\text{ mL}/50.0\text{ mL})]\} = 0.88 \text{ (or 88\% back extraction)}$$

c) The overall fraction of the compound in the original sample that will be present in the fresh portion of water after the back extraction is $(0.88)(0.993) = 0.87$ (or 87% overall back extraction).

29. Weak acids can be extracted from a sample that has been adjusted to a low pH. Weak bases can be extracted from a sample that has been adjusted to a high pH. Neutral compounds can be first extracted with either the weak acids or weak bases and then separated from these other components by adjusting the pH to a high value to back extract the weak acids or to a low value to back extract the weak bases.

30. Complexation formation can be used to alter the extent of an extraction when an added ligand forms a neutral complex with a metal ion, making it possible to extract this ion into an organic solvent. Oxine is one ligand that can be employed for this purpose.

31. The purity of an extracted solute will decrease when an increased number of extractions are carried out on a sample because the amount of other solutes that will be extracted will increase. The recovery of the solute will increase under the same conditions because it will also have a higher degree of extraction.

32. a) Compound A, 84.2% extraction; compound B, 4.6% extraction

    b) Compound A, 97.5% extraction ($n = 2$); 99.6% extraction ($n = 3$); 99.9% extraction ($n = 4$); Compound B, 8.95% extraction ($n = 2$); 13.1% extraction ($n = 3$); 17.1% extraction ($n = 4$)

    c) The recovery of compound A increases, but its purity decreases as more extractions are conducted on the sample.

33. a) Compound A, 84.2% extraction ($n = 1$); 97.5% extraction ($n = 2$); 99.6% extraction ($n = 3$); 99.9% extraction ($n = 4$); Compound B, 32.4% extraction ($n = 1$); 54.3% extraction ($n = 2$); 69.2% extraction ($n = 3$); 79.2% extraction ($n = 4$)

    b) Compound B now has a higher degree of extraction, which will lead to more coextraction with Compound A.

    c) If Compound B is a weak acid or weak base, changing D_c by altering the pH might be one possible option. A larger difference in D_c between Compounds A and B is needed to improve the extraction of Compound A if multiple extraction steps are used. Another possibility is to use a countercurrent extraction.

34. A countercurrent extraction is a special type of extraction that uses multiple portions of both the original sample solvent and extracting phase. In this system, one phase is kept in a series of tubes or containers, each of which contains a fixed amount of the second phase. After equilibration, each portion of the second phase is moved down one tube. The sample is applied to the first tube of this system. As the components of this sample distribute between the two phases and are moved with the upper second phase, they travel from one tube to the next based on their distribution constants. A countercurrent extraction offers the advantages of having higher recovery versus a single-step extraction and better purity versus a normal multistep extraction.

35. A Craig apparatus uses a series of glass tubes that can hold a fixed portion of one phase while allowing the simultaneous transfer of each top phase to the next tube. The result of this process is that solutes partition between the phases in each tube, but eventually travel to the end of the apparatus where they are collected, creating a countercurrent extraction.

36. Chromatography is a separation technique in which the components of a sample are separated based on how they distribute between two chemical or physical phases, one of which is stationary and the other of which travels through the separation system.

37. The three main parts of a chromatographic system are the mobile phase, stationary phase and support. The mobile phase is the phase that is flowing through the column and causes sample components to move toward the column's end. The stationary phase is the fixed phase that is coated or bonded within the column and that is responsible for delaying the movement of compounds as they travel through the column. The support is the material onto which the stationary phase is coated or attached.

38. A column is the tube or enclosed container that holds the stationary phase for a chromatographic system. A chromatograph is an instrument that is used to perform chromatography.

39. Like extractions, chromatography uses two phases (a mobile phase and stationary phase) for isolating one chemical from one another. However, chromatography is also based on the different rates of travel that substances have through the system.

40. These terms all refer to different types of chromatography based on the nature of the mobile phase that is present. If a gas is the mobile phase, the method is called gas chromatography. If the mobile phase is a liquid, the technique is known as liquid chromatography. If a supercritical fluid is the mobile phase, the method is known as supercritical fluid chromatography.

41. All chromatographic methods can be classified based on their separation mechanism and type of stationary phase. For instance, the use of underivatized solid particles as the stationary phase (and a separation mechanism based on adsorption) in GC or LC produces the methods of "gas-solid chromatography" and "liquid-solid chromatography". The use of a liquid coating as the stationary phase (and partitioning as the mechanism of separation) produces the methods of "gas-liquid chromatography" and "liquid-liquid chromatography".

42. If a chromatographic column is packed with support particles that contain the stationary phase, the method is called packed-bed chromatography. If the stationary phase is placed instead directly onto the interior wall of the column, the method is known as open-tubular chromatography. If the support and stationary phase are present on a flat plane, the method

is known as planar chromatography.

43. One of the strengths of chromatography is its ability to take the individual components of a sample and isolate these from each other, so they can be more easily identified or measured. Chromatography is often required in analytical methods that deal with complex samples. Such samples are a challenge for many detection techniques, which may work well for simple solutions but not for a mixture of many chemicals. At the same time, chromatography requires other techniques to help monitor the passage of chemicals through the column and to identify or measure these chemicals. The result is a combination of separation plus detection that makes chromatography valuable as a method for analyzing either simple or complex samples.

44. A chromatogram is a graph of the response in chromatography that is measured by a detector at the end of the column as a function of the time or volume of mobile phase that is needed for elution.

45. a) The void time is the time required for a totally nonbinding substance to travel through the column during chromatography.

b) The void volume is the volume of mobile phase that it takes to elute a totally nonretained substance from a chromatographic system.

c) The retention time is the average time required for a given retained substance to pass through a chromatographic system.

d) The retention volume is the average volume of mobile phase required to pass a given retained substance through a chromatographic system.

46. Band-broadening is a process that occurs as chemicals travel through a chromatographic

system or other separation device, in which the width of the region that contains each chemical gradually becomes broader.

The number, position, and size of peaks in a chromatogram provide a fingerprint pattern or chemical profile that is characteristic of the sample being analyzed. The identity of an eluting substance can be determined in many cases by comparing the retention times or retention volumes of this substance to values that are obtained for standards. The size of the peak can also be used with results for standards to determine the amount of the compound that is present in the sample.

47. Void volume = (0.45 min)(10.0 mL/min) = 4.5 mL

    Total retention volume for 2-propanol = (3.01 min)(10.0 mL/min) = 30.1 mL

48. At 0.050 mL/min:

    Void time = (4.5 mL)/(0.050 mL/min) = 90. min

    Retention time = (20.5 mL)/(0.050 mL/min) = 410 min

    At 0.50 mL/min:

    Void time = (4.5 mL)/(0.50 mL/min) = 9.0 min

    Retention time = (20.5 mL)/(0.50 mL/min) = 41 min

49. Based on plot of the relative peak height versus atrazine concentration for the standards and the measured peak height of 5.13 for the unknown, the atrazine concentration in the unknown is 2.41 µg/L. This concentration is below the allowable limit.

50. By using 1-propanol as an internal standard, a plot can be made of the areas for Methanol/1-Propanol versus methanol concentration. From this plot, the methanol

concentration in the unknown is found to be 450 mg/L.

51. a) The retention factor (k) is a measure of solute retention in chromatography that is calculated using $k = t'_R/t_M$ or $k = V'_R/V_M$, where t'_R is the adjusted retention time of the solute, t_M is the column void time, V'_R is the adjusted retention volume for the solute, and V_M is the column void volume. This factor is independent of column size and flow rate and can be directly related to the strength of the interactions that are occurring in the columns.

b) The adjusted retention time (t'_R) is a measure of solute retention in chromatography, as calculated from the retention time t_R through the relationship $t'_R = t_R - t_M$. This factor is similar in properties to the retention time, but it includes a correction for the column void time.

c) The adjusted retention volume (V'_R) is a measure of solute retention in chromatography that is calculated from the retention volume V_R through the relationship $V'_R = V_R - V_M$, where V_M is the column void volume. This factor is similar in properties to the retention volume, but it includes a correction for the column void volume.

52. a) Peak 1: retention time = 2.4 min, retention volume = (2.4 min)(1.0 mL/min) = 2.4 mL

   Peak 2: retention time = 4.6 min, retention volume = 4.6 mL

   Peak 3: retention time = 5.1 min, retention volume = 5.1 mL

   Peak 4: retention time = 5.7 min, retention volume = 5.7 mL

   Peak 5: retention time = 6.4 min, retention volume = 6.4 mL

   b) Peak 1: adjusted retention time = 2.4 min – 2.0 min = 0.4 min

   adjusted retention volume = 2.4 mL – 2.0 mL = 0.4 mL

   Peak 2: adjusted retention time = 2.6 min, adjusted retention volume = 2.6 mL

   Peak 3: adjusted retention time = 3.1 min, adjusted retention volume = 3.1 mL

Peak 4: adjusted retention time = 3.7 min, adjusted retention volume = 3.7 mL

Peak 5: adjusted retention time = 4.4 min, adjusted retention volume = 4.4 mL

c) Peak 1: retention factor = (2.4 min − 2.0 min)/(2.0 min) = 0.2

Peak 2: retention factor = (4.6 min − 2.0 min)/(2.0 min) = 1.3

Peak 3: retention factor = (5.1 min − 2.0 min)/(2.0 min) = 1.5$\underline{5}$ = 1.6

Peak 4: retention factor = (5.7 min − 2.0 min)/(2.0 min) = 1.8$\underline{5}$ = 1.8

Peak 5: retention factor = (6.4 min − 2.0 min)/(2.0 min) = 2.2

53. If the average mobile phase flow rate is decreased from 1.0 to 0.75 mL/min, the value for k will be unchanged, the value for t_R will increase by (1.0 mL/min)/(0.75 mL/min) = 1.33-fold, and the value for V_R will be unchanged.

If the column length is increased from 15 to 25 cm, the value for k will be unchanged, the value for t_R will increase by (25 cm/15 cm) = 1.67-fold, and the value for V_R will increase by 1.67-fold.

54. a) TNT, t'_R = 3.00 min, V'_R = 4.50 mL; RDX, t'_R = 4.15 min, V'_R = 6.22 mL; Tetryl, t'_R = 5.36 min, V'_R = 8.04 mL; HMX, t'_R = 6.35 min, V'_R = 9.52 mL

b) TNT, k = 1.50; RDX, k = 2.08; Tetryl, k = 2.68; HMX, k = 3.18

c) The total and adjusted retention times would all increase by 1.5-fold, but the retention factors would not be affected.

55. The retention of an analyte in chromatography is usually determined by the intermolecular forces that occur between this analyte, the stationary phase and mobile phase. The greater these interactions are for the analyte with the stationary phase versus the mobile phase, the stronger the analyte will be retained in a column and the larger its retention time or

retention volume will be.

56. a) $P_M \rightleftarrows P_S$

   where P = phenol, M = 2,3,4-trimethylpentane, S = 1,2,3-tris(2-cyanoethoxy)propane

   b) $$K_D = \frac{[P]_S}{[P]_M}$$

   c) $$k = \frac{[P]_S V_S}{[P]_M V_M} = K_D (V_S/V_M)$$

57. The separation of two chemicals by chromatography always requires that these chemicals have some difference in their retention, as produced when these chemicals have different equilibrium constants for their distribution between the mobile phase and stationary phase. This difference in retention is usually a result of differences in these equilibrium constants, with larger differences making it easier to resolve these chemicals. It is also possible in some cases to have differences in the phase ratio for two analytes, which will also contribute to a difference in retention.

58. If the phase ratio is the same for all three compounds, then their partition ratios will have the same relative sizes as their retention factors, giving relative values of 2.32:4.58:7.89 or (if normalizing and giving the solute a relative value of 1.00) 1.00:1.97:3:40.

59. If all other conditions are kept the same, the retention factor will increase by 2.2-fold, which is also the extent that the phase ratio is changing.

60. Band-broadening is a process that occurs as chemicals travel through a chromatographic system or other separation device, in which the width of the region that contains each chemical gradually becomes broader. If the chromatographic system produces narrow peaks, it is said to have a "high-efficiency" or "high-performance". If it gives rise to broad peaks, it is described as having a "low-efficiency" or "low-performance". It is always preferable to have a system that produces narrow peaks because these narrow peaks will make it easier to separate compounds that have small differences in their retention.

61. a) The term w_b represents the width of a peak at its baseline level. This factor is easy to measure but is highly dependent on the nature of the analyte and the separation conditions.
b) The term w_h represents the width of a peak at half-height. This factor is also easy to measure but is highly dependent on the nature of the analyte and the separation conditions.
c) The term σ is the standard deviation of a peak. This term is more difficult to measure exactly but can be related mathematically to the various band-broadening processes that are occurring in the columns.
d) The term N represents the number of theoretical plates. This value is easy to calculate and is a more uniform measure of band-broadening and efficiency than peak widths, but this value still depends on column length.
e) The term H is the height equivalent of a theoretical plate. This value can be calculated from N and is independent of column length. This term can also be related mathematically to the various band-broadening processes that are taking place in the column.

62. The expressions in Equation 20.21 include substitutions for the standard deviation in terms of the baseline and width and half-height width that are true for only a Gaussian peak. The relationship given in Equation 20.20 does not make this assumption and can be used for

either a Gaussian or non-Gaussian peak.

63. Peak 3: $N = 16\ (t_R/w_b)^2 = 16(4.04\ \text{min}/0.22\ \text{min})^2 = 53\underline{96} = 5400$

$H = L/N = (25\ \text{cm})/(53\underline{96}) = 0.0046\underline{3} = 0.0046\ \text{cm}$

Peak 4: $N = 16(4.36\ \text{min}/0.24\ \text{min})^2 = 52\underline{80} = 5300$

$H = L/N = (25\ \text{cm})/(52\underline{80}) = 0.0047\underline{3} = 0.0047\ \text{cm}$

Peak 5: $N = 16(5.03\ \text{min}/0.27\ \text{min})^2 = 55\underline{53} = 5600$

$H = L/N = (25\ \text{cm})/(55\underline{53}) = 0.0045\underline{0} = 0.0045\ \text{cm}$

Peak 6: $N = 16(7.97\ \text{min}/0.43\ \text{min})^2 = 54\underline{97} = 5500$

$H = L/N = (25\ \text{cm})/(54\underline{97}) = 0.0045\underline{4} = 0.0045\ \text{cm}$

64. If either the number of theoretical plates is increased or the plate height is decreased, the result would be sharper peaks and a better resolution between each peak if all other conditions are kept constant.

65. Peak tailing occurs when a chromatographic peak has a sharper front edge than back. Peak fronting occurs when a chromatographic peak has a sharper back edge than front. These types of non-ideal effects might occur when you are working with an overloaded column, a column that contains large empty spaces, an analyte that has slow binding or release from the stationary phase, or a column that has multiple types of sites that interact with analytes.

66. The *A/B* ratio (or "asymmetry factor") is a measure of the symmetry in a chromatographic peak. The *A/B* ratio is measured by first finding the baseline level for the peak along with the time (or volume) at which the peak's maximum response appears. Horizontal lines are drawn parallel to the baseline and at either one-half or one-tenth of the distance between the baseline and the peak maximum. Next, a vertical line is drawn down from the peak top,

and the widths of the line segments at one-half or one-tenth of the peak height are measured on either side of this vertical line. The width obtained between the vertical line and peak's trailing edge (the "A" distance) is then divided by the width between this vertical line and the peak's front edge (the "B" distance). For a Gaussian-shaped curve, the values for A and B will be the same, giving an A/B ratio of one. If peak tailing is present, an A/B ratio greater than one will be obtained. An A/B ratio less than one will be observed for a system with peak fronting.

67.   a) Peak tailing   b) Peak fronting   c) Gaussian peak

68. Using measurements at one-tenth of the total peak height, the A/B ratio for diquat is approximately 1.55 and the A/B ratio for paraquat is approximately 1.64. Both of these values indicate that peak tailing is present, as is also indicated by a close inspection of both peaks in the given set of chromatograms. The presence of this peak asymmetry will make it slightly harder to measure both of these analytes under such conditions due to overlap that will occur in the peaks for diquat and paraquat when both of these substances are present in the same sample.

69. a) Diffusion is a process by which a solute moves away from a region of high concentration to one of lower concentration. This process is important in affecting many types of band-broadening in chromatography.

b) Mass transfer is the movement of a solute from one region to another and is also an important process that affects chromatographic band-broadening.

c) A diffusion coefficient is a term used to describe the rate of a solute's diffusion. In chromatography, a large diffusion coefficient will result in faster movement of a solute between the mobile phase and stationary phase or other regions of the system.

70. a) Eddy diffusion is a band-broadening process that occurs whenever there are support particles within a column, as produced by the presence of the large number of flow paths around support particles.

b) Longitudinal diffusion refers to the broadening of a chemical's peak due to the diffusion of this chemical along the length of a chromatographic column or separation system.

c) Mobile phase mass transfer is a band-broadening process that results from the different rates of travel that a solute has across any given slice of a chromatographic column.

d) Stationary phase mass transfer is a band-broadening process that is related to the movement of chemicals between the stagnant mobile phase and stationary phase.

e) Stagnant mobile phase mass transfer is a band-broadening process that is related to the rate of diffusion or mass transfer of solutes as they go from the mobile phase outside the pores of the support to the mobile phase within the pores of the support or directly in contact with the support's surface.

f) Extra-column band-broadening is the broadening of a peak before or after it enters a chromatographic column.

71. The van Deemter equation is an equation used in chromatography to show the relationship between the plate height (H) for a system and the linear velocity (u). This equation has the general form $H = A + B/u + Cu$, where A, B, and C are constants of the system. Because these band-broadening processes are affected differently by a variation in flow rate, the overall result is a "U"-shaped curve for a plot of H versus u, in which there is an optimum linear velocity where H has its lowest value.

72. The optimum linear velocity in Figure 20.15 is at approximately 0.16 cm/s. Increasing from this linear velocity to 0.60 cm/s will result in a larger value for the plate height and a less efficient system. However, the time required for analytes to elute will also be shorter. It is possible to work at this higher linear velocity if sufficient efficiency is still present to provide adequate resolution for the desired separation.

73. A linear velocity of 12 cm/s is below the optimum linear velocity in the van Deemter curve, 20 cm/s is near the optimum, and 50 cm/s or 70 cm/s are above this optimum value.

74. The separation factor (α) is a measure of the relative difference in retention of two solutes as they pass through a column as determined by the ratio k_2/k_1, where k_1 is the retention factor for solute 1 and k_2 is the retention factor for solute 2. The separation factor becomes larger as the relative difference in retention increases between two peaks. This feature makes α useful in describing the effectiveness of a chromatographic separation. The separation factor can also indicate whether it is feasible to resolve two compounds by a given column, where a value greater than one is needed for a separation to occur.

75. The peak resolution (R_s) is a parameter used to describe the separation of two peaks in chromatography. The value of R_s for two adjacent peaks can be calculated through the following formula,

$$R_s = \frac{t_{R2} - t_{R1}}{(w_{b2} + w_{b1})/2}$$

where t_{R1} and w_{b1} are the retention time and baseline width (both in the same units of time) for the first eluting peak, and t_{R2} and w_{b2} are the retention time and baseline width of the second peak. An important advantage of using the peak resolution instead of the separation factor is that R_s considers both the difference in retention between two compounds (as represented by $t_{R2} - t_{R1}$) and the degree of band-broadening (as represented by w_{b1} and w_{b2}).

76. Separation factors:

    Peaks 1 vs. Peak 2, $\alpha = k_2/k_1 = 1.08$

    Peak 2 vs. Peak 3, $\alpha = 1.32$

    Peak 3 vs. Peak 4, $\alpha = 1.10$

    Peak 4 vs. Peak 5, $\alpha = 1.20$

    Peak 5 vs. Peak 6, $\alpha = 1.73$

    The two peaks with the lowest separation factor are peaks 1 and 2, followed closely by peaks 3 and 4. These would probably be the peaks that should be the focus of any further improvements that are required in this separation, if the separation factor is the only factor considered in this optimization process.

77. Peaks 1 and 2, $R_s = (3.31 \text{ min} - 3.14 \text{ min})/[(0.17 \text{ min} + 0.18 \text{ min})/2] = 0.97$

    Peaks 2 and 3, $R_s = 3.65$

    Peaks 3 and 4, $R_s = 1.39$

    Peaks 4 and 5, $R_s = 2.63$

    Peaks 5 and 6, $R_s = 8.4$

    Peaks 1 and 2 would again be the pair of peaks that should be focused on for further improvements during the optimization process, followed closely by peaks 3 and 4.

78. The "resolution equation" is an equation used to show the relationship among peak resolution, peak retention, efficiency, and selectivity in chromatography. One form of this equation is shown on the following page.

$$R_s = \frac{\sqrt{N}}{4} \cdot \frac{(\alpha - 1)}{\alpha} \cdot \frac{k}{(1+k)}$$

In this equation, R_s is the resolution, k is the retention factor for the second peak, α is the separation factor between the first and second peaks, and N is the number of theoretical plates for the column being used in this separation. This equation indicates that the degree of a separation in chromatography will be affected by changing three factors: 1) the extent of band-broadening in the column (N), 2) the overall degree of peak retention (k), and 3) the selectivity of the column's stationary phase in binding to one compound versus another (α).

79. a) An increase in the length of the column from 25.0 cm to 50.0 cm will give an increase in N by a factor of (50.0 cm/25.0 cm) = 2. Because the resolution is proportional to the square root of N, the value of R_s will increase from 0.96 to 0.96 $(2)^{1/2}$ = 1.3$\underline{6}$ = 1.4 if all other conditions remain constant.

b) In this case, the value of N is constant and the separation factor is unchanged, but the value of k is increased. The value of $k/(1 + k)$ for the second peak (1-acetophenone) increases from 2.2/(1 + 2.2) = 0.68$\underline{8}$ to 5.5/(1 + 5.5) = 0.846. The corresponding change in R_s is from 0.96 to 0.96 (0.84$\underline{6}$/0.68$\underline{8}$) = 1.1$\underline{8}$ = 1.2.

c) In this case, the separation factor is being changed, but the values of N and k in the resolution equation are constant. The separation factor term changes from {(2.2/2.1) − 1)}/(2.2/2.1) = 0.045$\underline{5}$ under the original conditions to {(2.2/2.0) − 1)}/(2.2/2.0) = 0.090$\underline{9}$ under the new conditions. This change will increase the resolution from 0.96 to 0.96 (0.90$\underline{9}$/0.045$\underline{5}$) = 1.9$\underline{2}$ = 1.9.

84. a) Response factors: f for butabarbital = 9.05/9.05 = 1.00; f for barbital = 9.01/9.05 = 0.996; f for pentobarbital = 8.24/9.05 = 0.910; f for secobarbital = 9.02/9.05 = 0.997; f for hexobarbital = 6.32/9.05 = 0.698.

b) The data obtained for the butabarbital standards gives a linear calibration curve with a best-fit equation of $y = 1.772\ x + 0.0581$. Based on the response factor determined from Part a), the response obtained for secobarbital would correspond to a response of (23.53)/(0.997) = 23.60 for the same concentration of butabarbital. Using this new value with the calibration curve for butabarbital, the effective concentration of secobarital in the same is estimated to be 13.3 mg/mL.

87. a) $t = (0.01\ \text{cm})^2/\{(2)(1\ \text{cm}^2/\text{s})\} = 0.00005$ s

b) $t = (0.01\ \text{cm})^2/\{(2)(10^{-5}\ \text{cm}^2/\text{s})\} = 5$ s

It can be seen from these results and those in Part a) that diffusion and mass transfer processes related to diffusion are much faster in gas chromatography than in liquid chromatography.

88. a) $R_s = (21.8\ \text{min} - 15.3\ \text{min})/\{(0.26\ \text{min} + 0.18\ \text{min})/2\} = 29.5$

$N = 5.545\ (21.8\ \text{min}/0.26\ \text{min})^2 = 39{,}000$ (for n-octane)

$= 5.545\ (15.3\ \text{min}/0.18\ \text{min})^2 = 40{,}000$ (for n-heptane)

$$SN = \frac{t_{R,(z+1)} - t_{R,z}}{w_{h,z} + w_{h,(z+1)}} - 1 = \frac{21.8\ \text{min} - 15.3\ \text{min}}{0.18\ \text{min} + 0.26\ \text{min}} - 1 = 13.8$$

b) $R_s = (43.5\ \text{min} - 30.5\ \text{min})/\{(0.36\ \text{min} + 0.25\ \text{min})/2\} = 42.6$

$N = 5.545\ (43.5\ \text{min}/0.36\ \text{min})^2 = 81{,}000$ (for n-octane)

$= 5.545\ (30.5\ \text{min}/0.25\ \text{min})^2 = 83{,}000$ (for n-heptane)

$$SN = \frac{t_{R,(z+1)} - t_{R,z}}{w_{h,z} + w_{h,(z+1)}} - 1 = \frac{43.5 \text{ min} - 30.5 \text{ min}}{0.25 \text{ min} + 0.36 \text{ min}} - 1 = 20.3$$

The number of plates has increased, which has lead to an increase in both the resolution and value of SN. The result is a method that can now be used to separate more peaks in a single run.

CHAPTER 21: GAS CHROMATOGRAPHY

1. Gas chromatography (GC) is a type of chromatography in which the mobile phase is a gas. GC is often used as a separation method for chemicals that are naturally volatile or that can be converted into a volatile form for analysis.

2. The first major component of a gas chromatograph is the gas source that supplies the mobile phase. The second part of the gas chromatograph is its injection system, which often consists of a heated loop or port into which the sample is placed and converted into a gaseous form. The third part of the system is the column, which contains the stationary phase and support material for the separation of components in a sample. This column is held in an enclosed area known as the column oven that maintains the temperature at a well-defined value. The fourth part of the GC system is a detector, and associated recording device, that monitors sample components as they leave the column.

3. a) A gas chromatogram is a plot of the detector response versus the time that has elapsed since sample injection onto a gas chromatographic system.

    b) A gas chromatograph is a system that is used to perform gas chromatography.

    c) A column oven is an enclosed area in a gas chromatographic system that is used to hold the column and maintain this column at a well-defined temperature.

4. GC is valuable tool for separating volatile chemicals prior to their measurement by an online detector or second analytical method. The location of a peak in a GC separation can be used to help in chemical identification, and the size of this peak can be compared to peaks for standards for use in measuring the amount of chemical that must have been present to

produce the peak.

5. A plot of the ratio of the relative peak area for MBTFA-morphine to MBTFA-naplorphine should be made versus the concentration of morphine in standards. The peak area ratio for the sample is then used with this plot to determine the concentration of morphine in the blood sample, giving a concentration of 54.5 ng/mL.

6. a) A plot made of the relative peak heights for toluene versus deuterated toluene on the y-axis and toluene concentration on the x-axis gives a linear relationship with a best-fit line of $y = 0.0033\,x + 0.0275$. The unknown sample gives a peak height ratio of $178/229 = 0.777$ under the same conditions, which would correspond to a toluene concentration of 227 ppb in the unknown sample.

b) The corresponding response in terms of toluene would be $88/0.71 = 124$, giving a peak height ratio of $124/238 = 0.52$. This would correspond to a concentration of 149 ppb for isopentane in the sample.

7. For it to be possible to examine a compound by gas chromatography, the injected analyte must be volatile and have good thermal stability.

8. All of these compounds have molar masses below 600 g/mol and boiling points below 500 °C at 1 atm, so it is possible (in theory) to analyze each by GC. However, some derivatization may be required for decanoic acid. Also, melene would be expected to have strong retention on a GC column due to its large size and high boiling point.

9. The approximate order of elution, from first to last, would be as follows: toluene, naphthalene, decanoic acid, and melene. This assumes that decanoic acid has sufficient

thermal stability for the GC analysis.

10. Thermal stability refers to the ability of a compound to be stable at a given temperature. This type of stability is important in GC because of the elevated temperatures that are often used in this method for elution.

11. Thermal stability can be tested by injecting a new analyte onto a GC system and seeing if this chemical produces a chromatogram that has a single, well-defined peak with good retention and detection properties. If thermal stability is a problem, such degradation can sometimes be minimized or eliminated by selecting proper sample injection conditions or by using chemical derivatization.

12. Derivatization is the process of altering the chemical structure of an analyte. Derivatization in GC typically involves replacing one or more polar groups on an analyte, such as an alcohol or amine group, with less polar groups. This change reduces the intermolecular interactions of the altered chemical, making it more volatile and easier to place into the gas phase. The same type of change also tends to make a compound more thermally stable.

13. A trimethylsilyl (TMS)-derivative is the product of a derivatization reaction that takes place between a chemical and TMS. This type of derivative tends to replace polar functional groups with a less polar group, making the resulting compound more volatile and often more thermally stable.

14. The conversion of the carboxylic acid in dicamba into a TMS derivative will produce a group that is less polar and that has less chance of undergoing hydrogen bonding. This process will take place through the general reaction that is shown in Equation 21.1, in which TMS reacts

with the OH in the carboxylic acid group of dicamba. Such a change should increase the volatility of dicamba in the derivatized form. This change should also increase the thermal stability of the dicamba derivative versus dicamba because a carboxylic acid is one type of group that often undergoes degradation at elevated temperatures.

15. The conversion of a carboxylic acid into a methyl ester produces a group that is less polar and that has less of a chance of undergoing hydrogen bonding. Such a change should increase the volatility of the chemical in the derivatized form. This change should also increase thermal stability because a carboxylic acid is one type of group that often undergoes degradation at elevated temperatures.

16. Chemical volatility is important in GC because a chemical must be able to enter the gas phase to be injected into a GC system. The more volatile a chemical is, the more time it will spend in the mobile phase and the less retained it will be on a GC column.

17. Homologs are compounds that have the same general structure but differ in the length of a single carbon chain. Volatility decreases and retention increases as we increase the size of the carbon chain in a group of homologs.

18. The approximate order of elution, from first to last, would be as follows:

    a) methanol, ethanol, 1-propanol, and 1-butanol

    b) benzene, toluene, and ethylbenzene

    c) butyric acid and caproic acid

19. Decreasing the column temperature leads to longer retention because this causes the injected analytes to be less volatile and spend less time in the mobile phase. Increasing the

temperature makes the analytes more volatile and causes them to pass through the column more quickly as they spend more time in the mobile phase.

20. The Kováts retention index is a measure of retention in gas chromatography, as determined by comparing a compound's retention on a given column to the retention seen under identical conditions for *n*-alkanes.

21. The following values can be calculated by using Equation 21.2.

    Kováts retention indices: benzene, $I = 639$; butanol, $I = 370$; nitropropane, $I = 246$; pyridine, $I = 478$; 2-pentanone, $I = 485$

22. a) The following values can be calculated by using Equation 21.2 for the new GC column.

    Kováts retention indices: benzene, $I = 721$; butanol, $I = 422$; nitropropane, $I = 327$; pyridine, $I = 499$; 2-pentanone, $I = 745$

    b) The values have increased in going from squalane to the new column, which is a general indication that the new column is more polar than squalene and tends to give higher retention.

23. A plot of $\log(t'_R)$ versus the number of carbon atoms in chain for the *n*-alkanes in Problem 22 gives a linear relationship with a best-fit line of $y = 0.1497\,x + 0.0013$. Based on this line, the values of $\log(t'_R)$ and t_R that would be expected for nonane are 1.35 and 23.3 min, while the values of $\log(t'_R)$ and t_R expected for decane are 1.50 and 32.5 min.

24. The low density of gases means that analytes in these gases can move about quickly by diffusion. This feature is important because most processes that cause peak broadening in chromatography will be reduced by the presence of fast diffusion, with longitudinal diffusion

being the main exception. The presence of less peak broadening, in turn, makes it easier for the column to discriminate between the analyte and other sample components. Low viscosity is another feature of gases that promotes high efficiency in GC. As the viscosity goes down for a mobile phase, it is possible to use a longer column, while still being with the allowable pressure range for the system.

25. The low viscosity of gases versus liquids makes it possible to use longer columns in GC compared to LC. A longer column will have a larger number of theoretical plates, which will increase the resolution of separations that are performed on that column.

26. a) $N = (2000 \text{ plates/m})(2 \text{ m}) = 4,000$

   $H = L/N = (2 \text{ m})/(4,000) = 0.0005 \text{ m (or 0.05 cm)}$

   b) $N = (3,200 \text{ plates/m})(20 \text{ m}) = 64,000$

   $H = (20 \text{ m})/(64,000) = 0.00031 \text{ m (or 0.031 cm)}$

   c) $N = (10,000 \text{ plates/m})(50 \text{ m}) = 500,000$

   $H = (50 \text{ m})/(500,000) = 0.0001 \text{ m (or 0.01 cm)}$

27. These values can be found by rearranging the relationship $N = 16 \, (t_R/w_b)^2$ for a Gaussian peak. Substituting in the known values for N and t_R results in the following calculated values for the baseline width (w_b): n-pentane, $w_b = 0.063$ min; toluene, $w_b = 0.21$ min; 1,2,4-trimethylbenzene, $w_b = 0.37$ min.

28. The role of the mobile phase in GC is to apply and transport compounds through the column.

29. The carrier gas is the mobile phase in gas chromatography. Examples of common carrier gases used in GC are hydrogen, helium, nitrogen, and argon.

30. The carrier gas should have high purity to avoid contamination or damage to the column and GC system. Impurities like water, oxygen, organic substances, and particulate matter can be removed by passing the carrier gas through a series of traps and filters before it enters the column.

31. It is important to control and/or monitor the pressure and flow rate of the carrier gas when using temperature programming (or flow programming) in GC. These factors not only affect the speed and efficiency of the separation, but they can also affect the response observed when using certain types of detectors (e.g., a thermal conductivity detector).

32. An isothermal method is a method that is performed at a constant temperature. An isothermal method works well if the sample is relatively simple or has only a few known compounds. The main strength of this approach is its simplicity in that no cooling down or re-equilibration period is needed between samples, which also helps minimize the time that elapses between sample injections.

33. The general elution problem is a problem that arises in chromatography and other separation methods when working with a complex sample, where it is often difficult to find a single set of conditions that can separate all sample components with adequate resolution and in a reasonable amount of time.

34. Gradient elution is an elution method in chromatography in which the separation conditions are changed during the analysis of a sample. A typical method using gradient elution will begin with conditions that allow early-eluting compounds to stay on the column longer, helping them to become better separated. The conditions are then changed over time to help

other compounds also elute with good resolution and within a satisfactory amount of time. As a result, this approach can be used to overcome the general elution problem.

35. Temperature programming is a method of performing gradient elution in gas chromatography, in which the temperature of the column is varied over time. This approach starts at a moderately low temperature, so that the most volatile substances in the sample are more strongly retained. The temperature is then gradually increased to allow other analytes in the sample to elute with reasonable retention times, while still being resolved from each other. The result is an improved separation for both weakly and strongly retained components in a complex sample.

36. A temperature program usually begins with an initial isothermal step at a relatively low column temperature. It is during this step that the sample is injected, allowing the most volatile compounds to interact with the column and to separate. The next step in the temperature program is known as the temperature ramp; it is during this step that analytes with intermediate or high boiling points are eluted from the column. A linear change in temperature over time is generally used because of its simplicity and ability to elute compounds with a large range of volatilities. After the temperature ramp has been completed, the third part of a temperature program is often an isothermal step in which the temperature is held for some period of time at the upper limit of the ramp. This step is useful in making sure that all analytes have time to elute and in ensuring that there are no low volatility substances that remain on the column from one sample injection to the next. The fourth step in the temperature program is a cooling-down period in which the column is returned back to its initial temperature.

37. A packed column is a chromatographic column that is filled with small support particles that act as an adsorbent or that are coated with the desired stationary phase. In GC, a packed column is made up of a glass or a metal tube that is usually 1–2 m long and a few millimeters in diameter.

38. Diatomaceous earth is a common support placed in packed GC columns. This material is formed from fossilized diatoms and mainly consists of silicon dioxide, or silica.

39. Packed GC columns, and their packed support particles, have a big surface area that can be used with large amounts of a stationary phase. This feature makes it possible to inject relatively large samples or quantities of chemicals onto such columns. A disadvantage of packed columns is that they often have lower efficiencies than columns based on open-tubular supports.

40. An open-tubular column (or "capillary column") is a tube that has a stationary phase coated on or attached to its interior surface. This type of column in GC will generally have a length between 10 and 100 m and an inner diameter of 0.1 to 0.75 mm.

41. a) A wall-coated open-tubular (WCOT) column, is an open-tubular column in GC in which a thin film of a liquid stationary phase is placed directly on the wall of the column. These columns are very efficient, but they have a small sample capacity as a result of their low surface area.

    b) A second type of open-tubular column is a support-coated open-tubular (SCOT) column, which has an interior wall that is coated with a thin layer of a particulate support, plus a thin film of a liquid stationary phase that is coated onto this support layer. This coating gives a

SCOT column a thicker layer of stationary phase than that found in a WCOT column, resulting in a less efficient column but one with a larger sample capacity.

c) A porous-layer open-tubular (PLOT) column is the third type of open-tubular column in GC. A PLOT column also contains a porous material that is deposited on the column's interior wall, but the surface of this material is now used directly as the stationary phase without any additional coating. This makes PLOT columns useful in the method of gas-solid chromatography.

42. A fused-silica open-tubular column (FSOT) is an open-tubular column in which the outer tube is made of fused-silica. This is the type of column that is often used in analytical applications of GC because of its high efficiency.

43. Gas-solid chromatography is a GC technique that involves the use of the same material as both the support and stationary phase, with retention being based on the adsorption of analytes to the support's surface. Supports used in this method include silica, alumina, and molecular sieves.

44. A molecular sieve is a porous material that is composed of a mixture of silica, alumina, water, and an oxide of an alkali or alkaline earth metal, such as sodium or calcium. When these materials are combined in a particular ratio, they produce a support with a series of pores with well-defined sizes and binding regions. This type of material is commonly used as a support and stationary phase in gas-solid chromatography.

45. Gas-liquid chromatography is a GC method in which a chemical coating or layer is placed onto the support and used as the stationary phase. The liquids that are used as stationary

phases in this method all have high boiling points and low volatilities, which allow them to stay within the column at the relatively high temperatures that are often used in GC for sample injection and elution. These liquids are also "wettable", which means they are easy to place onto a support in a thin, uniform layer.

46. The general structure of a polysiloxane consists of a backbone of silicon and oxygen atoms attached in long strings of Si–O–Si bonds. In GC, the total size of these chains varies from molecular masses of a few thousand up to over a million, which gives these polymers a low volatility. The remaining two bonds on each silicon atom are attached to side groups that can have a variety of structures. It is possible by altering the amount and type of these groups to produce stationary phases with a variety of polarities and specificities. Polysiloxanes also tend to have good temperature stability.

47. Column bleed is the loss of a stationary phase from a chromatographic column over time. This effect is caused when the stationary phase slowly vaporizes or breaks apart and leaves the column over time.

48. a) A bonded phase is a stationary phase that is covalently attached to a support material. This type of phase helps reduce the problem of column bleed by producing a more stable column.

b) A silanol group is a chemical group with the general formula –Si–OH and is located on the surface of a silica support. This type of group is often used to anchor a stationary phase to a support to reduce the problem of column bleed.

c) A cross-linked stationary phase is a stationary phase that has been cross-linked on a chromatographic support. The use of this type of stationary phase also helps reduce the problem of column bleed by providing a more stable column.

49. a) A thermal conductivity detector measures the ability of the eluting carrier gas and analyte mixture to conduct heat away from a hot-wire filament. This ability will vary as different analytes elute from the GC column.

b) A flame ionization detector (FID) detects organic compounds by measuring their ability to produce ions when they are burned in a flame. The flame in the FID is usually formed by burning the eluting compounds in a mixture of hydrogen and air. An additional amount of a make-up gas like nitrogen is sometimes combined with the carrier gas and hydrogen before entry into the flame to provide a steady, optimum flow rate for detection. Positively charged ions produced by the flame are collected by a negative electrode that surrounds the flame. As these ions are produced, they create a current at the electrode, thereby allowing the presence of the eluting compound to be detected.

50. a) A nitrogen-phosphorus detector is selective for the determination of nitrogen- or phosphorus-containing compounds. This detector is based on the measurement of ions that are produced from eluting compounds. These ions are generated by using thermal heating at or above a surface that can supply electrons to any electronegative species that surround it, forming negatively charged ions. This mechanism of ion formation is particularly efficient for nitrogen- or phosphorus-containing compounds.

b) An electron capture detector detects compounds that have electronegative atoms or groups in their structure, such as halogen atoms, nitro groups, polynuclear aromatic compounds, anhydrides, and conjugated carbonyl compounds, among others. This detector makes use of the capture of electrons by electronegative atoms or groups in the molecule.

These electrons are produced by a radioactive source, such as ^{3}H or ^{63}Ni, that emits beta particles (high energy electrons) as part of its decay process. As these particles are released, some of them will collide with the carrier gas. When these collisions occur, the carrier gas takes on some of the particle's energy and creates the release of a large number of secondary electrons at lower energies.

51. a) Parathion and b) imipramine

52. a) DDT and c) Freon-12

53. Gas chromatography/mass spectrometry is a method based on the combined use of gas chromatography with a mass spectrometer. If this approach is used to look at a broad range of ions, it can be used as a general detector. If it is used to look at just one or a few specific ions that are characteristic of a given analyte, it can be made to act as a selective detector.

54. A molecular ion is an ion that is produced from the intact structure of an analyte, such as through the loss of an electron or the addition a hydrogen ion. A fragment ion is an ion that is formed from only a portion of the original analyte. The molecular ion provides information on the molar mass of the analyte, while the fragment ions provide information on the analyte's structure.

55. Electron impact ionization (EI) is a method for the creation of ions in mass spectrometry by which analytes are passed through a beam of high-energy electrons. These electrons bombard some of the analyte molecules, causing electrons to be removed from the analyte and forming a molecular ion. Fragment ions are also often formed.

Chemical ionization (CI) is a method of ionization in mass spectrometry that makes use of a gas-phase acid to protonate the analyte and form a molecular ion with general structure MH^+. Both of these approaches are used to create ions for GC/MS. EI tends to be a harsher approach that results in more fragment ions, while CI tends to produce a greater amount of molecular ions.

56. A quadrupole mass analyzer uses a series of four parallel rods that are held at various well-defined potentials. Each opposing pair of rods is held at the same potential and any two neighboring rods are held at exactly opposite potentials. Over time, the potentials on these rods are continuously varied. This creates a varying electric field through which only ions with a certain mass-to-charge ratio will be able to pass. If these conditions are altered over time, ions with different mass-to-charge ratios can be collected and measured at different times.

57. a) A mass spectrum is a plot of the intensity of each ion that is detected over a range of mass-to-charge ratios.

b) A mass chromatogram is a plot of the number of ions that are measured at each elution time. A mass chromatogram can be constructed by using the intensities measured for all measured ions, or it can be prepared using only ions with particular mass-to-charge ratios.

c) The use of GC/MS as a general detector by collecting information on a wide range of ions is known as the full-scan mode of GC/MS. The full-scan mode is useful for looking at a broad range of compounds in a single analysis or for determining the identity of an unknown compound from its mass spectrum.

d) The use of GC/MS to collect information on only a few ions is called selected ion monitoring (SIM). In this approach, only a few ions characteristic of the compounds of interest are examined. SIM is employed when low detection limits are desired and when it is known in advance what compounds are to be analyzed and what types of ions they produce in the mass spectrometer.

58. a) GC/MS would be used to identify and confirm the presence of a wide range of possible performance-enhancing drugs.

b) The presence of halogen atoms in the structures of polychlorinated biphenyls would make ECD a good choice for the trace analysis of these chemicals in water samples.

c) A thermal conductivity detector could be used as a general detector that would be able to respond to the oxygen, carbon dioxide, and water that might be present as semitrace components in a sample of helium.

59. Gaseous samples can be injected into a GC system by passing the sample through a gas-tight valve or by using a gas-tight syringe to inject a known volume of gas into a GC system.

60. a) Direct injection uses a calibrated microsyringe to apply the desired volume of sample to the system. The microsyringe is passed through a gas-tight septum and into a heated chamber where the liquid sample and its contents are vaporized and swept by the carrier gas into the column. This approach allows essentially 100% transfer of the analytes into the column. The main limitation of direct injection is that it cannot be used with many open-tubular columns.

b) In split injection, a microsyringe is again used to inject a liquid sample into a GC system. As the sample is converted into a gas in the injection area, its vapors are divided so that only a small portion goes into the column. This procedure allows the volume of a sample that reaches the column to be greatly reduced. The result is a lower chance of overloading the column with the sample, especially for narrow-bore or moderate-bore open-tubular columns.

Split injection works well for relatively concentrated samples, but it can be difficult to use for trace analysis because much of the sample is discarded during the injection process. Another problem with split injection is that analytes with different volatilities may not have equal fractions that enter the column, which can create a variable recovery for analytes and can affect their measurement.

c) Splitless injection is also carried out with microsyringes and open-tubular columns, but the side vent of the injection system is now kept closed so that most of the injected and vaporized sample will go into the column. After the analytes have entered the column, the side vent is opened to flush out any undesired vapors from the injection chamber before the next sample is applied. Because little sample is lost during its injection, this approach is better to use than split injection when performing a trace analysis. However, splitless injection has a greater difficulty with column overloading by samples.

d) In cold on-column injection, a microsyringe containing the sample is passed through the "injector" and directly into the column or an uncoated precolumn. The region around the syringe is initially kept cool, so that the sample can be deposited in the column as a liquid film. This liquid is later heated, causing the analytes to enter the carrier gas to begin their passage through the column. Because the sample is applied directly to the column or precolumn in this method, essentially 100% of each analyte is available for measurement. This approach can have problems if there are any low volatility substances in the sample, which will accumulate at the top of the column or precolumn and alter the properties of the GC system over time.

61. Water binds strongly to many GC columns and can create problems with the long-term behavior and reproducibility of these columns. In addition, the water may contain dissolved solids, salts, or other nonvolatile compounds that are not suitable for injection onto a GC system. This problem can be avoided by first extracting the desired analytes into a nonaqueous solvent before they are injected onto a GC system.

62. Headspace analysis is based on the fact that volatile analytes in a liquid or solid sample will also be present in the vapor phase that is located above the sample. If a portion of this headspace is collected, it can be used to measure volatile analytes without interference from other, less volatile compounds that were in the original sample. This approach is often used to examine volatile organic compounds (VOCs) in water.

63. In the static method, the sample is placed into an enclosed container and its contents are allowed to distribute between the sample and its vapor phase. After equilibrium has been reached, a portion of the vapor phase is collected and injected onto a GC system. In the dynamic method, an inert gas is passed through the sample to carry away volatile compounds. This gas is then passed through a cold trap or solid adsorbent to collect and concentrate these volatile solutes for analysis. Although the dynamic technique requires more time and effort than the static method, it is more reproducible and allows better detection of compounds with relatively low volatilities.

64. A common way of dealing with analytes in a solid sample is to first extract the compounds of interest from the solid material, such as by using liquid–liquid extraction or supercritical fluid extraction. The extracted analytes are then placed into an organic solvent and treated as liquid samples, as described in the last section. It is also possible in some cases to examine

solid samples by using thermal desorption, in which a known quantity of the solid is placed into a chamber where the solid can be heated. As this solid is heated, its volatile components will enter the gas phase, allowing them to be trapped or placed into a GC system for testing.

65. a) Thermal desorption is a method that involves placing a known quantity of the solid into a chamber where the solid can be heated. As this solid is heated, its volatile components will enter the gas phase, allowing them to be trapped or placed into a GC system for testing.

b) Pyrolysis gas chromatography involves heating a solid sample in a controlled fashion to break the solid apart into smaller, more volatile chemical fragments.

c) In pyrolysis gas chromatography, the gas chromatogram that is created as volatile chemicals are released from a sample and separated by the GC column is known as a pyrogram. The pattern of eluting chemicals in the pyrogram forms a chemical fingerprint for the substance that is being tested, making it possible to characterize the structure of the original solid or to identify this solid by comparing it with known standard samples.

66. a) The extraction step is used to remove the analytes from the water, to possibly remove them from other interfering agents, and to provide a solvent that can later be used for direct injection onto the GC.

b) HDPE is a material that is nonvolatile and that cannot be examined directly by GC. Instead, this material is heated by pyrolysis, and its thermal degradation products are examined by GC to characterize the degree of branching that occurs within a preparation of this material.

c) Ethanol is volatile and easy to collect and analyze by headspace analysis. This approach also makes it possible to avoid injecting any water onto the GC system.

d) Use of cold on-column injection makes it possible to use a lower temperature during sample injection, which avoids problems due to the thermal degradation of endrin.

70. McReynolds constants for OV-17:

    Benzene $X' = 772 - 653 = 119$

    1-Butanol $Y' = 748 - 590 = 158$

    2-Pentanone $Z' = 789 - 627 = 162$

    1-Nitropropane $U' = 895 - 652 = 243$

    Pyridine $S' = 901 - 699 = 202$

71. a) The diglycerol column is the most polar, followed by Ethofat 60/60 and Igepal CO-630 (which have similar polarities), and OV-1 (the least polar stationary phase in this group).

b) The greatest retention for benzene (based on the X' values) will be on diglycerol, followed by Ethofat 60/40 and Igepal CO-630 (which are similar), and OV-1.

The greatest retention for 1-butanol (based on the Y' values) will be on diglycerol, followed by Ethofat 60/40 and Igepal CO-630 (which are similar), and OV-1.

The greatest retention for pyridine (based on the S' values) will be on diglycerol, followed by Ethofat 60/40 and Igepal CO-630 (which are similar), and OV-1.

The greatest retention for nitropropane (based on the U' values) will be on diglycerol, followed by Ethofat 60/40 and Igepal CO-630 (which are similar), and OV-1.

The greatest retention for 2-pentanone (based on the Z' values) will be on diglycerol, followed by Igepal CO-630, Ethofat 60/40, and OV-1.

72. a) This column should provide a general, relatively efficient system for separating a wide range of volatile compounds such as those that would be found in a mixture of C_1–C_{10}

saturated hydrocarbons.

b) An OV-1 column will provide a separation that is mainly based on volatility, which is appropriate for the separation and analysis of highly volatile chemicals such as VOCs in air.

c) A Carbowax 20M column will have a higher polarity than a DB-5 column, which will give the Carbowax 20M column higher retention for alcohols like methanol, ethanol, *n*-propanol, and 2-propanol. This higher retention should also result in higher resolution and a better separation.

CHAPTER 22: LIQUID CHROMATOGRAPHY

1. Liquid chromatography (LC) is a chromatographic technique in which the mobile phase is a liquid. This method is used to separate chemicals in liquid-based samples. This approach can be used for both identification and the measurement of chemicals based on the positions of their peaks and the sizes of these peaks when compared to those of standards for the same chemicals. LC is also often used as a sample pretreatment or separation method prior to the use of other analytical techniques.

2. A liquid chromatograph is a system that is used to perform LC. This type of system typically includes a support and stationary phase enclosed in a column and a liquid mobile phase that is delivered to the column by means of a pump. For analytical applications, an injection device is used to apply samples to the column, while a detector monitors and measures analytes as they leave the column. A collection device can also be placed after the column to capture analytes as they elute.

3. A liquid chromatogram is a plot of the detector response versus the time that has elapsed since sample injection onto a liquid chromatographic system. The retention time or volume of a peak in this plot can be employed to help identify a sample component, while the size of the peak can be related to the amount of this component in the sample.

4. The following results were obtained under the conditions used in this separation. Although the retention factors noted here are larger than those reported by the manufacturer of the C_{18} reversed-phase column, they do follow the same relative order as those reported for the standard test mixture and can be used to identify each of the components in this mixture.

$$t_M = 0.65 \text{ min}$$

Peak 1: Resorcinal, t_R = 1.05 min; k = (1.05 min – 0.65 min)/(0.65 min) = 0.62

Peak 2: Acetophenone, t_R = 2.10 min; k = (2.10 min – 0.65 min)/(0.65 min) = 2.23

Peak 3: Napthalene, t_R = 4.70 min; k = (4.70 min – 0.65 min)/(0.65 min) = 6.23

Peak 4: Anthracene, t_R = 9.60 min; k = (9.60 min – 0.65 min)/(0.65 min) = 13.8

5. a) Concentration of Ca^{2+} = 2.47 ppm

   b) This peak is due to Ca^{2+} because the retention factor for the peak in the unknown sample matches the retention factor for Ca^{2+} in the standard.

6. The first requirement that must be met before we can examine a chemical by LC is that it must be possible to place this chemical into a liquid that can be injected onto the column.

7. As was true in GC, the mobile phase in LC acts to carry solutes through the column. However, the greater density of a liquid versus a gas means that the liquid mobile phase in LC can also interact with the analytes and the stationary phase to affect analyte retention. This feature makes LC more flexible than GC when optimizing and controlling the degree of retention that is obtained in a separation.

8. a) In solvent A, k = 88.8; in solvent B, k = 29.2

   b) In solvent A, t_R = 119 min; in solvent B, t_R = 40.2 min

9. Diffusion is much slower in liquids than in gases, which is why LC tends to have greater band-broadening than GC at moderate to high flow rates. (Note: Longitudinal diffusion is greater in GC at low flow rates, which gives LC less band-broadening under such conditions.) The greater viscosity of liquids versus gases is also important to consider when

comparing LC with GC because the higher viscosity of liquids means that shorter columns are often needed in LC.

10. High-performance liquid chromatography (HPLC) is the modern instrumental form of LC. This method makes use of smaller and more efficient supports than classical LC. Because of this difference, HPLC often requires the use of high pressures and special pumps, detectors, and other instrumental components. The presence of a more efficient support in HPLC produces narrower peaks than classical LC, which provides better separations and lower limits of detection.

11. a) In a traditional porous particle, the mobile phase typically flows around, but not through, the particle. This situation means analytes must travel within the particle by means of diffusion, a relatively slow process that leads to significant band-broadening.
b) A perfusion particle contains larger pores that allow the mobile phase to pass both through and around the support particles. This type of flow decreases the average distance that solutes must diffuse to reach the stationary phase and results in less band-broadening.
c) A nonporous support has a support that consists of a solid core with the stationary phase located on the exterior. This is another way to decrease the average distance that a solute must diffuse to reach the stationary phase and to decrease band-broadening.
d) A monolithic support contains a porous support that is one continuous bed. This type of support has similar properties to a perfusion particle and also leads to a decrease in band-broadening compared to that of traditional porous particles.

12. Column 1, $N = 7980$, $H = 7.5$ μm; Column 2, $N = 5540$, $H = 18$ μm

Column 1 is more efficient than column 2.

13. a) $N = (12{,}000/15 \text{ cm})(25 \text{ cm}) = 20{,}000$

    b) $N = (25{,}000/15 \text{ cm})(5 \text{ cm}) = 8330$

    c) $N = (1000/15 \text{ cm})(25 \text{ cm}) = 1670$; $H = (15 \text{ cm}/1000) = 0.015 \text{ cm}$

    d) $N = (30{,}000/15 \text{ cm})(3 \text{ cm}) = 6000$; $H = (15 \text{ cm}/30{,}000) = 0.00050 \text{ cm}$

14. An LC separation is based on column chromatography when the support is held within an enclosed system such as a tube. It is also possible in LC to conduct a separation in which the stationary phase is placed onto a flat surface to perform "planar chromatography", such as by using a stationary phase that is present on a paper or on a plate used for thin-layer chromatography.

15. In LC the retention of solutes will depend on interactions involving *both* the mobile and stationary phases. This situation is different from that of GC, in that the mobile phase is used mainly to pass chemicals through the column and does not affect their retention.

16. A strong mobile phase in LC is a pure solvent or a solution that quickly elutes a retained analyte from a column. This situation is created when the analyte favors staying in the mobile phase versus the stationary phase, as occurs if the analyte is more soluble in the mobile phase or has only limited interactions with the stationary phase. A weak mobile phase in LC is a liquid that slowly elutes a retained analyte. This second situation occurs when the analyte has better solubility in the stationary phase than the mobile phase, or when the mobile phase promotes good interactions of the solute with the stationary phase.

17. Solvent programming is a type of gradient elution in which the composition of the mobile phase is changed over time.

    Isocratic elution is the use of a constant mobile phase composition for elution in chromatography.

18. a) Weak mobile phase = pH 8.2, 20 mM Tris buffer (solvent A); strong mobile phase = pH 8.2, 20 mM Tris plus 0.25 M NaCl (solvent B)

    b) The weak mobile phase is used to apply the sample to the column under conditions in which the desired analytes will tend to have their strongest retention. The strong mobile phase is later applied to elute these analytes from the column by decreasing their retention.

19. Adsorption chromatography is a chromatographic technique that separates solutes based on their adsorption to the surface of a support. In this case, the surface of the support acts as the stationary phase. Many of the supports used in gas-solid chromatography are also used in liquid-solid chromatography.

20. The process leading to retention in adsorption chromatography is shown below. This process involves the binding of an analyte (A) to the surface of a support and the competition of this analyte with the n moles of the mobile phase (M) for these binding sites.

$$A + n\, M\text{-Surface} \rightleftarrows A\text{-Surface} + n\, M$$

This model indicates that the retention of an analyte in adsorption chromatography will depend on the binding strength of A to the support and on the surface area of this support. This retention factor will also depend on how much mobile phase is displaced from the surface by A, the strength with which the mobile phase binds to the support, and the relative amount of mobile phase that is displaced by the analyte.

21. The elutropic strength is a measure of the strength with which a mobile phase is adsorbed to a solid support. A liquid with a large elutropic strength will strongly adsorb to the given

support, which will prevent the analyte from binding to the support. As a result, a liquid with a large elutropic strength will act as a strong mobile phase for that support because the presence of this liquid will cause the analyte to spend more time in the mobile phase and to elute quickly from the column.

22. Silica is the most popular support in adsorption chromatography. Because silica is polar in nature, it will most strongly retain polar compounds. A strong mobile phase for silica will also be polar. Alumina is another polar material that is used in adsorption chromatography. Like silica, alumina is a general-purpose support and can retain some polar solutes so strongly that they are irreversibly adsorbed onto its surface. Carbon-based materials are sometimes used as nonpolar supports in adsorption chromatography, giving columns that retain nonpolar solutes and that have a strong mobile phase that is nonpolar in nature.

23. The retention of solutes can be adjusted in adsorption chromatography by changing the elutropic strength of the mobile phase. The selectivity factor for this separation can be adjusted by keeping the elutropic strength essentially fixed by varying the solvents that are combined to reach this elutropic strength.

24. a) The approximate elutropic strength of hexane on silica is 0.

b) The elutropic strength of tetrahydrofuran on silica is 0.53.

c) Tetrahydrofuran is the stronger mobile phase on this stationary phase.

d) According to Figure 22.7, approximately 1.2% tetrahydrofuran must be added to hexane to change the elutropic strength of this mobile phase by 0.10 units, while about 3.2% is needed to change the elutropic strength by 0.20, and 18% is needed to change the elutropic strength by 0.30 units.

25. A mixture of 2% isopropyl alcohol plus 98% hexane or a mixture of 10% ethyl acetate plus 90% dichloromethane are two possibilities.

26. The relative low cost and widespread availability of supports like alumina and silica have made them popular as preparative tools by synthetic chemists to help purify new chemicals. Silica and alumina work well in separating chemicals that are present in an organic solvent, which will act as a weak mobile phase for these supports. Adsorption chromatography is especially useful in separating geometrical isomers and chemicals that belong to a given class of substances.

27. Partition chromatography is a liquid chromatographic technique in which solutes are separated based on their partitioning between a liquid mobile phase and a stationary phase coated on a solid support. This differs from adsorption chromatography, which uses analyte adsorption to a surface as the basis for a separation.

28. Normal-phase liquid chromatography (i.e., NPLC or normal-phase chromatography) is a type of partition chromatography that uses a polar stationary phase. Because NPLC has a polar stationary phase, it retains polar compounds most strongly. The weak mobile phase in NPLC is a nonpolar liquid, which is used as the injection solvent. A strong mobile phase is a polar liquid, such as water or methanol.

Reversed-phase liquid chromatography (i.e., RPLC or reversed-phase chromatography) is also a type of partition chromatography, but it uses a nonpolar stationary phase and retains nonpolar solutes the most strongly. The weak mobile phase in RPLC is a polar solvent, such as water, while the strong mobile phase is a less polar solvent.

29. Common stationary phases used in NPLC include aminopropyl, cyanopropyl and diol phases. The weak mobile phase in NPLC is a nonpolar liquid (for example, n-hexane or toluene), which is used as the injection solvent. A strong mobile phase is a polar liquid, such as water or methanol.

30. Common stationary phases used in RPLC are those that contain saturated C_8 or C_{18} chains. A polar solvent, such as water, will act as a weak mobile phase, while a less polar solvent (e.g., methanol or acetonitrile) will act as a strong mobile phase.

31. The retention of solutes in partition chromatography can be described by the solubility equilibrium given below, in which K_D is the distribution constant.

$$A_{\text{mobile phase}} \overset{K_D}{\rightleftarrows} A_{\text{stationary phase}} \qquad K_D = \frac{[A]_{\text{stationary phase}}}{[A]_{\text{mobile phase}}}$$

The distribution constant for the analyte in the mobile phase and the stationary phase of the column can be directly related to the solute's retention factor (k) as shown below,

$$k = K_D (V_s/V_M)$$

where V_s is the volume of stationary phase in the column, and V_M is the column void volume. This equation indicates the retention factor will increase in partition chromatography when we increase the tendency of this solute to enter the stationary phase (K_D) or increase the relative volume of stationary phase versus mobile phase in the column (V_s/V_M, the phase ratio).

32. A solvent polarity index is a measure of mobile phase strength in partition chromatography (e.g., in NPLC or RPLC). One advantage of using the polarity index to describe mobile phase strength in NPLC and RPLC is that P changes linearly as two different solvents are

mixed together. This makes this index useful in adjusting the mobile phase composition and analyte retention.

33. a) $P = 5.64$ b) $P = 4.31$

     c) $P_{tot} = 1.35$ d) $P_{tot} = 7.06$

34. a) $P_{tot} = \varphi_A P_A + \varphi_B P_B + \varphi_C P_C$

$$= (0.60)(5.64) + (0.20)(5.10) + (0.20)(10.2) = 6.4$$

     b) $P_{tot} = \varphi_A P_A + \varphi_B P_B$

$$6.4 = (\varphi_A)(5.64) + (1 - \varphi_A)(10.2) = 10.2 + (5.64 - 10.2)\varphi_A$$

$\varphi_A = 0.833$ $\varphi_B = 1 - \varphi_A = 1 - 0.833 = 0.167$

83.3% acetonitrile and 16.7% water

     c) $P_{tot} = \varphi_A P_A + \varphi_B P_B$

$$6.4 = (\varphi_A)(5.10) + (1 - \varphi_A)(10.2) = 10.2 + (5.10 - 10.2)\varphi_A$$

$\varphi_A = 0.745$ $\varphi_B = 1 - \varphi_A = 1 - 0.745 = 0.255$

74.5% methanol and 25.5% water

35. Bonded phases are commonly used in LC to reduce the problem of column bleeding. These bonded phases are now widely used in partition chromatography due to their better stability and efficiency when compared to liquid stationary phases.

36. Silica is often used as the support for NPLC or RPLC columns. To place bonded phases on this support, silanol groups on the surface of silica are treated with an organosilane that contains the desired stationary phase as a side chain.

37. Endcapping is the treatment of silica with a small organosilane to react with and cover silanol groups. When you are preparing a bonded-phase support with silica, it is important to react with, or cover, as many silanol groups as possible. If this is not done, the result is a support that has more than one way of interacting with analytes. These "mixed-mode" interactions can create broader peaks and give lower resolution. These effects can be minimized through endcapping by later reacting the silica with a small organosilane (such as trimethylchlorosilane) that can reach more silanol groups on the surface of silica.

38. a) The difference in the signs on the right reflects the difference in the polarities of the stationary phase in RPLC and NPLC.

   b) $P_{tot1} = (0.20)(3.92) + (0.80)(10.2) = 8.94$

   $P_{tot2} = (0.10)(3.92) + (0.90)(10.2) = 9.57$

   $\log(12.5/k_2) = (8.94 - 9.57)/2$

   $k_2 = 25.8$

   c) $\log(12.5/5.0) = (8.94 - P_{tot2})/2$

   $P_{tot2} = 8.14 = (x)(3.92) + (1 - x)(10.2)$, where x is the volume fraction of isopropanol

   Rearranging and solving for x gives $x = 0.328 = 0.33$. This result means a mixture of approximately 33% isopropanol plus 67% water is needed to provide the desired degree of retention.

39. 2,4-D is a weak acid that will have its neutral, acidic form as the principle species at a low pH. This species is what leads to the higher retention on a RPLC column at an acidic pH versus a neutral pH. This same pH dependence will also lead to a decrease in retention for 2,4-D on an anion-exchange column as the pH is decreased.

40. NPLC has similar applications to those listed earlier for adsorption chromatography with silica or alumina. These applications typically involve the use of NPLC for separating analytes in organic solvents and chemicals that contain polar functional groups. RPLC is by far the most popular type of partition chromatography and LC. The fact that the weak mobile phase for RPLC is a polar solvent like water is another valuable feature that allows aqueous-based samples to be injected directly onto a reversed-phase column. This feature makes RPLC popular for examining clinical, biological, and environmental samples.

41. Ion-exchange chromatography is a liquid chromatographic technique in which solutes are separated by their adsorption onto a support containing fixed charges on its surface.

42. Anion-exchange chromatography is a type of ion-exchange chromatography that uses a positively charged group to separate negative ions.

    Cation-exchange chromatography is a type of ion-exchange chromatography that uses a negatively charged group to separate positive ions.

43. a) Cation-exchange chromatography

    b) Anion-exchange chromatography

    c) Under these conditions, the amino acids will have a net negative charge; use anion-exchange chromatography

44. A typical ion-exchange reaction and equilibrium expression is shown below for the competition of a sample anion (A^-) and a competing anion (C^-) for a positively charged ion-exchange site on a support.

$$A^- + Support^+(C^-) \rightleftarrows Support^+(A^-) + C^-$$

$$K_{A,C} = \frac{[\text{Support}^+(A^-)][C^-]}{[A^-][\text{Support}^+(C^-)]}$$

This reaction and equilibrium expression are both similar to those shown in Equation 22.9 for the binding of a sample cation to a support with negatively charged groups in the presence of a competing cation.

45. A selectivity coefficient ($K_{A,C}$) is a type of equilibrium constant used to describe an ion-exchange reaction (see answer to previous problem for an example).

46. Factors that will affect the retention of charged analytes on an ion-exchange column include: a) the nature and accessibility of the ion-exchange groups on the support, b) the type and concentration of analyte ions, and c) the nature and concentration of the competing ions in the mobile phase. The pH of the mobile phase will also be important if we have ion-exchange sites, analytes, or competing ions that are weak acids or bases because a change in pH may affect the charges on such agents.

47. Silica can be used as a support for ion-exchange chromatography if it has been modified to contain charged groups on its surface. Another support that is commonly used in ion-exchange chromatography for small inorganic and organic ions is *polystyrene*, which is prepared by polymerizing styrene in the presence of divinylbenzene. Carbohydrate-based gels are another common type of support used in ion-exchange chromatography. These materials are especially useful in the separation of biological compounds, which can have strong, undesirable binding to organic polymer resins like polystyrene.

48. A weak mobile phase in ion-exchange chromatography is a solution with a low ionic strength, leading to the strong retention of sample ions. A strong mobile phase in ion-

exchange chromatography is usually a solution that contains a high concentration of competing ions. The retention of analytes in this method will also be affected by the type of competing ion and type of ion-exchange sites that are present. In addition, the pH of the mobile phase can be adjusted to alter retention if you are working with analytes, competing ions, or exchange sites that are weak acids or weak bases. Adding a complexing agent to the mobile phase can also affect the charge of an analyte and alter its retention.

49. Ion-exchange is frequently employed in removing certain types of ions from samples or solutions. Ion-exchange chromatography has also been used for many years as a preparative tool in biochemistry for purifying proteins, peptides, and nucleotides. In addition, ion-exchange supports are frequently employed for concentrating small, inorganic and organic ions from samples like food, environmental samples, and commercial products to help analyze trace metals or ionic contaminants. Another application of ion-exchange chromatography is in the direct separation and analysis of samples through the use of a special type of ion-exchange chromatography known as ion chromatography.

50. Suppressor ion chromatography is a type of ion chromatography in which the use of a second column or membrane separator (of opposite charge to the first ion-exchange column) replaces competing ions that have high conductivity with ions that have a lower conductivity (see description of Figure 22.13 in the text for an example).

51. Size-exclusion chromatography is a liquid chromatographic technique that separates substances according to differences in their size, such as through their ability to enter the pores of a solid support.

52. All analytes in size-exclusion chromatography will elute with a retention volume that is between the volume of mobile phase that is outside of all the support pores (known as the "excluded volume") and the true void volume of the column.

53. The ideal support in size-exclusion chromatography consists of a porous material that does not interact directly with the injected solute. An important feature of all the supports is its pore size, which will determine the size of compounds that it can be used to separate.

54. Because retention is based on whether a solute can enter the pores of the support or not, the mobile phase does not affect retention and there is no weak or strong mobile phase in size-exclusion chromatography.

55. Gel filtration chromatography is a type of size-exclusion chromatography that uses an aqueous mobile phase. Gel permeation chromatography is a type of size-exclusion chromatography that uses an organic mobile phase.

56. As a preparative tool, size-exclusion chromatography is often used with biological samples to remove small solutes from large agents like proteins. It can also be used to transfer large analytes from one solution to another or to remove salts from a sample. In analytical applications, size-exclusion chromatography is frequently employed in the separation of biomolecules and polymers. This approach can also be utilized in estimating the molar mass of an analyte like a protein or the distribution of molar masses in a polymer.

57. K_o = (25.2 mL – 14.1 mL)/(29.3 mL – 14.1 mL) = 0.730

58. a) V_M = 11.00 mL, V_e = 19.01 mL

　　b) K_o = (16.75 mL – 11.00 mL)/(19.01 mL = 11.00 mL) = 0.718

c) A plot of log(MW) versus K_o gives an approximately linear response in between V_M and V_e. Based on the value of K_o from Part b) and this linear region, the molar mass of the known protein is 69,000 g/mol.

59. Affinity chromatography is a liquid chromatographic method that is based on biologically-related interactions. This method makes use of the selective, reversible interactions that characterize most biological systems.

60. a) An affinity ligand is the immobilized molecule that is used in affinity chromatography as the stationary phase.

    b) "Immobilization method" is a term used in affinity chromatography to describe the method by which the affinity ligand is attached to the support.

    c) A high specificity ligand is a ligand that binds to one, or just a few, closely related compounds.

    d) A general ligand is a ligand that can bind to a group of structurally related compounds.

    e) An application buffer in affinity chromatography is the weak mobile phase for this method.

    f) The elution buffer is the strong mobile phase in affinity chromatography and acts to readily remove the analyte from the affinity ligand.

61. In the "on/off" elution method of affinity chromatography, the sample is first applied to the column in the presence of an application buffer. Because of the strong and selective nature of most biological interactions, the affinity ligand will bind to the analyte of interest during this step, while allowing most other sample components to pass through as a nonretained peak. After these nonretained components have been washed from the column, a separate

elution buffer is applied to release the retained analyte. This analyte is detected as it leaves the column or is collected for later use. The column and affinity ligand are then placed back into the original mobile phase, allowing them to be regenerated prior to the injection of the next sample.

62. The retention of analyte (A) in an affinity column can be described by a complexation reaction in which A combines with the affinity ligand (L) to form the complex (A–L),

$$A + L \underset{}{\overset{K_A}{\rightleftarrows}} A\text{-}L \qquad K_A = \frac{[A\text{-}L]}{[A][L]}$$

where K_A is the association equilibrium constant for the formation of complex A–L. If a 1:1 complex is formed between A and L, the retention factor for A on the affinity column can be related to K_A and the amount of affinity ligand by the following equation,

$$k = K_A (m_L/V_M)$$

where m_L in this equation is the total moles of active ligand sites in the column and V_M is the void volume of the column. This last equation indicates that the retention factor in affinity chromatography will depend both on the strength of binding between the analyte and ligand (K_A) and the concentration of the available binding sites for an analyte on the affinity ligand (m_L/V_M).

63. a) $k = K_A\, m_L/V_M = (4.0 \times 10^8\, M^{-1})(150 \times 10^{-9}\, \text{mol})/(1.5 \times 10^{-3}\, \text{L}) = 4.0 \times 10^4$

b) $t_M = (1.5\, \text{mL})/(1.0\, \text{mL/min}) = 1.5\, \text{min}$

$t_R = (k + 1)\, t_M = (4.0 \times 10^4 + 1)(1.5\, \text{min}) = 60{,}000\, \text{min or } 1000\, \text{h}$

64. $k = (3\, \text{min} - 1.5\, \text{min})/(1.5\, \text{min}) = 1.0$

$k = K_A\, m_L/V_M$

$$1.0 = K_A (150 \times 10^{-9} \text{ mol})/(1.5 \times 10^{-3} \text{ L}) \quad \text{or} \quad K_A = 1.0 \times 10^4 \, M^{-1}$$

65. A chiral separation refers to the use of chromatography to separate and examine the individual forms of a chiral compound. One way this type of separation can be accomplished is by using a stationary phase that is also chiral and that is able to interact to different degrees with the various individual forms of a chiral analyte. This type of separation is important in pharmaceutical analysis because many drugs exist in several chiral forms.

66. Carbohydrate gels like agarose or cellulose are commonly used with affinity ligands for the purification of biological molecules. Silica can be used with affinity ligands by first converting this support into a diol-bonded phase or other form that has low nonspecific binding for most biological agents.

67. A weak mobile phase in affinity chromatography is one that allows strong binding between the analyte and affinity ligand (i.e., the "application buffer"). This weak mobile phase is usually a solvent that mimics the pH, ionic strength, and polarity of the affinity ligand in its natural environment. A strong mobile phase in affinity chromatography is a solvent that can readily remove the analyte from the affinity ligand (i.e., the "elution buffer").

68. Biospecific elution is an elution technique used in affinity chromatography in which a competing agent is added to the mobile phase to displace the analyte from the affinity ligand. Nonspecific elution is an elution technique used in affinity chromatography in which the pH, ionic strength, or polarity of the mobile phase are altered to lower the association equilibrium constant for the analyte–ligand interaction.

69. A refractive-index (RI) detector is a detector used in LC that measures the ability of the mobile phase and analytes to refract or bend light. In one design for this type of detector, light from a visible light source is passed through two flow cells, one containing mobile phase eluting from the column and the other containing a reference solution. These flow cells are at an angle to one another, which causes the light to be bent at their interface if there is any difference in the content and refractive index of their solutions. As analytes elute from the column, the refractive index of the solution in the sample flow cell will be different from that in the reference flow cell. This difference in refractive index causes the light beam to be bent and produces a response at the detector.

   A key advantage of an RI detector is that it will respond to any compound that has a different refractive index from the mobile phase, provided that enough solute is present to give a measurable signal. This makes an RI detector useful in work where an analyte cannot be easily measured by other devices or where the nature or properties of an analyte are not yet known. One disadvantage of an RI detector is that it does not have limits of detection as low as absorbance detectors or many other HPLC detectors. In addition, its signal is sensitive to changes in the mobile phase composition and temperature, making the RI detector difficult to use with gradient elution.

70. The simplest type of absorbance detector for HPLC is a fixed-wavelength absorbance detector, which is set to always monitor a specific wavelength. A more complex design is the variable-wavelength absorbance detector, which allows the monitored wavelength to be varied over a wide range. A photodiode-array detector is an absorbance detector that uses an array of small detector cells to simultaneously measure the change in absorbance at many wavelengths. This array makes it possible to record an entire spectrum for a compound as it

elutes from a column, which can be valuable in identifying overlapping peaks. The cost increases as you go from a fixed-wavelength detector to a variable-wavelength detector or photodiode-array detector, but the amount of information that can be acquired also increases.

71. a) A fluorescence detector measures the ability of chemicals to absorb and emit light at a particular set of wavelengths. Because these wavelengths are characteristic of a given chemical, this method can provide a signal that has a low background and is reasonably specific for the analyte of interest. This type of detector can be used to examine chemicals that are naturally fluorescent or that can be converted to a fluorescent derivative.

    b) An evaporative light-scattering detector measures the degree of light scattering that is produced for any solute that is less volatile than the mobile phase as the mobile phase leaving the column is converted into a spray of small droplets and the solvent in these droplets evaporates. This is a general detector that is capable of examining analytes that often cannot be detected by absorbance, such as lipids and carbohydrates.

    c) A conductivity detector can monitor ionic compounds by measuring the ability of the mobile phase and its contents to conduct a current when placed in an electrical field.

    d) An electrochemical detector can be used to measure the ability of an analyte to undergo either oxidation or reduction at an electrode and given potential.

72. Electrospray ionization (ESI) is an ionization method used in mass spectrometry (MS), in which the sample is placed into a solvent and sprayed from a highly charged needle (3–5 kV). The solvent in the charged droplets evaporates away quickly, giving smaller droplets with an excess of positive or negative charge that will eventually cause the droplet to divide and molecules in the droplet to be desorbed as ions that enter the gas phase. This ionization

method is often used in combining LC with MS by ionizing analytes, such as proteins and peptides, as they leave an LC column.

73. a) Both chromatograms are based on the measurement of ion intensity at a given elution time. They are different in that one case uses ions with specific mass-to-charge ratios for the measurement, while the other uses the overall intensity of ions that cover a broad range of mass-to-charge ratios.

b) If a specific set of ions are measured with mass-to-charge ratios that are specific for a given analyte, ESI mass spectrometry could be used as a specific means for detection. If the overall intensity of many types of ions is measured, then this approach would act more as a general detector.

74. An injection valve is often used in HPLC to introduce a sample into the system (see Figure 22.23 for a typical valve that is utilized for this purpose).

75. In a reciprocating pump, a rotating cam is used to move a piston in and out of a solvent chamber; the movement of the piston causes the mobile phase to flow into the chamber and out toward the column. A syringe pump consists of a syringe in which a plunger is pressed in by a motor, creating a flow of the mobile phase out of the syringe.

76. Because analytes examined by HPLC are injected as a solution onto the column, these analytes must first be placed into a liquid that is compatible with the mobile phase used at the beginning of the separation method. One or more pretreatment steps, such as an extraction, may be needed to dissolve and transfer these substances into an appropriate solvent. If solid matter is present in the sample, this matter would be removed by centrifugation or filtration prior to injection to avoid clogging within the column and chromatographic system.

Derivatization may be used to improve the response of an analyte in devices such as fluorescence or electrochemical detectors or used to improve the separation of the solute from other sample components.

77. Precolumn derivatization takes place before the sample is injected and can be used to either change the analyte's retention on a column or the ability this analyte has to respond on a given detector. Postcolumn derivatization occurs after the analyte has eluted from a column and is used only to alter the detection of the analyte.

79. a) $R_F = D_s/D_f = (3.47 \text{ cm})/(4.50 \text{ cm}) = 0.77$

    b) The most likely candidate is phenobarbital, which has the same R_F value as the unknown compound.

80. Band 1: $R_F = D_s/D_f = (1.23 \text{ cm})/(5.23 \text{ cm}) = 0.235$

    $k = (1 - R_F)/R_F = (1.000 - 0.235)/(0.235) = 3.25$

    Band 2: $R_F = D_s/D_f = (1.86 \text{ cm})/(5.23 \text{ cm}) = 0.356$

    $k = (1 - R_F)/R_F = (1.000 - 0.356)/(0.356) = 1.81$

    Band 3: $R_F = D_s/D_f = (2.59 \text{ cm})/(5.23 \text{ cm}) = 0.495$

    $k = (1 - R_F)/R_F = (1.000 - 0.495)/(0.495) = 1.82$

81. a) Ion-exchange chromatography would be used for the separation of these ions.

    b) The analysis of morphine and its metabolites in serum samples would be performed by RPLC, which will provide a general separation based on polarity and will allow an aqueous sample to be injected onto the column.

c) Isolation of a specific bacterial protein from a cell culture could be accomplished by various means, but affinity chromatography using a ligand for this protein would be one relatively easy approach for obtaining a simple and selective separation.

d) The use of ion chromatography and an anion-exchange analytical column would be a logical choice for the analysis of nitrate and nitrite ions in drinking water.

CHAPTER 23: ELECTROPHORESIS

1. Electrophoresis is a technique in which solutes are separated by their different rates of migration in an electric field. It is a common method used to separate solutes that differ in their charge-to-size ratios.

2. Zone electrophoresis is an electrophoretic method that uses small amounts of sample to allow analytes to be separated into narrow bands or zones. Moving boundary electrophoresis is an electrophoretic method that produces a series of moving boundaries between regions that contained different mixtures of analytes. Of these two methods, zone electrophoresis is more common in modern laboratories.

3. a) Migration distance is the distance that an analyte travels on a support in a given amount of time, such to describe separations based on gel electrophoresis.

   b) Migration time is the time required for an analyte to travel a given distance, such as used to describe separations based on capillary electrophoresis.

   c) An electropherogram is a plot of detector response versus migration time in electrophoresis.

4. $v = (35 \text{ cm})/(5.63 \text{ min}) = 6.22 \text{ cm/min}$

   Distance in 2.5 min = $(6.22 \text{ cm/min})(2.5 \text{ min}) = 15.5\underline{5} \text{ cm} = 15.6 \text{ cm}$

5. $v = (3.2 \text{ cm})/(30 \text{ min}) = 0.10\underline{7} \text{ cm/min} = 0.11 \text{ cm/min}$

   At 200 V, v will increase by (200 V/100 V) = 2-fold to $0.21\underline{3} \text{ cm/min} = 0.21 \text{ cm/min}$

   Time to travel 10 cm at 200 V = $(10 \text{ cm})/(0.21\underline{3} \text{ cm/min}) = 46.\underline{9} = 47 \text{ min}$

6. The overall rate of travel of a charged solute in electrophoresis will depend on two opposing forces. The first of these forces is the attraction of a charged solute toward the electrode of opposite charge. The second force acting on the solute is resistance to its movement, as created by the surrounding medium. When an electric field is applied, a solute will accelerate toward the electrode of opposite charge until these two forces become equal in size (although opposite in direction). At this point, a steady-state situation is produced, in which the solute begins to move at a constant velocity.

7. Electrophoretic mobility (μ) is a constant used in electrophoresis to relate the velocity of migration (v) for a charged solute to the strength of the applied electric field (E), where $v = \mu E$. The value of μ is equal to $z/(6 \pi r \eta)$, where z is the charge on the solute, r is the solvated radius of the ion, and η is the viscosity of the medium.

8. $v = (21.5 \text{ cm})/(8.31 \text{ min}) = 2.59 \text{ cm/min}$

    $\mu = (L_d/t_m)/(V/L)$

    $= \{(21.5 \text{ cm})/(8.31 \text{ min})\}/\{10.0 \text{ kV}/25.0 \text{ cm}) = 6.47 \text{ cm}^2/\text{kV} \cdot \text{min}$

9. The migration velocity would increase by (15.0 kV)/(10.0 kV) = 1.5-fold, going from 2.59 cm/min to (2.59 cm/min)(1.5) = 3.88 cm/min.

    The migration time will decrease by 1.5-fold from 8.31 min to (8.31 min)/(1.5) = 5.54 min.

    There will be no change in the electrophoretic mobility.

10. A change in pH can affect the migration of weak acids or weak bases. Complexation or solubility reactions can also affect the observed migration of an analyte.

11. The main species for dicamba throughout the given pH range is the conjugate base, A^-. The pK_a of dicamba is sufficiently low such that the relative amount of this conjugate base versus dicamba is not significantly altered by the change in pH. However, the second pK_a for DCSA does occur in the range over which the pH is being altered. This means the relative amounts of the acid–base forms of DCSA are changing, which alters the apparent mobility that is observed for this compound.

12. β-Cyclodextrin is a chiral agent that can form reversible complexes with some drugs. If the two different forms of a chiral drug have different degrees of binding to this ligand, this will affect their observed mobilities to different extents. The results would be a chiral separation.

13. Electroosmosis refers to the movement of the running buffer in electrophoresis, as caused by the presence of fixed charged in the system and the creation of an electrical double layer at the support's surface. The presence of electrosmosis will affect the overall mobility observed for all solutes migrating in the system.

14. Electroosmotic mobility is the observed mobility due to electroosmosis in electrophoresis. The value of μ_{eo} depends on such factors as the size of the electric field, the type of running buffer that is being employed, and the type of charge that is present on the support.

15. $\mu_{eo} = (L_d/t_m)/(V/L)$

    $= \{(30.0 \text{ cm})/(1.52 \text{ min})\}/\{30.0 \text{ kV}/50.0 \text{ cm}\} = 32.9 \text{ cm}^2/\text{kV} \cdot \text{min}$

16. $\mu_{eo} = (L_d/t_m)/(V/L)$ or $t_m = (L_d/\mu_{eo})/(V/L)$

    $t_m = \{(0.020 \text{ m})/(8.3 \times 10^{-10} \text{ m}^2/\text{V} \cdot \text{s})\}/(20{,}000 \text{ V}/0.025 \text{ m}) = 30 \text{ s (or 0.50 min)}$

17. a) $N = 59,200$ b) $H = 5.9 \times 10^{-4}$ cm c) $N = 6400$

18. a) $R_s = 1.94$ b) $R_s = 2.00$

19. Longitudinal diffusion occurs in electrophoresis when a solute diffuses away from the center of its band along the direction of travel, causing this band to broaden over time and to become less concentrated. One factor that affects the extent of this band-broadening is the "size" of the diffusing solute, or its solvated radius. Because larger analytes have slower diffusion, they will be less affected by longitudinal diffusion than will smaller substances. The rate of this diffusion will also decrease as we increase the viscosity of the running buffer or lower the temperature of the system. One way we can minimize the effects of longitudinal diffusion in electrophoresis is to have an analyte move through a porous support. If the pores of this support are sufficiently small, they will inhibit the movement of analytes due to diffusion and help provide narrower bands.

20. Joule heating is a process that is caused by heating that occurs whenever an electric field is applied to the system. According to Ohm's law, placing a voltage across a medium with a resistance requires that a current is present to maintain this voltage across the medium. As current flows through the system, heat is generated. As heat is produced, the temperature of the electrophoretic system will begin to rise. This rise in temperature will increase longitudinal diffusion and lead to increased band-broadening. In addition, if the heat is not distributed uniformly throughout the electrophoretic system, the temperature will not be the same throughout the system. An uneven temperature will lead to regions with different densities (causing mixing) and different rates of diffusion, which results in even more band-broadening.

21. One way Joule heating can be decreased is by using a lower voltage for the separation. An alternative approach is to use more efficient cooling for the system, which would allow higher voltages to be used and provide shorter separation times. Another possibility is to add a support to the electrophoretic system that minimizes the effects of Joule heating due to uneven heat distribution and density gradients in the running buffer. Another factor that affects Joule heating is the ionic strength of the running buffer. A lower ionic strength for this buffer will lower heat production because at low ionic strengths there are fewer ions in this buffer.

22. a) A plot of I versus V should result in a linear relationship with a slope of resistance R if all other conditions are held constant.

    b) Joule heating can lead to such deviations.

23. Eddy diffusion in electrophoresis can contribute to band-broadening if a support is used to minimize the effects of Joule heating, a situation that creates multiple flow paths for analytes through the support.

24. Wick flow is a source of band-broadening in gel electrophoresis due to evaporation of solvent from the wicks, such as in the presence of Joule heating. As this solvent is lost, it is replenished by the flow of more solvent through the wicks and from the buffer reservoirs. This flow leads to a net movement of buffer from each reservoir toward the center of the support. The rate of this flow depends on the rate of solvent evaporation, so it will increase with the use of a high voltage or a high current.

25. Gel electrophoresis is an electrophoretic method that is performed by applying a sample to a gel support, which is then placed into an electric field. The position of a sample band versus that of standards can be used in identification, and the size of the band can be used to measure the amount of analyte that is present.

26. Based on a calibration curve made with the above standards and the result for the unknown, the amount of protein in the unknown sample is 9.1 ng.

27. The locations of the bands can be used to determine which types of proteins are present in the sample. The size of these bands, as indicated in the densitometer scan, can be used to determine the relative amount of protein in each of these bands.

28. Some typical systems for carrying out gel electrophoresis are shown in Figure 23.7, which may have a support that is held in either a vertical or horizontal position. This support contains a running buffer with ions that carry a current through the support when an electric field is applied. To replenish this buffer and its components as they move through the support or evaporate, the ends of the support are placed in contact with two reservoirs that contain the same buffer solution and the electrodes. Once samples have been placed on the support, the electrodes are connected to a power supply and used to apply a voltage across the support. This electric field is passed through the system for a given amount of time, causing the sample components to migrate. After the electric field has been turned off, the gel is removed and examined to locate the analyte bands.

29. Cellulose acetate, filter paper, and starch are useful supports for work with relatively small molecules, like amino acids and nucleotides. Electrophoresis involving large molecules,

such as DNA, can be carried out on agarose. Polyacrylamide is often used in work with proteins.

30. The samples in gel electrophoresis are applied to small wells that are made in the gel during its preparation. A common approach to create narrow sample bands is to employ two types of gels in the system: a "stacking gel" and a "running gel". The running gel is the support used for the electrophoretic separation of substances in the sample. The stacking gel has a lower degree of cross-linking (giving it larger pores) and is located on top of the running gel. The stacking gel is also the section of the support in which the sample wells are located. After a sample has been placed in the wells and an electric field has been applied, analytes will travel quickly through the stacking gel until they reach its boundary with the running gel. These substances will then travel much more slowly, allowing other parts of the sample to catch up and to form a narrower, more concentrated band at the top of the running gel.

31. a) A densitometer is a scanning device that is used for direct detection on gels by using absorbance measurements.

    b) Coomassie Brilliant Blue is a common stain used to see protein bands in gel electrophoresis.

    c) Silver staining uses silver nitrate to detect low concentration proteins.

    d) Blotting is an approach for detection in gel electrophoresis during which a portion of an analyte band is transferred to a support, such as nitrocelluose, where the analytes are reacted with a labeled agent.

32. A Southern blot is a blotting method used to detect specific sequences of DNA, based on the binding of these sequences to an added, known sequence of DNA that is labeled with a

radioactive tag or with a label that can undergo chemiluminescence. A northern blot is a blotting method that is used to detect specific sequences of RNA through their binding with a labeled DNA probe. A western blot is a blotting method used to detect specific proteins based on the transfer of these proteins onto a support like nitrocellulose or nylon, followed by the treatment of this support with labeled antibodies that can specifically bind the proteins of interest.

33. Matrix-assisted laser desorption/ionization time-of-flight mass spectrometry (MALDI-TOF MS) is a type of mass spectrometry in which a special matrix capable of absorbing light from a laser is used for chemical ionization. This approach is often used to determine the molar mass of bands in gel electrophoresis, such as for proteins, thus making it possible to identify the contents of this band. MALDI-TOF MS can also be used to look at peptides, polysaccharides, nucleic acids, and some synthetic polymers.

34. The proteins in a sample are first denatured and their disulfide bonds broken through the use of a reducing agent, converting the proteins into a set of single-stranded polypeptides. These polypeptides are then treated with sodium dodecyl sulfate (SDS), a surfactant that coats each protein and forms roughly linear rods that have an exterior layer of negative charge. The result for a mixture of proteins is a series of rods with different lengths, but similar charge-to-mass ratios. Next, these protein rods are passed through a porous polyacrylamide gel in the presence of an electric field. The negative charges on these rods (from the SDS coating) causes them to all move toward the positive electrode, while the pores of the gel allow small rods to travel more quickly than large rods to this electrode. The different distances of travel on this gel in a given period of time can be compared to the migration distances for protein

standards on the same gel to obtain the molar mass of an unknown protein.

35. A plot of log(MW) versus migration distance results in an approximately linear relationship with the following best-fit line: $y = -0.555\,x + 5.41$. Using this best-fit line, the approximate molecular weight of the unknown protein is 52 kDa.

36. Using the best-fit line from the previous problem gives the following estimated migration distances.

    18.5 kDa protein: 2.06 cm

    40.2 kDa protein: 1.45 cm

    91.8 kDa protein: 0.81 cm

37. Isoelectric focusing (IEF) is an electrophoretic method used to separate for zwitterions (e.g., proteins or peptides) based on their isoelectric points by having these compounds migrate in an electric field across a pH gradient.

38. Ampholytes are a mixture of small zwitterions that are used in isoelectric focusing to produce a stable pH gradient.

39. Two-dimensional electrophoresis (2-D electrophoresis) is a method in which two different types of electrophoresis are performed on a single sample. The first of these separations is usually based on isoelectric focusing and the second on sodium dodecyl sulfate polyacrylamide gel electrophoresis (SDS-PAGE). The use of two different separation approaches in this method makes it possible to obtain separations of complex mixtures of analytes such as proteins or peptides.

40. Capillary electrophoresis (CE) is a type of electrophoresis that is performed using a narrow capillary that is filled with a running buffer. The use of narrow bore tubes provides efficient removal of Joule heating by allowing this heat to be quickly dissipated to the surrounding environment. This removal of heat helps to decrease band-broadening and provides much more efficient and faster separations than gel electrophoresis.

41. A plot of the peak height for nitrate versus the internal standard is made with the concentration of nitrate on the x-axis. Based on this plot and the results for the unknown, the amount of nitrate in the sample is 4.3 mg/L.

42. A plot of the peak area ratio for the peptide compared to fluorescein is first made versus the concentration of the peptide. The result is a linear relationship with a best-fit line of $y = 0.230\,x + 0.200$. The peak area ratio for the unknown sample is $4098/556 = 7.37$, which can be used with the best-fit line to obtain an estimated concentration of 31 nM for the peptide in the unknown sample.

43. One reason that CE is more efficient than gel electrophoresis is that Joule heating is greatly reduced as a source of band-broadening. Also, capillary electrophoresis is often used with no gel or support present, which eliminates eddy diffusion and secondary interactions with the support (other than the capillary wall). The result is that longitudinal diffusion now becomes the main source of band-broadening.

44. If it is assumed that longitudinal diffusion is the only major source of band-broadening present in this system, the number of theoretical plates can be estimated by using Equation 23.8.

$$N = \frac{\mu V L_d}{2DL} = \frac{(1.58 \text{ cm}^2/\text{kV} \cdot \text{min})(20.0 \text{ kV})(33.0 \text{ cm})}{2(3.0 \times 10^{-5} \text{ cm}^2/\text{s})(40.0 \text{ cm})} = 434{,}500$$

The migration time can be found by using Equation 23.9 and solving for t_m.

$$\mu = (L_d/t_m)/(V/L) \quad \text{or} \quad t_m = L_d/\{(V/L)\mu\}$$

$$t_m = (33.0 \text{ cm})/\{(1.58 \text{ cm}^2/\text{kV} \cdot \text{min})(20.0 \text{ kV}/40.0 \text{ cm})\} = 41.8 \text{ min}$$

45. a) $N = 16 (t_m/w_b)^2$ (from Chapter 22)

       $= 16 \{(14.8 \text{ min})(60 \text{ s/min})/(26 \text{ s})\}^2$

       $= 18{,}700$

b) $\mu = (L_d/t_m)/(V/L)$ (in the absence of any electroosmotic flow)

       $= \{(38.0 \text{ cm})/(14.8 \text{ min})\}/\{15.0 \text{ kV}/42.5 \text{ cm})= 7.27 \text{ cm}^2/\text{kV} \cdot \text{min}$

$$N = \frac{\mu V L_d}{2DL} = \frac{(7.27 \text{ cm}^2/\text{kV} \cdot \text{min})(15.0 \text{ kV})(38.0 \text{ cm})}{2D(42.5 \text{ cm})} = 18{,}700$$

Rearranging the above expression for N and solving for the diffusion coefficient gives $D = 2.6 \times 10^{-3}$ cm^2/min (or 4.3×10^{-5} cm/s). It is assumed in this calculation that longitudinal diffusion is the only significant source of band-broadening.

46. The main components of a CE system include a power supply and electrodes for applying the electric field, two containers that create a contact between these electrodes and the solution within the capillary, an on-line detector, and a means for injecting samples onto the capillary. Because these instruments can use voltages up to 25–30 kV, they include safety features that protect the user from the high voltage region and that can turn off this voltage when the system is opened for maintenance or for inserting samples and reagents.

47. An uncoated silica capillary can lead to a significant amount of flow due to electroosmosis when you are working at a neutral or basic pH, as is caused by deprotonation of the silica's surface silanol groups. This electroosmosis will cause all analytes, regardless of their charge, to travel in the same direction through the capillary. This effect means that a sample containing many types of ions can be injected at one end of the capillary (at the positive electrode), with electroosmosis then carrying these through to the other end (to the negative electrode) and past an online detector.

48. The normal polarity mode is a method used in CE during which a sample is injected at one end of the capillary (the positive electrode when using an uncoated silica capillary) and carried by electroosmosis to a detector near the other end of the capillary (the negative electrode, in this case). The reversed polarity mode is a method used in capillary electrophoresis during which a sample is injected at one end of the capillary (the negative electrode when using an uncoated silica capillary) and migrates against electroosmosis to a detector near the other end of the capillary (the positive electrode, in this case).

49. The small volume of a CE capillary and the high efficiency of CE both require that a small injection volume is used in CE. Although special techniques must be used to attain these small injection volumes, this feature also makes it possible to work with samples that have only small volumes of material that are available for analysis.

50. Hydrodynamic injection is an injection technique employed in CE which uses a difference in pressure to deliver a sample to the capillary. Electrokinetic injection is an injection technique employed in CE in which an electric field is applied across the capillary, allowing electroosmostic flow and the electrophoretic mobility of the analytes to cause them to enter

51. Sample stacking is a method for concentrating samples and providing narrow analyte bands in capillary electrophoresis. This approach makes use of a sample that has a lower ionic strength (and lower conductivity) than that of the running buffer.

52. Selective detection methods that are used for capillary electrophoresis include electrochemical and fluorescence detection, as well as mass spectrometry. Ultraviolet-visible absorbance, conductance detection and mass spectrometry are often employed in CE for more general detection. Many of these methods are also used in liquid chromatography.

53. Placing the migration times for the proteins into Equation 23.10, along with the measured peak areas, gives the following results.

    Protein 1: $A_{c,1} = (3430)/(20.3 \text{ min}) = 169$

    Protein 2: $A_{c,2} = (1235)/(24.5 \text{ min}) = 50.4$

    If these proteins have a similar response at the detector, then there is 169/50.4 = 3.35-fold more of the first isoform versus the second isoform.

54. If the applied voltage is increased from 15.0 kV to 20.0 kV (a 1.33-fold increase), there will be a corresponding 1.33-fold decrease in both the migration time and the apparent peak area. The new peak area under these conditions will be (11,250 units)/(1.33) = 8,460.

55. Laser-induced fluorescence (LIF) is a method that employs a laser to excite a fluorescent compound, allowing the detection of this agent through its subsequent emission of light. This approach can provide low detection limits for analytes that are fluorescent or that have been

converted to fluorescent derivatives.

56. Capillary sieving electrophoresis is a type of capillary electrophoresis in which an agent is included in the separation that can separate analytes based on their size. One way to perform capillary sieving electrophoresis is to place a porous gel in the capillary, like the polyacrylamide gels employed in SDS-PAGE. A second approach is to add to the running buffer a large polymer that can entangle with analytes and alter their rate of migration.

57. A normalized migration time of 2.65 in Figure 23.18 corresponds to a log(MW) value of approximately 5.04, or MW = $10^{5.04}$ = 110,000 Da.

58. a) Electrokinetic chromatography is a type of capillary electrophoresis in which a charged agent is placed into the running buffer to interact with analytes.

b) Micellar electrokinetic chromatography (MEKC) is a subset of electrokinetic chromatography that employs micelles as running buffer additives.

c) A micelle is a particle formed by the aggregation of a large number of surfactant molecules, such as sodium dodecyl sulfate (SDS).

d) The critical micelle concentration is the threshold concentration of a surfactant above which the surfactant molecules come together to form micelles.

59. Capillary isoelectric focusing (CIEF) is the use of isoelectric focusing in a capillary electrophoresis system. One way CIEF can be conducted is by placing the electrodes in contact with two different electrolyte solutions: 1) the "catholyte", which is a basic solution by the cathode and 2) the "anolyte", which is an acidic solution located by the anode. The capillary contains a mixture of ampholytes that will create a pH gradient when an electric

field is applied between these electrodes. A coated capillary is also used in this case to minimize or eliminate electroosmotic flow. When a sample is injected onto this system, its zwitterions will migrate until they reach a region where the pH is equal to their pI. Once these bands have formed, they are pushed through the capillary and past the detector by applying pressure to the system.

60. Affinity capillary electrophoresis (ACE) is a type of capillary electrophoresis in which biologically-related agents are placed as additives in the running buffer. One common use of ACE is in the separation of chiral analytes through the use of binding agents like cyclodextrins or proteins. This method can also be used in clinical and pharmaceutical assays and for the study of biological interactions.

62. a) In the absence of electroosmotic flow:

$$v = \mu V/L$$
$$= (2.50 \text{ cm}^2/\text{kV} \cdot \text{min})(15.0 \text{ kV})/(30.0 \text{ cm})$$
$$= 1.25 \text{ cm/min}$$

$$t_m = (L_d L)/(\mu V)$$
$$= (25.0 \text{ cm})(30.0 \text{ cm})/\{(2.50 \text{ cm}^2/\text{kV} \cdot \text{min})(15.0 \text{ kV})\}$$
$$= 20.0 \text{ min}$$

b) In the presence of electroosmotic flow:

$$v = (\mu + \mu_{osm}) V/L$$
$$= (2.50 \text{ cm}^2/\text{kV} \cdot \text{min} + 4.10 \text{ cm}^2/\text{kV} \cdot \text{min})(15.0 \text{ kV})/(30.0 \text{ cm})$$
$$= 3.30 \text{ cm/min}$$

$$t_m = (L_d\, L)/\{(\mu + \mu_{osm})V\}$$

$$= (25.0\text{ cm})(30.0\text{ cm})/\{(2.50\text{ cm}^2/\text{kV}\cdot\text{min} + 4.10\text{ cm}^2/\text{kV}\cdot\text{min})(15.0\text{ kV})\}$$

$$= 7.58 \text{ min}$$

c) In the absence of electroosmotic flow:

$$v = \mu\, V/L = (-2.50\text{ cm}^2/\text{kV}\cdot\text{min})(15.0\text{ kV})/(30.0\text{ cm})$$

$$= -1.25 \text{ cm/min}$$

$$t_m = (L_d\, L)/(\mu\, V) = (25.0\text{ cm})(30.0\text{ cm})/\{(-2.50\text{ cm}^2/\text{kV}\cdot\text{min})(15.0\text{ kV})\}$$

$$= -20.0 \text{ min}$$

(Note: This "negative" time simply indicates that the anion is traveling in the wrong direction and will not enter the capillary for separation and detection in the absence of electroosmotic flow.)

In the presence of electroosmotic flow:

$$v = (\mu + \mu_{osm})\, V/L$$

$$= (-2.50\text{ cm}^2/\text{kV}\cdot\text{min} + 4.10\text{ cm}^2/\text{kV}\cdot\text{min})(15.0\text{ kV})/(30.0\text{ cm})$$

$$= 0.80 \text{ cm/min}$$

$$t_m = (L_d\, L)/\{(\mu + \mu_{osm})V\}$$

$$= (25.0\text{ cm})(30.0\text{ cm})/\{(-2.50\text{ cm}^2/\text{kV}\cdot\text{min} + 4.10\text{ cm}^2/\text{kV}\cdot\text{min})(15.0\text{ kV})\}$$

$$= 31.3 \text{ min}$$